Principles of
Statistical Radiophysics 3

S. M. Rytov Yu. A. Kravtsov
V. I. Tatarskii

Principles of Statistical Radiophysics 3

Elements of Random Fields

With 36 Figures

Springer-Verlag Berlin Heidelberg New York
London Paris Tokyo

Professor Sergei M. Rytov
Corresp. Member of the USSR Academy of Sciences,
Department of General Physics and Astronomy, USSR Academy of Sciences,
SU-117901 Moscow, USSR

Professor Yurii A. Kravtsov
General Physics Institute, USSR Academy of Sciences,
SU-117942 Moscow, USSR

Professor Valeryan I. Tatarskii
Corresp. Member of the USSR Academy of Sciences, Inst. of Atmospheric Physics,
USSR Academy of Sciences, Pyzhevsky Per., SU-109017 Moscow, USSR

Translator

Alexander P. Repyev
Samarkandsky Boulevard 13-3-63, SU-109507 Moscow, USSR

Title of the original Russian edition: *Vvedenie v statisticheskuyu radiofiziku II,* sluchainuie polya
2. revised and enlarged edition

ISBN-13:978-3-642-72687-3 e-ISBN-13:978-3-642-72685-9
DOI: 10.1007/978-3-642-72685-9

Library of Congress Cataloging-in-Publication Data. (Revised for volume 3) Rytov, S.M., 1908-. Principles of statistical radiophysics. Translation of: Vvedenie v statisticheskuiu radiofiziku. Bibliography: v. 1, p. 245–248. Includes index. Contents: 1. Elements of random process theory–3. Elements of random fields. 1. Radio waves–Mathematics. 2. Stochastic processes. I. Kravtsov, Iurii Aleksandrovich. II. Tatarskii, V. I. (Valer'ian Il'ich). I. Title. QC661.R9213 1987 537.5'34 87-4644

Softcover reprint of the hardcover 1st edition 1989

Typesetting: Macmillan India Ltd., India
2157/3150-543210

Foreword

Principles of Statistical Radiophysics is concerned with the theory of random functions (processes and fields) treated in close association with a number of applications in physics. Primarily, the book deals with radiophysics in its broadest sense, i.e., viewed as a general theory of oscillations and waves of any physical nature[1]. This translation is based on the second (two-volume) Russian edition. It appears in four volumes:

1. Elements of Random Process Theory
2. Correlation Theory of Random Processes
3. Elements of Random Fields
4. Wave Propagation Through Random Media.

The four volumes are, naturally, to a large extent conceptually interconnected (being linked, for instance, by cross-references); yet for the advanced reader each of them might be of interest on its own. This motivated the division of the *Principles* into four separate volumes.

The text is designed for graduate and postgraduate students majoring in radiophysics, radio engineering, or other branches of physics and technology dealing with oscillations and waves (e.g., acoustics and optics). As a rule, early in their career these students face problems involving the use of random functions. The book provides a sound basis from which to understand and solve problems at this level. In addition, it paves the way for a more profound study of the mathematical theory, should it be necessary[2]. The reader is assumed to be familiar with probability theory.

In progressing from one volume to the next, the reader will see that the physical problems under consideration become more complex and the mathematical machinery more involved. This results quite naturally from the course-oriented origin of the book, based as it is on summarized lecture notes. Their extensive teaching experience has convinced the authors that such an approach is preferable to a more uniform presentation. Each chapter is followed by a number of problems worked out in detail. The problems not only serve as exercises, but in many cases contain additional theoretical material and further literature references.

S.M. Rytov
Yu.A. Kravtsov
V.I. Tatarskii

Moscow, March 1986

[1] It should be noted that certain questions in statistical radiophysics are not covered in this book. For example, it is neither concerned with quantum radiophysics and quantum electronics, nor with statistical phenomena related to the propagation of waves in nonlinear media, see [1].

[2] Treatments of the mathematical theory are given, for instance, in [2–9], and more specialized applications of the theory to radiophysics and engineering are to be found in [10–19].

Preface

Unlike the preceding two volumes devoted to random functions of *one* variable, Volume 3 is concerned with the more general case of multi-variate functions, i.e., random fields. The variables are usually time t and spatial coordinates $\boldsymbol{r} = \{x, y, z\}$. Some problems only involve subsets of these variates: either the number of spatial coordinates is less than three (like for fields on a surface or on a line), or there is no time dependence (static fields) or deterministic processes are treated (say, monochromatic fields).

S. M. Rytov
Yu. A. Kravtsov

Moscow, December 1988

Contents

1. Fundamentals

In the present volume and in [1.1] we for the most part use the *correlation theory* of random wave fields. The emphasis here is therefore on this theory and on spectral representations of fields. As we pass from random processes to multivariate random functions (fields), we will find that the required generalizations are mathematically obvious. However, it would still be expedient to start off with definitions of some basic generalized concepts, see Sect. 1.1.

Up until Sect. 1.6 we consider the special case of constant (time-independent) random fields, which from the very beginning are assumed to be complex (Sect. 1.2). As was noted in [1.2, 3] correlation theory is of especial interest for *stationary* processes. The same is true of fields, the analog of stationarity here being the spatial *homogeneity* of a field, i.e., homogeneity in spatial coordinates (Sect. 1.3).

A novel property of the field (as compared with the univariate random function, or process) is its possible anisotropy, i.e., the dependence of its statistical characteristics on the direction in space.

For physical applications, the requirement that a field be homogeneous not infrequently turns out to be too stringent, as does the requirement that a process be stationary. That is why certain deviations from stationarity i.e., processes with stationary increments and quasi-stationary processes (Sects. II.4.2, 3)* are of especial interest for the theory of random processes. For random fields, similar generalizations are desirable and possible. These are *locally homogeneous* (Sect. 1.4) and *quasi-homogeneous* (Sect. 1.5) fields.

Section 1.6 considers spectral representations in the general case, where random fields depend both on spatial coordinates and on time. These will prove useful later in the treatment.

The concluding section (Sect. 1.7) is purely mathematical in nature, and it is only for this reason that it has been included in this chapter. It is devoted to another, more general way of describing random functions than that we have used so far—the so-called functional description of random fields in terms of *characteristic functionals*. This technique is applied in some exercises at the end of the chapter. Specifically, it is used in Exercise 1.22 to give a consistent derivation of an equation for the probability density of states of a nonlinear dynamic system exposed to delta-correlated random actions (the Einstein–Fokker–Planck equation). This result has already been used (without proof) in Sect. I.6.4.

---

* (Labels starting with a Roman number refer to the respective volume of the Principles of Statistical Radiophysics, e.g. (I.2.1) stands for Eq. (2.1) of Vol. 1.)

1.1 Basic Definitions

Volumes I and II [1.2, 3], almost without exception, deal with univariate random functions $\xi(\alpha)$, the variable α in the majority of applications being time t. If $\xi(\alpha, \beta, \ldots)$ is a multivariate random function, it is usually referred to as the *random field* (in the space of variates $\alpha, \beta, \ldots$). From the very beginning we will confine ourselves to the case of the four variables $\alpha, \beta \ldots$: the time t, and the coordinates of a point in space $\boldsymbol{r} = (x, y, z)$. We can, of course, refer to $\xi(t, \boldsymbol{r})$ as a field in a four-dimensional space, but in nonrelativistic problems it is more customary and convenient to speak about a variable (time dependent) field in a three-dimensional space (x, y, z).

The random field can also be described by N functions (or components) $\xi^{(i)}(t, \boldsymbol{r})$ $(i = 1, 2, \ldots, N)$, when it is then called the N-component random field. In the case of processes with only one argument (time t), we used the term N-variate random process or function. But now we have to distinguish between the number of variables t, x, y, z and the number of components of the field itself. Mathematically, the components $\xi^{(i)}(t, \boldsymbol{r})$ may be arbitrary and may even have different dimensions. For example, the density fluctuations of a liquid, ϱ, pressure p, temperature T, and velocity $\boldsymbol{v}$ together form a six-component random field. But physically, of special interest are, of course, those cases where the components $\xi^{(i)}(t, \boldsymbol{r})$ have the same dimensions and exhibit the same behavior under orthogonal transformations of coordinates in the space x, y, z, i.e., they are the components of a tensor. Given this approach, it is more convenient in the above example to speak about *four* fields: three scalar fields (ϱ, p, and T) and one vector field ($\boldsymbol{v}$).

A number of definitions and properties introduced and established in earlier volumes for random functions of a single variable can be generalized in an obvious manner to multivariate fields and specifically to space-time fields that vary with x, y, z, and t. These generalizations also concern the question of what is required in order to give a *complete definition* of a random field.

Let us denote a point in a four-dimensional world by $Q = (t, \boldsymbol{r})$. A one-component random field $\xi(Q)$ is said to be defined *completely* if all its n-variate, or n-point $(n = 1, 2, \ldots)$ probability densities are known. That is, for any number n of arbitrarily selected points Q, the functions are given by

$$\begin{aligned} & w_n\{\xi_1, \ldots, \xi_n\} d\xi_1 \ldots d\xi_n \\ & \quad = P\{\xi_\nu \leqslant \xi(Q_\nu) < \xi_\nu + d\xi_\nu \quad (\nu = 1, 2, \ldots, n)\} \ , \end{aligned} \tag{1.1}$$

where $P\{\xi \leqslant \xi(Q) < \xi + d\xi\}$ is the probability that the random variable $\xi(Q)$ will assume a value lying in the interval $\{\xi, \xi + d\xi)$[1]. Similarly, an N-component field $\xi^{(i)}(Q)$ is statistically defined (described) completely by a collection of nN-variate

[1] In [1.2, 3] we used only this way of defining a random function, which by no means suggests that it is the only one possible (see Sect. 1.7).

probability densities

$$w_{nN}(\xi_1^{(1)}, \xi_2^{(1)}, \dots, \xi_n^{(N)}) d\xi_1^{(1)} d\xi_2^{(1)} \dots d\xi_n^{(N)}$$
$$= P\{\xi_\nu^{(i)} \leqslant \xi^{(i)}(Q_\nu) < \xi_\nu^{(i)} + d\xi_\nu^{(i)}, \quad \nu = 1, 2, \dots, n, \quad i = 1, 2, \dots, N\} \ . \tag{1.2}$$

Clearly, these probability densities have to be subject to the conditions of non-negativity, symmetry, hierarchy, and normalization (Sect. I.4.1).

The notion of statistical homogeneity can also quite naturally be generalized to random fields. A one-component random field is called *homogeneous* (in the strict sense), i.e., stationary[2] in t and homogeneous in x, y, z, if all the n-variate probability densities w_n are invariant under the translation $Q \to Q + \delta Q$.

$$P\{\xi_\nu \leqslant \xi(Q_\nu + \delta Q) < \xi_\nu + d\xi_\nu, \quad \nu = 1, 2, \dots, n\}$$
$$= P\{\xi_\nu \leqslant \xi(Q_\nu) < \xi_\nu + d\xi_\nu, \quad \nu = 1, 2, \dots, n\} \ . \tag{1.3}$$

With an N-component random field $\xi^{(i)}$, if all the nN-variate probability densities w_{nN} are invariant under the above translation δQ, then the fields $\xi^{(i)}$ $(i = 1, 2, \dots, N)$ are said to be *homogeneous and homogeneously connected* (in the Q-space).

The notion of *spatial homogeneity* for random fields (the invariance of w_n under translation $\boldsymbol{r} \to \boldsymbol{r} + \delta\boldsymbol{r}$) is a natural generalization of the notion of stationarity for random functions of time (Sect. I.4.3). But the multidimensionality of the space of the variables x, y, z, t opens up new possibilities, namely: in some of the variables the field may be homogeneous, and in the others, inhomogeneous. For example, along with fields homogeneous in the Q-space, there are stationary but spatially inhomogeneous fields, or homogeneous nonstationary fields. In problems on waves, there are many fields that are homogeneous in certain surfaces only, say, in a plane or sphere.

Given the multivariate probability densities, we can work out the moments of any order for the random field ξ. In the general case, these moments will be functions of coordinates: $Q_\nu = (t_\nu, \boldsymbol{r}_\nu)$. In many cases we are interested in the lowest moments (of first and second order), which are used in the correlation theory of random fields. The basic concepts in this theory are the same as in the general correlation theory of random functions.

The *mean* of the random field $\xi(Q)$ (first moment) is computed using the univariate probability density $w_1(\xi)$[3]

$$\langle \xi(Q) \rangle = \int \xi w_1(\xi) d\xi \ . \tag{1.4}$$

The fluctuational part of the random field ξ, i.e., $\xi - \langle \xi \rangle$, will again be labelled by a

---

[2] Homogeneity in time is usually termed stationarity.

[3] Statistical averaging is denoted here by angle brackets, but where it is more convenient, we will use a bar as in [1.2, 3], so that $\langle \xi \rangle = \bar{\xi}$.

tilde above ξ:

$$\tilde{\xi} \equiv \xi - \langle \xi \rangle \equiv \xi - \bar{\xi} \ . \tag{1.5}$$

The *mixed moment* of the second order, B_ξ, is calculated with the help of the bivariate probability density $w_2(\xi_1, \xi_2)$

$$B_\xi(Q_1, Q_2) \equiv \langle \xi(Q_1)\xi(Q_2) \rangle = \iint \xi_1 \xi_2 w_2(\xi_1, \xi_2) d\xi_1 d\xi_2 \ , \tag{1.6}$$

as is the *covariance* (or rather, *autocovariance*)

$$\begin{aligned}\psi_\xi(Q_1, Q_2) &\equiv \langle \tilde{\xi}(Q_1)\tilde{\xi}(Q_2) \rangle \\ &= \iint (\xi_1 - \langle \xi_1 \rangle)(\xi_2 - \langle \xi_2 \rangle) w_2(\xi_1, \xi_2) d\xi_1 d\xi_2 \\ &= B_\xi(Q_1, Q_2) - \langle \xi(Q_1) \rangle \langle \xi(Q_2) \rangle \ .\end{aligned} \tag{1.7}$$

For random fields with zero mean, the functions $B_\xi(Q_1, Q_2) = \langle \xi(Q_1)\xi(Q_2) \rangle$ and $\psi_\xi(Q_1, Q_2)$ coincide. The *variance* of the random field $D[\xi]$, i.e., the mean square of the fluctuations is

$$D[\xi] \equiv \sigma_\xi^2(Q) \equiv \langle \tilde{\xi}^2(Q) \rangle = \langle [\xi(Q) - \langle \xi(Q) \rangle]^2 \rangle = \psi_\xi(Q, Q) \ . \tag{1.8}$$

In correlation theory, a multicomponent field $\xi^{(i)}(Q)$ is characterized by the means $\langle \xi^{(i)}(Q) \rangle$ and the matrix of second-order moments

$$B_{ik}(1, 2) = \langle \xi^{(i)}(1)\xi^{(k)}(2) \rangle \ , \tag{1.9}$$

or, equivalently, by the *correlation matrix*

$$\psi_{ik}(1, 2) = \langle \tilde{\xi}^{(i)}(1)\tilde{\xi}^{(k)}(2) \rangle = B_{ik}(1, 2) - \langle \xi^{(i)}(1) \rangle \langle \xi^{(k)}(2) \rangle \ . \tag{1.10}$$

Here 1 and 2 are a short-hand notation for the arguments Q_1 and Q_2. The diagonal elements of the correlation matrix are the *covariances* $\psi_{kk}(1, 2)$, and the non-diagonal elements the *mutual covariances* $\psi_{ik}(1, 2)$ of the fields $\xi^{(i)}$ and $\xi^{(k)}$.

Apart from the real random fields $\xi(Q)$ discussed above, we will also deal with the *complex* fields

$$\zeta(Q) = \xi(Q) + i\eta(Q) \ , \tag{1.11}$$

where $\xi = \mathrm{Re}\{\zeta\}$ and $\eta = \mathrm{Im}\{\zeta\}$ are real functions in the Q-space. A complex random field is completely defined statistically by the $2n$-variate probability densities $w_{2n}(\xi_1, \ldots, \xi_n; \eta_1, \ldots, \eta_n)$ $(n = 1, 2, \ldots)$, which are used to express the probabilities

$$P\{\xi_\mu \leqslant \xi(Q_\mu) < \xi_\mu + d\xi_\mu, \quad \eta_\nu \leqslant \eta(Q_\nu) < \eta_\nu + d\eta_\nu \quad (\mu, \nu = 1, 2, \ldots, n)\} \ ,$$

which is analogous to (1.1). In terms of w_{2n} we can express any moment of the complex random field ζ and, in particular, its lowest moments. These latter are of especial interest for the problems to be dealt with below, as it is only rarely that we will venture beyond the confines of the correlation theory of random fields.

1.2 Spatial Covariances of Complex Random Fields

To begin with, consider random fields that are dependent only on space coordinates, i.e., time-independent ones. We will be looking at time-dependent fields in Sect. 1.6.

Let a one-component complex random field ξ vary with $\boldsymbol{r}$ alone

$$\zeta(\boldsymbol{r}) = \xi(\boldsymbol{r}) + \mathrm{i}\eta(\boldsymbol{r}) \ .$$

By definition, the covariance of ζ is

$$\begin{aligned}\psi_\zeta(\boldsymbol{r}_1, \boldsymbol{r}_2) &\equiv \langle \tilde{\zeta}(\boldsymbol{r}_1)\tilde{\zeta}^*(\boldsymbol{r}_2)\rangle \\ &= \langle \zeta(\boldsymbol{r}_1)\zeta^*(\boldsymbol{r}_2)\rangle - \langle \zeta(\boldsymbol{r}_1)\rangle\langle \zeta^*(\boldsymbol{r}_2)\rangle \ .\end{aligned} \tag{1.12}$$

Putting here $\boldsymbol{r}_1 = \boldsymbol{r}_2 = \boldsymbol{r}$, we obtain at $\boldsymbol{r}$ the variance

$$\sigma_\zeta^2(\boldsymbol{r}) \equiv \langle |\tilde{\zeta}(\boldsymbol{r})|^2\rangle = \sigma_\xi^2 + \sigma_\eta^2 = \psi_\zeta(\boldsymbol{r}, \boldsymbol{r}) \ . \tag{1.13}$$

Many features of the spatial covariance $\psi_\zeta(\boldsymbol{r}_1, \boldsymbol{r}_2)$ have their respective counterparts in the theory of random processes (Sects. I.4.1; I.4.4, 5, and II.1.1). We will now summarize these features.

The covariance $\psi_\zeta(\boldsymbol{r}_1, \boldsymbol{r}_2)$ is *Hermitian*

$$\psi_\zeta(\boldsymbol{r}_1, \boldsymbol{r}_2) = \psi_\zeta^*(\boldsymbol{r}_2, \boldsymbol{r}_1) \ , \tag{1.14}$$

as follows from the definition (1.12). In the special case of the *real* random field ($\zeta = \zeta^*$) the covariance is *symmetrical*

$$\psi_\zeta(\boldsymbol{r}_1, \boldsymbol{r}_2) = \psi_\zeta(\boldsymbol{r}_2, \boldsymbol{r}_1) \ . \tag{1.15}$$

Further, the squared modulus of $\psi_\zeta(\boldsymbol{r}_1, \boldsymbol{r}_2)$ never exceeds the product of the variances $\sigma_\zeta^2(\boldsymbol{r}_1)$ and $\sigma_\zeta^2(\boldsymbol{r}_2)$

$$|\psi_\zeta(\boldsymbol{r}_1, \boldsymbol{r}_2)|^2 \leqslant \sigma_\zeta^2(\boldsymbol{r}_1)\sigma_\zeta^2(\boldsymbol{r}_2) = \psi_\zeta(\boldsymbol{r}_1, \boldsymbol{r}_1)\psi_\zeta(\boldsymbol{r}_2, \boldsymbol{r}_2) \ . \tag{1.16}$$

It follows from (1.16), specifically, that the *correlation coefficient* (or the *normalized covariance*) defined as

$$K_\zeta(\boldsymbol{r}_1, \boldsymbol{r}_2) = \frac{\psi_\zeta(\boldsymbol{r}_1, \boldsymbol{r}_2)}{\sigma_\zeta(\boldsymbol{r}_1)\sigma_\zeta(\boldsymbol{r}_2)} = \frac{\psi_\zeta(\boldsymbol{r}_1, \boldsymbol{r}_2)}{\sqrt{\psi_\zeta(\boldsymbol{r}_1, \boldsymbol{r}_1)\psi_\zeta(\boldsymbol{r}_2, \boldsymbol{r}_2)}} \tag{1.17}$$

has the absolute value that does not exceed unity

$$|K_\zeta(\boldsymbol{r}_1, \boldsymbol{r}_2)| \leqslant 1 \ . \tag{1.18}$$

The quantity $\langle|\int\tilde{\zeta}(\boldsymbol{r})u(\boldsymbol{r})d^3r|^2\rangle$ is nonnegative, hence the covariance $\psi_\zeta(\boldsymbol{r}_1,\boldsymbol{r}_2)$ is *positive definite*, i.e.,

$$\iint\limits_V \psi_\zeta(\boldsymbol{r}_1,\boldsymbol{r}_2)u(\boldsymbol{r}_1)u^*(\boldsymbol{r}_2)d^3r_1d^3r_2 \geqslant 0 \ , \tag{1.19}$$

where $u(\boldsymbol{r})$ is an arbitrary complex function and V is an arbitrary integration region, for which the integral (1.19) does exist.

The *mutual* covariance of two complex fields $\mu(\boldsymbol{r})$ and $v(\boldsymbol{r})$ is

$$\psi_{\mu v}(\boldsymbol{r}_1,\boldsymbol{r}_2) = \langle\tilde{\mu}(\boldsymbol{r}_1)\tilde{v}^*(\boldsymbol{r}_2)\rangle \ . \tag{1.20}$$

The squared modulus of $\psi_{\mu v}$ is not larger than the product of the variances σ_μ^2 and σ_v^2

$$|\psi_{\mu v}(\boldsymbol{r}_1,\boldsymbol{r}_2)|^2 \leqslant \sigma_\mu^2(\boldsymbol{r}_1)\sigma_v^2(\boldsymbol{r}_2) \ , \tag{1.21}$$

which is analogous to (1.16). In consequence, the *mutual correlation coefficient*

$$K_{\mu v}(\boldsymbol{r}_1,\boldsymbol{r}_2) = \frac{\psi_{\mu v}(\boldsymbol{r}_1,\boldsymbol{r}_2)}{\sigma_\mu(\boldsymbol{r}_1)\sigma_v(\boldsymbol{r}_2)} \tag{1.22}$$

has a modulus that is not larger than unity: $|K_{\mu v}| \leqslant 1$. It is of significance that, unlike the covariances, the mutual covariance $\psi_{\mu v}$ is neither Hermitian (1.14) nor positive definite (1.19). Rather, it has the property $\psi_{\mu v}^*(\boldsymbol{r}_1,\boldsymbol{r}_2) = \psi_{v\mu}(\boldsymbol{r}_2,\boldsymbol{r}_1)$.

Besides $\psi_\zeta(\boldsymbol{r}_1,\boldsymbol{r}_2)$, it is sometimes convenient to introduce one more covariance of the complex field

$$\tilde{\psi}_\zeta(\boldsymbol{r}_1,\boldsymbol{r}_2) = \langle\tilde{\zeta}(\boldsymbol{r}_1)\tilde{\zeta}(\boldsymbol{r}_2)\rangle \ . \tag{1.23}$$

Earlier (Sect. II.1.1) it was termed the *second* covariance. Unlike $\psi_\zeta(\boldsymbol{r}_1,\boldsymbol{r}_2)$ given by (1.12), the random field $\tilde{\zeta}(\boldsymbol{r}_2)$ enters into (1.22) without its complex conjugate. The second covariance is symmetrical with respect to its arguments $[\tilde{\psi}_\zeta(\boldsymbol{r}_2,\boldsymbol{r}_1) = \tilde{\psi}_\zeta(\boldsymbol{r}_1,\boldsymbol{r}_2)]$, and for a real field ζ it clearly coincides with $\psi_\zeta(\boldsymbol{r}_1,\boldsymbol{r}_2)$.

It is possible to express the covariances

$$\psi_\xi(\boldsymbol{r}_1,\boldsymbol{r}_2) = \langle\tilde{\xi}(\boldsymbol{r}_1)\tilde{\xi}(\boldsymbol{r}_2)\rangle, \quad \psi_\eta(\boldsymbol{r}_1,\boldsymbol{r}_2) = \langle\tilde{\eta}(\boldsymbol{r}_1)\tilde{\eta}(\boldsymbol{r}_2)\rangle$$

and mutual covariance

$$\psi_{\xi\eta}(\boldsymbol{r}_1,\boldsymbol{r}_2) = \langle\tilde{\xi}(\boldsymbol{r}_1)\tilde{\eta}(\boldsymbol{r}_2)\rangle = \psi_{\eta\xi}(\boldsymbol{r}_2,\boldsymbol{r}_1)$$

of the real and imaginary parts of the complex field $\zeta = \xi + \mathrm{i}\eta$ in terms of $\psi_\zeta(\boldsymbol{r}_1,\boldsymbol{r}_2)$ and $\tilde{\psi}_\zeta(\boldsymbol{r}_1,\boldsymbol{r}_2)$.

Note that, generally speaking, $\psi_{\xi\eta}$ is not symmetrical with respect to $\boldsymbol{r}_1$ and $\boldsymbol{r}_2$. Using the obvious relationships

$$\begin{aligned}\psi_\zeta(\boldsymbol{r}_1,\boldsymbol{r}_2) &= \psi_\xi(\boldsymbol{r}_1,\boldsymbol{r}_2)+\psi_\eta(\boldsymbol{r}_1,\boldsymbol{r}_2)-\mathrm{i}\psi_{\xi\eta}(\boldsymbol{r}_1,\boldsymbol{r}_2)+\mathrm{i}\psi_{\eta\xi}(\boldsymbol{r}_1,\boldsymbol{r}_2) \ , \\ \tilde{\psi}_\zeta(\boldsymbol{r}_1,\boldsymbol{r}_2) &= \psi_\xi(\boldsymbol{r}_1,\boldsymbol{r}_2)-\psi_\eta(\boldsymbol{r}_1,\boldsymbol{r}_2)+\mathrm{i}\psi_{\xi\eta}(\boldsymbol{r}_1,\boldsymbol{r}_2)+\mathrm{i}\psi_{\eta\xi}(\boldsymbol{r}_1,\boldsymbol{r}_2) \ ,\end{aligned} \tag{1.24}$$

we arrive at the following expressions for the components of the correlation matrix of the *real* fields $\xi(\boldsymbol{r})$ and $\eta(\boldsymbol{r})$:

$$\begin{aligned}
\psi_\xi(\boldsymbol{r}_1,\boldsymbol{r}_2) &= \tfrac{1}{2}\mathrm{Re}\{\psi_\zeta(\boldsymbol{r}_1,\boldsymbol{r}_2)+\tilde{\psi}_\zeta(\boldsymbol{r}_1,\boldsymbol{r}_2)\} \ ,\\
\psi_\eta(\boldsymbol{r}_1,\boldsymbol{r}_2) &= \tfrac{1}{2}\mathrm{Re}\{\psi_\zeta(\boldsymbol{r}_1,\boldsymbol{r}_2)-\tilde{\psi}_\zeta(\boldsymbol{r}_1,\boldsymbol{r}_2)\} \ ,\\
\psi_{\xi\eta}(\boldsymbol{r}_1,\boldsymbol{r}_2) &= \tfrac{1}{2}\mathrm{Im}\{\tilde{\psi}_\zeta(\boldsymbol{r}_1,\boldsymbol{r}_2)-\psi_\zeta(\boldsymbol{r}_1,\boldsymbol{r}_2)\} \ ,\\
\psi_{\eta\xi}(\boldsymbol{r}_1,\boldsymbol{r}_2) &= \tfrac{1}{2}\mathrm{Im}\{\tilde{\psi}_\zeta(\boldsymbol{r}_1,\boldsymbol{r}_2)+\psi_\zeta(\boldsymbol{r}_1,\boldsymbol{r}_2)\} \ .
\end{aligned} \tag{1.25}$$

Equations (1.25) differ from their companion expressions (II.1.3) only in that they contain the radius-vectors $\boldsymbol{r}_1$ and $\boldsymbol{r}_2$ instead of times t and t'.

In the case of the *statistically homogeneous* (in the wide sense) random field ζ, the invariance under the translation $\boldsymbol{r}\to\boldsymbol{r}+\delta\boldsymbol{r}$ only holds for the mean and the second moments. That is to say,

$$\langle\zeta(\boldsymbol{r})\rangle = \langle\zeta(\boldsymbol{r}+\delta\boldsymbol{r})\rangle \ , \tag{1.26}$$

$$\psi_\zeta(\boldsymbol{r}_1,\boldsymbol{r}_2) = \psi_\zeta(\boldsymbol{r}_1+\delta\boldsymbol{r},\boldsymbol{r}_2+\delta\boldsymbol{r}) \ . \tag{1.27}$$

Condition (1.26) means that $\langle\zeta\rangle = \mathrm{const}$. Putting $\delta\boldsymbol{r} = -\boldsymbol{r}_2$ in (1.27), we find that $\psi_\zeta(\boldsymbol{r}_1,\boldsymbol{r}_2) = \psi_\zeta(\boldsymbol{r}_1-\boldsymbol{r}_2, 0)$, i.e., the covariance of a statistically homogeneous field is dependent only on the *difference* $\boldsymbol{r}_1-\boldsymbol{r}_2$, and not on $\boldsymbol{r}_1$ and $\boldsymbol{r}_2$ separately. Instead of $\psi_\zeta(\boldsymbol{r}_1-\boldsymbol{r}_2, 0)$ we will write simply $\psi_\zeta(\boldsymbol{r})$

$$\psi_\zeta(\boldsymbol{r}) = \langle\tilde{\zeta}(\boldsymbol{r}_1)\tilde{\zeta}^*(\boldsymbol{r}_1-\boldsymbol{r})\rangle \ . \tag{1.28}$$

The second covariance $\tilde{\psi}_\zeta(\boldsymbol{r})$ is written in much the same way (only without the complex conjugate).

The variance of a statistically homogeneous field is constant, i.e.,

$$\sigma_\zeta^2 \equiv \langle|\tilde{\zeta}(\boldsymbol{r})|^2\rangle = \psi_\zeta(0) \ , \tag{1.29}$$

and the general properties (1.14), (1.16) and (1.19) of the covariance become

$$\psi_\zeta(\boldsymbol{r}) = \psi_\zeta^*(-\boldsymbol{r}) \ , \tag{1.30}$$

$$|\psi_\zeta(\boldsymbol{r})| \leqslant \sigma_\zeta^2 \ , \tag{1.31}$$

$$\int_V d^3r_1 \int_V d^3r_2\, \psi_\zeta(\boldsymbol{r}_1-\boldsymbol{r}_2)u(\boldsymbol{r}_1)u^*(\boldsymbol{r}_2) \geqslant 0 \ . \tag{1.32}$$

They are similar in form to respective properties of stationary random processes. According to (1.30),

$$\mathrm{Re}\{\psi_\zeta(-\boldsymbol{r})\} = \mathrm{Re}\{\psi_\zeta(\boldsymbol{r})\}, \quad \mathrm{Im}\{\psi_\zeta(-\boldsymbol{r})\} = -\mathrm{Im}\{\psi_\zeta(\boldsymbol{r})\} \ ,$$

i.e., the real part of ψ_ζ is even in $\boldsymbol{r}$, and the imaginary part is odd. If $\psi_\zeta(\boldsymbol{r})$ is even, then it is automatically real.

The second covariance, being symmetrical with respect to its arguments, is even for homogeneous fields: $\tilde{\psi}_\zeta(-\boldsymbol{r}) = \tilde{\psi}_\zeta(\boldsymbol{r})$. For the particular case where not only the second, but also the first covariances are even, expressions (1.25) take the form

$$\psi_\xi(\boldsymbol{r}) = [\psi_\zeta(\boldsymbol{r}) + \mathrm{Re}\{\tilde{\psi}_\zeta(\boldsymbol{r})\}]/2, \quad \psi_\eta(\boldsymbol{r}) = [\psi_\zeta(\boldsymbol{r}) - \mathrm{Re}\{\tilde{\psi}_\zeta(\boldsymbol{r})\}]/2 \ ,$$
$$\psi_{\xi\eta}(\boldsymbol{r}) = \psi_{\eta\xi}(\boldsymbol{r}) = \mathrm{Im}\{\tilde{\psi}_\zeta(\boldsymbol{r})\}/2 \ . \tag{1.33}$$

It follows that the real and imaginary parts ξ and η of the complex statistically homogeneous field $\zeta = \xi + i\eta$ are statistically homogeneous and, moreover, homogeneously connected.

Statistically homogeneous fields, with $\psi_\zeta(\boldsymbol{r})$ and $\tilde{\psi}_\zeta(\boldsymbol{r})$ dependent *only on the magnitude* (but *not on the direction*) of the vector $\boldsymbol{r} = \boldsymbol{r}_1 - \boldsymbol{r}_2$ connecting the points $\boldsymbol{r}_1$ and $\boldsymbol{r}_2$

$$\psi_\zeta(\boldsymbol{r}) = \psi_\zeta(r), \quad \tilde{\psi}_\zeta(\boldsymbol{r}) = \tilde{\psi}_\zeta(r) \ , \tag{1.34}$$

where $r = |\boldsymbol{r}| = \sqrt{x^2 + y^2 + z^2}$, are said to be *statistically isotropic*. It is impossible to indicate a complete analog for isotropic fields in the theory of random processes. A limited parallel, however, can be drawn with stationary real processes whose covariance is even: $\psi(\tau) = \psi(|\tau|)$.

The covariance of isotropic random fields is always even and hence always real. It follows from (1.33) that if the field ζ is isotropic, then the fields $\xi = \mathrm{Re}\{\zeta\}$ and $\eta = \mathrm{Im}\{\zeta\}$ are isotropic and isotropically interconnected.

Examples of the covariance of a homogeneous isotropic random field ζ are the covariance in the form of the Gaussian curve

$$\psi_\zeta(r) = \sigma_\zeta^2 \mathrm{e}^{-r^2/2l^2} \ , \tag{1.35}$$

and the exponential covariance

$$\psi_\xi(r) = \sigma_\zeta^2 \mathrm{e}^{-r/l} \ . \tag{1.36}$$

In both cases l is the *correlation radius* of the field, i.e., the distance $r = |\boldsymbol{r}_1 - \boldsymbol{r}_2|$, over which $\psi_\zeta(\boldsymbol{r})$ is about halved as compared with the variance σ_ζ^2. For fields with a covariance of arbitrary form, the effective (or integral) correlation radius is normally given by

$$l_{\mathrm{ef}} = \frac{1}{\sigma_\zeta^2} \int_0^\infty \psi_\zeta(r) dr = \int_0^\infty K_\zeta(r) dr \ . \tag{1.37}$$

For the Gaussian covariance (1.35): $l_{\mathrm{ef}} = l(\pi/2)^{-1/2}$, and for the exponential covariance (1.36) $l_{\mathrm{ef}} = l$.

Some covariances have a certain space scale, but no effective correlation radius, since the integral in (1.37) diverges. This is the case, for instance, with the covariance of the form

$$\psi_\zeta(r) = \sigma_\zeta^2 (1 + r^2/l^2)^{-\mu}, \quad \mu < \tfrac{1}{2} \ .$$

Although the notion of the correlation radius (like the correlation time for random processes) is often useful, it does not permit of a universal definition that might be suitable for arbitrary fluctuations. Not infrequently, covariances have several characteristic space scales. Examples include a fast oscillating covariance with a slowly varying envelope, and the covariances of anisotropic fields to be discussed below.

The covariances of statistically homogeneous, but *anisotropic*, fields vary not only with the magnitude but also with the direction of $\boldsymbol{r} = \boldsymbol{r}_1 - \boldsymbol{r}_2$, as in the following two examples:

$$\psi_\zeta(\boldsymbol{r}) = \psi_\zeta[(\alpha x + \beta y + \gamma z)/l] \ , \tag{1.38}$$

$$\psi_\zeta(\boldsymbol{r}) = \psi_\zeta\left(\frac{|x|}{a}, \frac{|y|}{b}, \frac{|z|}{c}\right) . \tag{1.39}$$

It is quite obvious that anisotropic fields have no counterparts in the theory of random processes. Distances at which the values of the anisotropic random field become uncorrelated vary with the direction. So, in the anisotropic Gaussian covariance

$$\psi_\zeta(\boldsymbol{r}) = \sigma_\zeta^2 \exp\left[-\frac{1}{2}\left(\frac{x^2}{a^2} + \frac{y^2}{b^2} + \frac{z^2}{c^2}\right)\right] , \tag{1.40}$$

which is a special case of (1.39), a, b, and c characterize the scale of the spatial correlation in the x, y, and z-directions. The scale l for covariances of the form (1.38) characterizes the correlation radius in the direction perpendicular to the plane $\alpha x + \beta y + \gamma z = 0$, whereas in the plane itself (and those parallel to it) correlation occurs up to infinity.

We sometimes speak of the *anisomerism* of random fields rather than of statistical anisotropy. The term "anisomeric fluctuations" is convenient in some problems of electrodynamics and elasticity dealing with statistically *anisotropic fluctuations* of parameters in *anisotropic media*, but we will make little use of this term.

Along with moments of the first and second order, we can also introduce higher-order moments of the complex random field ζ. The $(m+n)$-point moment $B_{m,n}$, or the moment of the $(m+n)$th order, is normally defined as follows:

$$B_{m,n}(1, 2, \ldots, m+n) = \langle \zeta(1) \ldots \zeta(m)\zeta^*(m+1) \ldots \zeta^*(m+n)\rangle \ . \tag{1.41}$$

The field ζ appears here m times, and its complex conjugate ζ^*, n times. To simplify notation, instead of $\boldsymbol{r}_j$ we again write j alone. So we write, say, $\zeta(m)$, instead of $\zeta(\boldsymbol{r}_m)$, and so forth. The second moments B and $\tilde{B}$ thus become

$$B(1,2) = B_{1,1}(1,2), \quad \tilde{B}(1,2) = B_{2,0}(1,2) \ . \tag{1.42}$$

Some simple properties of $B_{m,n}$ follow directly from the definition. It is thus obvious that $B_{n,n}(1, \ldots, n; n, \ldots, 1) \geqslant 0$. Permutations within the groups of arguments $(1, \ldots, m)$ and $(m+1, \ldots, m+n)$ do not change $B_{m,n}$, and complex conjugations bring about the following changes in the arrangement of the arguments and subscripts:

$$[B_{m,n}(1, \ldots, m+n)]^* = B_{n,m}(m+n, \ldots, 1) \ . \tag{1.43}$$

This identity holds for the special cases of Hermitian B and symmetric $\tilde{B}$.

1.3 Spatial Spectral Representations of Homogeneous Random Fields

Let us formally represent the fluctuational component of the univariate random field $\tilde{\zeta}(\boldsymbol{r})$ as the three-dimensional Fourier integral

$$\tilde{\zeta}(\boldsymbol{r}) = \int_{-\infty}^{\infty} \zeta(\boldsymbol{k}) e^{i\boldsymbol{k}\cdot\boldsymbol{r}} d^3k \ . \tag{1.44}$$

Here $\zeta(\boldsymbol{k})$ is the spatial spectral amplitude (or $\boldsymbol{k}$-amplitude, for short) of ζ that is given in terms of $\tilde{\zeta}(\boldsymbol{r})$ using the inverse Fourier transform

$$\zeta(\boldsymbol{k}) = \frac{1}{(2\pi)^3} \int_{-\infty}^{\infty} \tilde{\zeta}(\boldsymbol{r}) e^{-i\boldsymbol{k}\cdot\boldsymbol{r}} d^3k \ . \tag{1.45}$$

The shorter notation $\zeta_{\boldsymbol{k}}$ is sometimes used instead of $\zeta(\boldsymbol{k})$.

We referred to the spectral expansion (1.44) as formal because, for a non-decreasing (at infinity) function such as the *homogeneous* random field $\tilde{\zeta}(\boldsymbol{r})$, the Fourier transform $\zeta(\tilde{\boldsymbol{k}})$ does not exist. For the spectral representations of such random fields to have their conventional mathematical meaning, they should be supposed to be homogeneous in a large, but finite volume V outside which they vanish fast enough. The transition to infinite V should only be made at the final stage, i.e., after the statistical averaging. Here we will not, however, resort to such procedures (as in a similar situation with spectral representations of stationary random processes), but will instead assume that the integrals (1.44, 45) exist in the sense of probabilistic convergence, namely in the mean square (Sect. II.1.3).

From (1.44) the covariance of the one-component random field $\psi_\zeta(\boldsymbol{r}_1 - \boldsymbol{r}_2)$ is

$$\psi_\zeta(\boldsymbol{r}_1 - \boldsymbol{r}_2) = \iint_{-\infty}^{\infty} \langle \zeta(\boldsymbol{k}')\zeta^*(\boldsymbol{k}'')\rangle e^{i(\boldsymbol{k}'\boldsymbol{r}_1 - \boldsymbol{k}''\boldsymbol{r}_2)} d^3k'\, d^3k'' \ . \tag{1.46}$$

The spectral amplitudes $\zeta(\boldsymbol{k}')$ and $\zeta^*(\boldsymbol{k}'')$ appear to be delta-correlated in $\boldsymbol{k}$. In fact, by (1.45),

$$\langle \zeta(\boldsymbol{k}')\zeta^*(\boldsymbol{k}'')\rangle = \frac{1}{(2\pi)^6} \iint_{-\infty}^{\infty} \langle \tilde{\zeta}(\boldsymbol{r}_1)\tilde{\zeta}^*(\boldsymbol{r}_2)\rangle e^{-i(\boldsymbol{k}'\cdot\boldsymbol{r}_1 - \boldsymbol{k}''\cdot\boldsymbol{r}_2)} d^3r_1\, d^3r_2 \ .$$

We here pass to new variables of integration $\boldsymbol{r} = \boldsymbol{r}_1 - \boldsymbol{r}_2$ and $\boldsymbol{R} = (\boldsymbol{r}_1 + \boldsymbol{r}_2)/2$, for which $d^3r_1\, d^3r_2 = d^3r d^3R$. If we consider that

$$\langle \tilde{\zeta}(\boldsymbol{r}_1)\tilde{\zeta}(\boldsymbol{r}_2)\rangle = \psi_\zeta(\boldsymbol{r}_1 - \boldsymbol{r}_2) = \psi_\zeta(\boldsymbol{r}) \quad \text{and}$$

$$\boldsymbol{k}'\cdot\boldsymbol{r}_1 - \boldsymbol{k}''\cdot\boldsymbol{r}_2 = (\boldsymbol{k}' - \boldsymbol{k}'')\cdot\boldsymbol{R} + (\boldsymbol{k}' + \boldsymbol{k}'')\cdot\boldsymbol{r}/2 \ ,$$

and integrate with respect to $\boldsymbol{R}$, we will get

$$\langle \zeta(\boldsymbol{k}')\zeta^*(\boldsymbol{k}'')\rangle = \Phi_\zeta(\boldsymbol{k}')\delta(\boldsymbol{k}' - \boldsymbol{k}'') \ , \quad \text{where} \tag{1.47}$$

$$\Phi_\zeta(\boldsymbol{k}) = \frac{1}{(2\pi)^3} \int_{-\infty}^{\infty} \psi_\zeta(\boldsymbol{r}) e^{-i\boldsymbol{k}\cdot\boldsymbol{r}} d^3r \ . \tag{1.48}$$

Substituting (1.47) into (1.46), we arrive at

$$\psi_\zeta(\boldsymbol{r}) = \int_{-\infty}^{\infty} \Phi_\zeta(\boldsymbol{k}) \mathrm{e}^{\mathrm{i}\boldsymbol{k}\cdot\boldsymbol{r}} d^3k \ . \tag{1.49}$$

The function $\Phi_\zeta(\boldsymbol{k})$ is termed the *spatial spectral density* (***k**-density*) or, for short, the *spatial spectrum* of the random homogeneous field $\zeta(\boldsymbol{r})$. It is analogous to the spectral density (ω-density) $g(\omega)$ in the theory of stationary random processes. Equation (1.48) is a generalization of the Wiener–Khintchine theorem (Sect. II.1.4) to random fields. It will be noticed that it is sometimes convenient to utilize one- and two-dimensional as well as three-dimensional spectral representations (see Exercises 1.2–4). It is easily seen that the spatial spectrum $\Phi_\zeta(\boldsymbol{k})$ is a *real* and *nonnegative* quantity. That it is real follows from (1.30), and nonnegative from the fact that the covariance is positive definite.

Unlike $\Phi_\zeta(\boldsymbol{k})$, in the general case the *mutual* spatial spectrum

$$\Phi_{\mu\nu}(\boldsymbol{k}) = \frac{1}{(2\pi)^3} \int_{-\infty}^{\infty} \psi_{\mu\nu}(\boldsymbol{r}) \mathrm{e}^{-\mathrm{i}\boldsymbol{k}\cdot\boldsymbol{r}} d^3r \tag{1.50}$$

corresponding to the mutual covariance $\psi_{\mu\nu}(\boldsymbol{r})$ of two complex fields $\mu(\boldsymbol{r})$ and $\nu(\boldsymbol{r})$ is a complex function.

For even covariances $\psi_\zeta(-\boldsymbol{r}) = \psi_\zeta(\boldsymbol{r})$, in particular for the covariance of a real or isotropic field, the spectral density is even as well: $\Phi_\zeta(-\boldsymbol{k}) = \Phi_\zeta(\boldsymbol{k})$. In that case, $\psi_\zeta(\boldsymbol{r})$ and $\Phi_\zeta(\boldsymbol{k})$ are related by the Fourier cosine transform

$$\psi_\zeta(\boldsymbol{r}) = \int_{-\infty}^{\infty} \Phi_\zeta(\boldsymbol{k}) \cos \boldsymbol{k}\cdot\boldsymbol{r}\, d^3k \ , \tag{1.51}$$

$$\Phi_\zeta(\boldsymbol{k}) = \frac{1}{(2\pi)^3} \int_{-\infty}^{\infty} \psi_\zeta(\boldsymbol{r}) \cos \boldsymbol{k}\cdot\boldsymbol{r}\, d^3r \ . \tag{1.52}$$

In the special case of an *isotropic* homogeneous random field with $\psi_\zeta(\boldsymbol{r}) = \psi_\zeta(r)$, the spatial spectral density is only dependent on the *magnitude* of $\boldsymbol{k}$. In order to see this, we pass in (1.18) to the polar coordinates (r, θ, φ) with the polar axis directed along $\boldsymbol{k}$. Then, $\boldsymbol{k}\cdot\boldsymbol{r} = kr\cos\theta$, and we get

$$\begin{aligned} \Phi_\zeta(\boldsymbol{k}) &= \frac{1}{(2\pi)^3} \int_0^{2\pi} d\varphi \int_0^{\pi} \sin\theta\, d\theta \int_0^{\infty} \psi_\zeta(r) \mathrm{e}^{-\mathrm{i}kr\cos\theta} r^2\, dr \\ &= \frac{1}{2\pi^2 k} \int_0^{\infty} \sin kr\, \psi_\zeta(r)\, r\, dr = \Phi_\zeta(k) \ . \end{aligned} \tag{1.53}$$

The fact that Φ_ζ only varies with k enables (1.49) and (1.51) to be written in a simpler form, namely in terms of a single integral with respect to k. Introducing spherical coordinates in the space of wave vectors $\boldsymbol{k}$ (with the polar axis directed along $\boldsymbol{r}$) and integrating with respect to the angle variables, we get

$$\psi_\zeta(r) = \frac{4\pi}{r} \int_0^{\infty} \sin kr\, \Phi_\zeta(k)\, k\, dk \ . \tag{1.54}$$

By way of example we will furnish the value of the spatial spectrum $\Phi_\zeta(k)$ of a homogeneous and isotropic random field with the Gaussian covariance (1.35)

$$\Phi_\zeta(k) = (2\pi)^{-3/2} \sigma_\zeta^2 l^3 e^{-1/2k^2l^2} , \tag{1.55}$$

and the exponential covariance (1.36)

$$\Phi_\zeta(k) = \frac{\sigma_\zeta^2 l^3}{\pi(1+k^2l^2)^2} . \tag{1.56}$$

These expressions are obtainable using any of the equations (1.48), (1.52), or (1.53), the last being the simplest.

It is seen from these examples that the "uncertainty" (width) of the spectrum Δk is inversely proportional to the correlation radius of the random field: $\Delta k \sim 1/l$. But l characterizes the "uncertainty" (width) Δr of the covariance, which differs markedly from zero only for $r \lesssim \Delta r \sim l$. As a result, for isotropic random fields we may formulate the *uncertainty* (i.e., width) *relation*

$$\Delta k \Delta r \geqslant 1 , \tag{1.57}$$

which is an analog of the relation $\Delta\omega\Delta t \gtrsim 1$ for random processes (Sect. II.2.2). According to (1.57), short-correlated fields correspond to wide spatial spectra (Δk is large), whereas when correlation radii are large the spectrum is narrow.

With anisotropic fields the uncertainties (correlation radii) in various directions are dissimilar, and instead of one inequality (1.57) we now have three inequalities:

$$\Delta k_x \Delta r_x \gtrsim 1, \quad \Delta k_y \Delta r_y \gtrsim 1, \quad \Delta k_z \Delta r_z \gtrsim 1 . \tag{1.58}$$

These relations can be illustrated by the spatial spectrum corresponding to the Gaussian covariance (1.40)

$$\Phi_\zeta(\boldsymbol{k}) = \frac{\sigma_\zeta^2 abc}{(2\pi)^{3/2}} \exp\left(-\frac{1}{2}(a^2k_x^2 + b^2k_y^2 + c^2k_z^2)\right) . \tag{1.59}$$

The widths of the spectrum along the k_x, k_y, k_z axes are inversely proportional to a, b, and c, respectively, i.e., to the space scale of the covariance (1.40) along the x, y, and z-axes.

For $\psi_\zeta(\boldsymbol{r})$, i.e., the integral on the right side of (1.51) and (1.54), to exist, the spectral density $\Phi_\zeta(\boldsymbol{k})$ must decrease with k faster than k^{-3}. On the other hand, as $k \to 0$, $\Phi_\zeta(\boldsymbol{k})$ may have a power singularity of the form $k^{-\alpha}$ with $\alpha < 3$.

The existence condition for a covariance is closely related to the requirements to be fulfilled by the spectrum of the *differentiable* random fields. As in the theory of random processes (Sect. I.4.6), we can see that the necessary and sufficient condition for the existence (in the mean square, of course) of the first spatial derivatives, e.g., $\nabla\zeta(\boldsymbol{r})$, reduces to the existence of the quantity

$$\langle \nabla\zeta(\boldsymbol{r}_1) \, \nabla\zeta^*(\boldsymbol{r}_2) \rangle = -\Delta\psi_\zeta(\boldsymbol{r}_1 - \boldsymbol{r}_2) . \tag{1.60}$$

Substituting here the spectral expansion (1.49) and setting $\boldsymbol{r}_1 = \boldsymbol{r}_2 = \boldsymbol{r}$, we obtain the existence condition for $\nabla\zeta$ in the form

$$\langle|\nabla\zeta(\boldsymbol{r})|^2\rangle = -\Delta\psi_\zeta(0) = \int_{-\infty}^{\infty} k^2\Phi_\zeta(\boldsymbol{k})\,d^3k < \infty \ . \tag{1.61}$$

As $k \to 0$, this condition permits of power singularities of the form $\Phi_\zeta(\boldsymbol{k}) \sim k^{-\alpha}$ of a higher order than is required for $\psi_\zeta(\boldsymbol{r})$ to exist: (1.61) converges if $\alpha < 5$, whereas (1.49) converges only if $\alpha < 3$. At the same time, (1.61) is a more stringent requirement on the rate at which $\Phi_\zeta(k)$ decreases as $k \to \infty$. It is necessary that $\Phi_\zeta(\boldsymbol{k})$ fall off with k as $k^{-\mu}$ for $\mu > 5$ [for $\psi_\zeta(\boldsymbol{r})$ to exist requires only that $\mu > 3$]. The latter condition is not satisfied, for example, by a random field with the exponential covariance (1.36), since its spectral density (1.56) decreases at infinity insufficiently fast (as k^{-4}). Consequently, the field is not differentiable.

Yet more rigid requirements are imposed on $\Phi_\zeta(\boldsymbol{k})$ when $k \to \infty$ by the existence of nth derivatives of the random field. It follows from the condition $\langle|\nabla^n\zeta(r)|^2\rangle < \infty$ that the following inequality must be obeyed:

$$\int k^{2n}\Phi_\zeta(\boldsymbol{k})\,d^3k < \infty \ , \tag{1.62}$$

so that with increasing k the spectrum $\Phi_\zeta(\boldsymbol{k})$ must decay faster than k^{-3-2n}. Clearly, $\Phi_\zeta(\boldsymbol{k})$ of an infinitely differentiable field should decrease as $k \to \infty$ faster than any negative power of k, e.g., in an exponential manner. An example is a random field with the Gaussian covariance (1.35).

But an example of a spectrum that decays slower than an exponential, but faster than $1/k^m$, is the function

$$\Phi_\zeta(k) = C\exp[-(\ln k)^2] = Ck^{-\ln k}$$

for which at any finite m

$$\int_0^\infty k^m\Phi_\zeta(k)\,dk = \sqrt{\pi}\exp\left[\frac{(m+1)^2}{4}\right] < \infty \ .$$

1.4 Locally Homogeneous Random Fields

A complex random field $\zeta(\boldsymbol{r})$ can be characterized not only by the covariance (1.12), but also by the so-called *structure function*, which is the mean square of the modulus of the *increment* of the fluctuational component $\tilde{\zeta}(\boldsymbol{r})$ of $\zeta(\boldsymbol{r})$

$$D_\zeta(\boldsymbol{r}_1, \boldsymbol{r}_2) = \langle|\tilde{\zeta}(\boldsymbol{r}_1) - \tilde{\zeta}(\boldsymbol{r}_2)|^2\rangle \ . \tag{1.63}$$

Clearly, at $\boldsymbol{r}_1 = \boldsymbol{r}_2$ the structure function vanishes, i.e., $D_\zeta(\boldsymbol{r}_1, \boldsymbol{r}_1) = 0$.

If $D_\zeta(\boldsymbol{r}_1, \boldsymbol{r}_2)$ and the increment of the mean field $f_\zeta(\boldsymbol{r}_1, \boldsymbol{r}_2) = \langle\zeta(\boldsymbol{r}_1)\rangle - \langle\zeta(\boldsymbol{r}_2)\rangle$ vary with the difference $\boldsymbol{r}_1 - \boldsymbol{r}_2$ only, i.e.,

$$D_\zeta(\boldsymbol{r}_1, \boldsymbol{r}_2) = D_\zeta(\boldsymbol{r}_1 - \boldsymbol{r}_2), \quad f_\zeta(\boldsymbol{r}_1, \boldsymbol{r}_2) = f_\zeta(\boldsymbol{r}_1 - \boldsymbol{r}_2) \ , \tag{1.64}$$

then such fields are called *locally homogeneous*. The notion of locally homogeneous fields was introduced by *Kolmogorov* [1.4, 5]. The local homogeneity conditions impose certain constraints on the moments of the *increments* of $\zeta(r)$, i.e., the *differences* of values of the field at two points $\boldsymbol{r}_1$ and $\boldsymbol{r}_2$, and not on the moments of these individual values $\zeta(\boldsymbol{r}_1)$ and $\zeta(\boldsymbol{r}_2)$. By virtue of the complete analogy of these fields to random processes with stationary (first) increments (Sect. II.4.2), we might also refer to a locally homogeneous field as a *random field with homogeneous increments*, but this name has not been taken up in the literature.

The requirement that $f_\zeta(\boldsymbol{r}_1, \boldsymbol{r}_2) = \langle\zeta(\boldsymbol{r}_1)\rangle - \langle\zeta(\boldsymbol{r}_2)\rangle$ be homogeneous can only be met if the mean field varies linearly with $\boldsymbol{r}$: $\langle\zeta(\boldsymbol{r})\rangle = \langle\boldsymbol{a}\rangle\boldsymbol{r} + \text{const}$, where $\boldsymbol{a}$ is an arbitrary vector that is not necessarily random. In fact, for $\langle\zeta(\boldsymbol{r})\rangle$ varying linearly with $\boldsymbol{r}$ we have

$$f_\zeta(\boldsymbol{r}_1, \boldsymbol{r}_2) = \langle\boldsymbol{a}\rangle\cdot(\boldsymbol{r}_1 - \boldsymbol{r}_2) = f_\zeta(\boldsymbol{r}_1 - \boldsymbol{r}_2) \ .$$

If the field has the covariance $\psi_\zeta(\boldsymbol{r}_1, \boldsymbol{r}_2) = \langle\tilde{\zeta}(\boldsymbol{r}_1)\,\tilde{\zeta}^*(\boldsymbol{r}_2)\rangle$, then, by (1.63),

$$D_\zeta(\boldsymbol{r}_1, \boldsymbol{r}_2) = \psi_\zeta(\boldsymbol{r}_1, \boldsymbol{r}_1) + \psi_\zeta(\boldsymbol{r}_2, \boldsymbol{r}_2) - \psi_\zeta(\boldsymbol{r}_1, \boldsymbol{r}_2) - \psi_\zeta^*(\boldsymbol{r}_1, \boldsymbol{r}_2) \ . \tag{1.65}$$

Specifically, for the homogeneous field $\zeta(r)$ we have from (1.65)

$$D_\zeta(\boldsymbol{r}) = 2[\psi_\zeta(0) - \psi_\zeta(\boldsymbol{r})] \ . \tag{1.66}$$

An important advantage of the structure function is its ability to retain its validity in the general case where the covariance does not exist. The situation here is as with random processes (stationary processes are a particular case of processes with stationary increments). The analogy becomes still more complete in the case of *locally homogeneous isotropic fields*. These are random fields whose $D_\zeta(\boldsymbol{r})$ and $f_\zeta(\boldsymbol{r})$ vary only with the magnitude of $\boldsymbol{r} = \boldsymbol{r}_1 - \boldsymbol{r}_2$, i.e.,

$$D_\zeta(\boldsymbol{r}) = D_\zeta(r), \quad f_\zeta(\boldsymbol{r}) = f_\zeta(r) \ . \tag{1.67}$$

The second of these conditions is met if $\langle\boldsymbol{a}\rangle = 0$, i.e., if $\boldsymbol{a}$ is isotropically distributed in space. In applications, in particular to the theory of atmospheric turbulence, $\boldsymbol{a}$ is often taken to be zero. In any case, the mean of the locally homogeneous isotropic field is constant: $\langle\zeta\rangle = \text{const}$. For the homogeneous isotropic random field, (1.66) takes the form

$$D_\zeta(r) = 2[\psi_\zeta(0) - \psi_\zeta(r)] \ , \tag{1.68}$$

where $D_\zeta(r)$ and $\psi_\zeta(r)$ are real. If $\psi_\zeta(r)$ vanishes as $r \to \infty$, then $D_\zeta(\infty) = 2\psi_\zeta(0)$, which enables $\psi_\zeta(r)$ to be expressed through $D_\zeta(r)$

$$\psi_\zeta(r) = [D_\zeta(\infty) - D_\zeta(r)]/2 \ . \tag{1.69}$$

Significantly, the structure function is not affected by spatially extended fluctuations of ζ, i.e., by those with the characteristic dimension $l \gg r = |\boldsymbol{r}_1 - \boldsymbol{r}_2|$. Since the perturbations due to such fluctuations are essentially the same at points $\boldsymbol{r}_1$ and $\boldsymbol{r}_2$, the difference $\tilde{\zeta}(\boldsymbol{r}_1) - \tilde{\zeta}(\boldsymbol{r}_2)$, and hence their contribution to $D_\zeta(r)$, is small. But the covariance accounts for fluctuations of any scale to the same extent. It is for this reason that the use of the structure functions instead of the covariance appears to be physically justified in those cases where large-scale fluctuations of the field do not influence the phenomena of interest. This by no means implies that such fluctuations do not exist. On the contrary, their contribution to the resulting fluctuation perturbations may even be large, but for the phenomena observed they are immaterial.

An example is the statistical theory of developed turbulence [1.6,7], i.e., turbulent fluid motion with eddies that vary widely in size l. The most interesting (from the view-point of their inherent regularities) are eddies whose dimensions are markedly smaller than those of the largest ones (L_0, the so-called *outer scale of turbulence*) produced by the primary flow, and still strongly influenced by its geometry and kinematics. It is for the subrange $l \ll L_0$ that Kolmogorov introduced the notion of the local homogeneity of the random field. He also suggested that the field can be described statistically by a function D_ζ that simply excludes large-scale inhomogeneities from consideration.

Thus, the homogeneity assumption for D_ζ is less restrictive than a similar assumption for ψ_ζ (homogeneous fields), in that it allows a wider class of random fields (locally homogeneous fields) to be covered.

Let us now look at the spatial spectral representations of locally homogeneous fields. By analogy with the spectral representations of the processes with stationary random increments (Sect. II.4.2) we have

$$D_\zeta(\boldsymbol{r}) = 2 \int_{-\infty}^{\infty} \Phi_\zeta(\boldsymbol{k})(1-\cos \boldsymbol{k}\cdot\boldsymbol{r})\, d^3k \; . \tag{1.70}$$

The integral (1.70) converges as $k \to 0$ if the spectrum $\Phi_\zeta(k)$ has a power singularity of the form $k^{-\alpha}$ where $\alpha < 5$, whereas $\psi_\zeta(r)$ only exists for $\alpha < 3$.

For locally homogeneous fields the inversion formula for (1.70) is not as readily derived as for homogeneous fields. It is necessary first to differentiate (1.70) with respect to $\boldsymbol{r}$ and then to carry out the inverse Fourier transformation. We thus obtain

$$\Phi_\zeta(\boldsymbol{k}) = \frac{1}{16\pi^3 k^2} \int_{-\infty}^{\infty} \boldsymbol{k} \cdot \nabla D_\zeta(\boldsymbol{r}) \sin \boldsymbol{k}\cdot\boldsymbol{r}\, d^3r \; . \tag{1.71}$$

For the isotropic field we pass in (1.70, 71) to spherical coordinates and take into account the property $\boldsymbol{k}\, \nabla D_\zeta(r) = \boldsymbol{k}\boldsymbol{r} D'(r)/r$ (the prime signifies a derivative with respect to r)

$$D_\zeta(r) = 8\pi \int_0^{\infty} \left(1 - \frac{\sin kr}{kr}\right) \Phi_\zeta(k)\, k^2\, dk \; , \tag{1.72}$$

$$\Phi_\zeta(k) = \frac{1}{4\pi^2 k^3} \int_0^{\infty} (\sin kr - kr \cos kr)\, D_\zeta'(r)\, dr \; . \tag{1.73}$$

As an example of a locally homogeneous and isotropic field consider the spatial fluctuations of the electric permittivity ε in a turbulent atmosphere. The structure function of these fluctuations

$$D_\varepsilon(r) = \langle |\tilde{\varepsilon}(\boldsymbol{r}_1) - \tilde{\varepsilon}(\boldsymbol{r}_2)|^2 \rangle$$

at sufficiently large r obeys the "two-thirds law" due to *Kolmogorov* and *Obukhov* (see for example [1.6])

$$D_\varepsilon(r) \approx C_\varepsilon^2 r^{2/3}, \quad r \gg l_0 , \tag{1.74}$$

where C is the so-called *structure constant*, and l_0 is the *inner scale of turbulence*[4]. But for small r, i.e., for $r \ll l_0$, the structure function varies as the square of r:

$$D_\varepsilon(r) \approx C_\varepsilon^2 l_0^{-4/3} r^2, \quad r \ll l_0 . \tag{1.75}$$

The inner scale l_0 enters into the equations (1.74, 75) in such a manner that the values of both asymptotic expressions are equal at $r = l_0$. The corresponding curves are given in Fig. 1.1 by broken lines, whereas the real behavior of the structure function $D_\varepsilon(r)$ is shown by a continuous line.

A spectrum fitting the real behavior of $D_\varepsilon(r)$ can be selected from the following considerations. For the power structure function

$$D(r) = C^2 r^\mu, \quad 0 < \mu < 2 , \tag{1.76}$$

the spatial spectral density is [1.6]

$$\Phi(k) = \frac{\Gamma(\mu+2)}{4\pi^2} \sin\left(\frac{\pi\mu}{2}\right) C^2 k^{-(\mu+3)} , \tag{1.77}$$

so that to the "two-thirds law" ($\mu = 2/3$) corresponds the spectrum

$$\Phi_\varepsilon(k) = \frac{\Gamma(\frac{8}{3})\sin \pi/3}{4\pi^2} C_\varepsilon^2 k^{-11/3} = 0.033\, C_\varepsilon^2 k^{-11/3} . \tag{1.78}$$

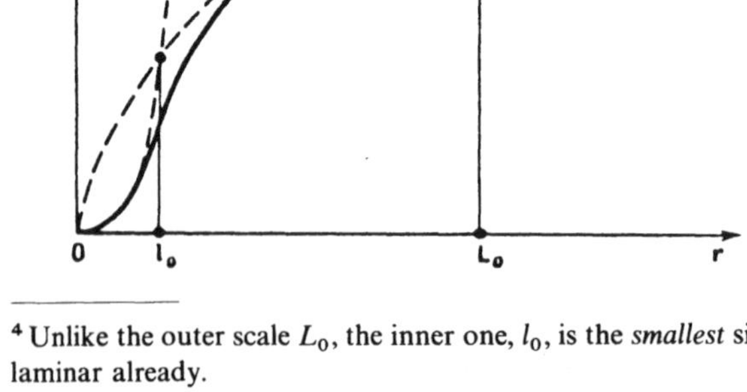

Fig. 1.1. The structure function of the fluctuations of the electric permittivity of a turbulent atmosphere (*solid line*) and its asymptotic behavior at short distances (quadratic dependence), in the intermediate region (two-thirds law), and at infinity

[4] Unlike the outer scale L_0, the inner one, l_0, is the *smallest* size of eddies. Perturbations with $l < l_0$ are laminar already.

To obtain the square-law behavior of $D_\varepsilon(r)$ at small $r \ll l_0$ we should suppress the spectral density (1.78) at large $k \gg 1/l_0$. Physically, this operation reflects the dissipation of eddies due to viscosity, when their scale becomes small $(1/k \ll l_0)$. The density (1.77) can be suppressed at large k, say, by inserting the factor $\exp(-k^2/\kappa_m^2)$, i.e., by letting

$$\Phi_\varepsilon(k) = 0.033 C_\varepsilon^2 k^{-11/3} \mathrm{e}^{-k^2/\kappa_m^2} \; . \tag{1.79}$$

The resulting structure function $D_\varepsilon(r)$ will exhibit the desired asymptotic behavior (1.74, 75) if the cut-off parameter κ_m is taken to be equal to $5.92/l_0$. The function adequately describes a turbulent field at values of r that do not exceed the outer scale of turbulence L_0. But in fact, for $r \gg L_0$ the structure function "saturates" (see Fig. 1.1) and tends to a finite value that is conveniently written as

$$D_\varepsilon(\infty) = C_\varepsilon^2 L_0^{2/3} \; . \tag{1.80}$$

If L_0 is introduced in such a way, then the curve (1.74) crosses the level $D_\varepsilon = D_\varepsilon(\infty)$ just at $r = L_0$. From (1.68) the limiting value of (1.80) is $2\sigma_\varepsilon^2$, i.e., twice the mean square fluctuation: $C_\varepsilon^2 L_0^{2/3} = 2\sigma_\varepsilon^2$.

As a result, it becomes possible to approximate $D_\varepsilon(r)$ in individual intervals by the following functions:

$$D(r) = \begin{cases} C_\varepsilon^2 l_0^{-4/3} r^2 = 2\sigma_\varepsilon^2 r^2 l_0^{-4/3} L_0^{-2/3} \; , & r \ll l_0 \; , \\ C_\varepsilon^2 r^{2/3} = 2\sigma_\varepsilon^2 (r/L_0)^{2/3} \; , & l_0 \ll r \ll L_0 \; , \\ C_\varepsilon^2 L_0^{2/3} = 2\sigma_\varepsilon^2 \; , & L_0 \ll r \; . \end{cases} \tag{1.81}$$

That the structure function is limited as $r \to \infty$ can also be deduced from (1.79) by substituting $(k^2 + \kappa_0^2)^{-11/6}$ for $k^{-11/3}$, where $\kappa_0 = 2\pi/L_0$ is the wave number corresponding to the outer scale of turbulence. In so doing, the spectral density will be

$$\Phi_\varepsilon(k) = 0.033 C_\varepsilon^2 (k^2 + \kappa_0^2)^{-11/6} \mathrm{e}^{-k^2/\kappa_m^2} \; , \tag{1.82}$$

hence (1.70) and (1.72) will be finite at any r.

It goes without saying that the fluctuations ε in a real-world turbulent atmosphere behave in a more complex way than is suggested by the above approximations. Nonetheless, in many radiophysical problems, (1.82) describes the spatial spectrum of turbulent fluctuations ε well enough.

A systematic exposition of the issues of statistical hydrodynamics is to be found, for instance, in the monographs [1.6, 7].

1.5 Quasi-Homogeneous Fields

Unfortunately, we cannot always apply the idealization of real inhomogeneous fields using the locally homogeneous fields just considered which describe a rather specific kind of inhomogeneity. Therefore, other special cases of perturbations also become important. Included under this heading are, in particular, fields whose perturbations are in a sense spatially *slow*, or rather smooth, and otherwise arbitrary. These are the so-called statistically *quasi-homogeneous* fields, i.e., those with smoothly varying statistical characteristics: means, variances, etc. Obviously, these are a spatial analog to the quasi-stationary random processes treated in Sect. II.4.3.

According to (1.17), in the general case, the covariance is related to the correlation coefficient K_ζ by

$$\psi_\zeta(\boldsymbol{r}_1, \boldsymbol{r}_2) = \sigma_\zeta(\boldsymbol{r}_1)\,\sigma_\zeta(\boldsymbol{r}_2)\,K_\zeta(\boldsymbol{r}_1, \boldsymbol{r}_2) \ .$$

In terms of $\boldsymbol{r} = \boldsymbol{r}_1 - \boldsymbol{r}_2$ and $\boldsymbol{R} = (\boldsymbol{r}_1 + \boldsymbol{r}_2)/2$, this relation takes the form

$$\psi_\zeta(\boldsymbol{r}_1, \boldsymbol{r}_2) = \sigma_\zeta(\boldsymbol{R} + \boldsymbol{r}/2)\cdot\sigma_\zeta(\boldsymbol{R} - \boldsymbol{r}/2)\,K_\zeta(\boldsymbol{r}, \boldsymbol{R}) \tag{1.83}$$

If l_ζ is the correlation radius of the field ζ [i.e., the space scale of $K_\zeta(\boldsymbol{r}, \boldsymbol{R})$ in $\boldsymbol{r} = \boldsymbol{r}_1 - \boldsymbol{r}_2$], and L is the characteristic space scale (in $\boldsymbol{R}$) of the variance σ_ζ^2, the mean field $\langle\zeta\rangle$ and the correlation coefficient K_ζ, then for quasi-homogeneous fields we have

$$l_\zeta \ll L \ . \tag{1.84}$$

By virtue of (1.84) we can substitute σ_ζ^2 for $\sigma_\zeta(\boldsymbol{R} + \boldsymbol{r}/2)\,\sigma_\zeta(\boldsymbol{R} - \boldsymbol{r}/2)$ in (1.83). This change is undoubtedly justified within the sphere of radius $r \lesssim l_\zeta$; the value of the factor at K_ζ elsewhere is immaterial, since $K_\zeta(\boldsymbol{r}, \boldsymbol{R}) \to 0$ for $r \gg l_\zeta$. We thus arrive at the following expression for the covariance of the quasi-homogeneous field:

$$\psi_\zeta(\boldsymbol{r}, \boldsymbol{R}) = \sigma_\zeta^2(\boldsymbol{R})\,K_\zeta(\boldsymbol{r}, \boldsymbol{R}) \ . \tag{1.85}$$

Because quasi-homogeneity is a deviation from strict homogeneity, the spectral amplitudes $\zeta(\boldsymbol{k})$ are no longer delta-correlated, just like ω-amplitudes in the theory of quasi-stationary random processes (Sect. II.4.3).

An examination of the spectral representation of these processes (Sect. II.4.3) revealed, however, a physical significance to the *instantaneous frequency spectrum* $g(\omega, t)$ produced by decomposing the covariance $\psi(\tau, t)$ in the "fast" variable $\tau = t_1 - t_2$. The same is true of the quasi-homogeneous field. We write the Fourier transform only in $\boldsymbol{r} = \boldsymbol{r}_1 - \boldsymbol{r}_2$

$$\psi_\zeta(\boldsymbol{r}, \boldsymbol{R}) = \int_{-\infty}^{\infty} \Phi_\zeta(\boldsymbol{k}, \boldsymbol{R})\,\mathrm{e}^{\mathrm{i}\boldsymbol{k}\cdot\boldsymbol{r}}\,d^3k \ . \tag{1.86}$$

The quantity

$$\Phi_\zeta(\boldsymbol{k}, \boldsymbol{R}) = \frac{1}{(2\pi)^3}\int_{-\infty}^{\infty} \psi_\zeta(\boldsymbol{r}, \boldsymbol{R})\,\mathrm{e}^{-\mathrm{i}\boldsymbol{k}\cdot\boldsymbol{r}}\,d^3r \tag{1.87}$$

may be treated as the *local* spatial spectral density that varies slowly from one point $\boldsymbol{R}$ to another (with a characteristic scale $L \gg l_\zeta$).

As to fields with both kinds of violation of spatial homogeneity, local homogeneity and quasi-homogeneity, they may be called *locally quasi-homogeneous* and formally defined as *fields whose first increments are quasi-homogeneous.* An example of a locally quasi-homogeneous (isotropic) field is that of the fluctuations of the dielectric permittivity $\tilde{\varepsilon}(\boldsymbol{r})$ in a turbulent atmosphere that is smoothly inhomogeneous in height z [1.6].

1.6 Space-Time Spectral Representations of Random Fields

Up to this point, we have not concerned ourselves with the time dependence of random fields in order to bring out the features due to their dependence on several spatial coordinates, which are not inherent in univariate random functions.

Of course, if we take into consideration the variation of random fields with time, we will have a variety of combinations of spatial and temporal statistical properties of the field $\zeta(t, \boldsymbol{r})$ (stationarity in t, homogeneity in $\boldsymbol{r}$ and various types of deviation from stationarity and homogeneity) already mentioned in Sect. 1.1. The formal use of the four-variate (space-time) harmonic expansions of $\zeta(t, \boldsymbol{r})$ presents no problem, so that the appropriate generalizations hardly need a detailed treatment. We will only present the most important equations and expressions, which we will need later in the book.

For the fluctuations of the one-component random field, $\tilde{\zeta}(t, \boldsymbol{r})$, the four-variate spectral expansion is

$$\tilde{\zeta}(t, \boldsymbol{r}) = \int\int \zeta(\omega, \boldsymbol{k}) e^{i(\boldsymbol{k}\cdot\boldsymbol{r} - \omega t)} d\omega\, d^3 k \ , \tag{1.88}$$

where $\zeta(\omega, \boldsymbol{k})$ is the $\omega\boldsymbol{k}$-amplitude (or rather, the amplitude density) of the field, which is sometimes denoted as $\zeta_{\omega\boldsymbol{k}}$. As opposed to the case in [1.3], where harmonic expansions of processes contain the temporal factor $\exp(i\omega t)$, in the present volume and in [1.1] the time dependence will be taken care of by the factor $\exp(-i\omega t)$, since the spatial harmonic factor $\exp(i\boldsymbol{k}\cdot\boldsymbol{r})$ is more convenient when dealing with wave phenomena than $\exp(-i\boldsymbol{k}\cdot\boldsymbol{r})$. In appropriate complex expressions from [1.3] we will need to substitute $-i\omega$ for $i\omega$.

The spectral amplitudes $\zeta(\omega', \boldsymbol{k}')$ and $\zeta(\omega'', \boldsymbol{k}'')$ of a *homogeneous stationary* random field are delta-correlated both in ω, and in $\boldsymbol{k}$

$$\langle \zeta(\omega', \boldsymbol{k}')\, \zeta^*(\omega'', \boldsymbol{k}'') \rangle = G_\zeta(\omega', \boldsymbol{k}')\, \delta(\omega' - \omega'')\, \delta(\boldsymbol{k}' - \boldsymbol{k}'') \ . \tag{1.89}$$

Here $G_\zeta(\omega, \boldsymbol{k})$ is the space-time spectrum of the field (or $\omega\boldsymbol{k}$-density), which is always nonnegative. The spectrum is sometimes represented in the form $G_\zeta(\omega, \boldsymbol{k}) = |\zeta_{\omega\boldsymbol{k}}|^2$, but it should be remembered that it is just another symbol for the factor multiplying the delta-functions. The space-time covariance $\Psi_\zeta(\tau, \boldsymbol{r})$ is connected with the $\omega\boldsymbol{k}$-density $G_\zeta(\omega, \boldsymbol{k})$ by the Fourier transform

$$\Psi_\zeta(\tau, \boldsymbol{r}) \equiv \langle \tilde{\zeta}(t_1, \boldsymbol{r}_1) \tilde{\zeta}^*(t_1 - \tau, \boldsymbol{r}_1 - \boldsymbol{r}) \rangle = \int\int_{-\infty}^{\infty} G_\zeta(\omega, \boldsymbol{k}) e^{i(\boldsymbol{k}\cdot\boldsymbol{r} - \omega\tau)} d\omega\, d^3 k \tag{1.90}$$

(the generalized Wiener–Khintchine theorem). The inverse Fourier transform is

$$G_\zeta(\omega, \boldsymbol{k}) = \frac{1}{(2\pi)^4} \iint\limits_{-\infty}^{\infty} \Psi_\zeta(\tau, \boldsymbol{r}) e^{-i(\boldsymbol{k}\cdot\boldsymbol{r} - \omega\tau)} d\tau \, d^3 r \ . \tag{1.91}$$

In the case of an even covariance, $\Psi(-\tau, -\boldsymbol{r}) = \Psi(\tau, \boldsymbol{r})$, the exponential function in these expansions may be replaced by $\cos(\boldsymbol{k}\cdot\boldsymbol{r} - \omega\tau)$. We can express both the space spectrum $\Phi_\zeta(\boldsymbol{k})$ and the temporal (frequency) spectrum $g_\zeta(\omega)$ of the random field in terms of the $\omega\boldsymbol{k}$-density $G_\zeta(\omega, \boldsymbol{k})$:

$$\Phi_\zeta(\boldsymbol{k}) = \int\limits_{-\infty}^{\infty} G_\zeta(\omega, \boldsymbol{k}) d\omega = \int\limits_{-\infty}^{\infty} \Psi_\zeta(0, \boldsymbol{r}) e^{-i\boldsymbol{k}\cdot\boldsymbol{r}} d^3 r \ , \tag{1.92}$$

$$g_\zeta(\omega) = \int\limits_{-\infty}^{\infty} G_\zeta(\omega, \boldsymbol{k}) d^3 k = \int\limits_{-\infty}^{\infty} \Psi_\zeta(\tau, 0) e^{i\omega\tau} d\tau \ . \tag{1.93}$$

As an example, consider the field with so-called "frozen" inhomogeneities, i.e., a field where all the temporal variations of $\zeta(t, \boldsymbol{r})$ are determined only by the space perturbations $v(\boldsymbol{r})$ travelling at a constant speed $\boldsymbol{v}$. If the homogeneous random field $v(\boldsymbol{r})$ at rest is described by the space covariance $\psi_v(\boldsymbol{r})$, then for nonrelativistic velocities[5] $\zeta(t, \boldsymbol{r}) = v(\boldsymbol{r} - \boldsymbol{v}t)$, and then

$$\Psi_\zeta(\tau, \boldsymbol{r}) = \psi_v(\boldsymbol{r} - \boldsymbol{v}\tau) \ . \tag{1.94}$$

Substituting (1.94) into (1.91) gives the following simple relationship between the $\omega\boldsymbol{k}$-density $G_\zeta(\omega, \boldsymbol{k})$ of the "frozen" field $\zeta(t, \boldsymbol{r})$ and the space spectral density $\Phi_v(k)$ of the field $v(\boldsymbol{r})$ at rest:

$$G_\zeta(\omega, \boldsymbol{k}) = \Phi_v(\boldsymbol{k}) \delta(\omega - \boldsymbol{k}\cdot\boldsymbol{v}) \ . \tag{1.95}$$

Using (1.92) and (1.95), we can easily show that

$$\Phi_\zeta(\boldsymbol{k}) = \Phi_v(\boldsymbol{k}) \ , \tag{1.96}$$

i.e., that the space spectra of the fields $\zeta(t, \boldsymbol{r})$ and $v(\boldsymbol{r})$ coincide. As regards the temporal spectrum of fluctuations of $\zeta(t, \boldsymbol{r})$, we find from (1.93) and (1.95)

$$\begin{aligned} g_\zeta(\omega) &= \int\limits_{-\infty}^{\infty} dk_\parallel \int\limits_{-\infty}^{\infty} d^2 k_\perp \Phi_v(k_\parallel, \boldsymbol{k}_\perp) \delta(\omega - k_\parallel v) \\ &= \frac{1}{v} \int\limits_{-\infty}^{\infty} \Phi_v\left(\frac{\omega}{v}, \boldsymbol{k}_\perp\right) d^2 k_\perp \ , \end{aligned} \tag{1.97}$$

where $k_\parallel$ and $\boldsymbol{k}_\perp$ are respectively the longitudinal and transverse (relative to the velocity $\boldsymbol{v}$) components of the wave vector $\boldsymbol{k}$. Expression (1.97) is simplified in the case of the isotropic field $v(\boldsymbol{r})$, when $\Phi_v(k_\parallel, \boldsymbol{k}_\perp) = \Phi_v(\sqrt{k_\parallel^2 + k_\perp^2})$. Passing in (1.97) to

---

[5] For peculiarities of the relativistic case see Exercise 1.8.7.

the polar coordinates and introducing instead of $k_\perp$ a new variable of integration $\kappa = \sqrt{(\omega/v)^2 + k_\perp^2}$, we arrive at

$$g_\zeta(\omega) = \frac{2\pi}{v} \int\limits_{|\omega|/v}^{\infty} \Phi_v(\kappa)\kappa d\kappa \ . \tag{1.98}$$

Differentiating (1.98) with respect to ω, we can readily express the space spectrum of the frozen isotropic field $\Phi_\zeta(k) = \Phi_v(k)$ in terms of the frequency spectrum $g_\zeta(\omega)$

$$\Phi_\zeta(k) = -\frac{v^2}{2\pi k} \frac{dg_\zeta(\omega)}{d\omega}\bigg|_{\omega = kv} \ . \tag{1.99}$$

Equations of the form (1.98, 99) are often used with fields that can, to a sufficient accuracy, be viewed as frozen.

What has been said above about the spectral representation of the one-component field can easily be generalized to the n-component field $\zeta^{(i)}(t, \boldsymbol{r})$ ($i = 1, 2, \ldots, n$). The simplest properties are characteristic of the *homogeneous and homogeneously connected* (in the wide sense) fields in the space $Q = (t, \boldsymbol{r})$.

In the case of the *vector* field $\zeta^{(i)}$, when coordinates are subjected to orthogonal transformations the elements of the correlation matrix $\Psi_{ij}^{(\zeta)}$ are treated as the components of an n-dimensional tensor of the second rank, i.e., we can speak about the *correlation tensor* of a vector field. For homogeneous isotropic vector fields a number of elegant theorems have been proved, some of which are formulated in Exercises 1.8.8–10.

The multicomponent random field $\zeta^{(i)}(t, \boldsymbol{r})$, represented as a Fourier integral, has the form analogous to (1.88), and for the homogeneous stationary field the following equality is met:

$$\langle \zeta_i(\omega', \boldsymbol{k}')\zeta_j^*(\omega'', \boldsymbol{k}'')\rangle = G_{ij}^{(\zeta)}(\omega', \boldsymbol{k}')\delta(\omega' - \omega'')\delta(\boldsymbol{k}' - \boldsymbol{k}'') \ , \tag{1.100}$$

which leads naturally to the generalization of the Wiener–Khintchine theorem to multicomponent space-time fields

$$\Psi_{ij}^{(\zeta)}(\tau, \boldsymbol{r}) = \iint\limits_{-\infty}^{\infty} G_{ij}^{(\zeta)}(\omega, \boldsymbol{k})\mathrm{e}^{\mathrm{i}(\boldsymbol{k}\cdot\boldsymbol{r} - \omega t)}d\omega\, d^3k \tag{1.101}$$

with the inverse transformation

$$G_{ij}^{(\zeta)}(\omega, \boldsymbol{k}) = \frac{1}{(2\pi)^4} \iint\limits_{-\infty}^{\infty} \Psi_{ij}^{(\zeta)}(\tau, \boldsymbol{r})\mathrm{e}^{-\mathrm{i}(\boldsymbol{k}\cdot\boldsymbol{r} - \omega t)}d\tau\, d^3r \ . \tag{1.102}$$

The correlation matrix of the multicomponent field $\Psi_{ij}^{(\zeta)}$ is positive definite. It follows that the diagonal components of the matrix of the $\omega\boldsymbol{k}$-density are real and nonnegative: $G_{ii}^{(\zeta)}(\omega, \boldsymbol{k}) \geqslant 0$. However, the nondiagonal elements are generally complex.

The following point should be stressed. Generally speaking, the integration region with respect to ω and $\boldsymbol{k}$ in the Fourier expansions of the fields themselves and of their covariances (1.90) and (1.101) is four-dimensional. But given that the fields under consideration satisfy certain dynamic equations (wave equations in particular) in a *homogeneous stationary* medium *free of sources* (i.e., the equations themselves are homogeneous), then ω and $\boldsymbol{k}$ are no longer independent, but subject to the so-called *dispersion relation*

$$\Delta(\omega, \boldsymbol{k}) = 0 \ . \tag{1.103}$$

This equation follows from the requirement that the plane monochromatic wave $\exp(\mathrm{i}\boldsymbol{k}\cdot\boldsymbol{r} - \mathrm{i}\omega t)$ be a *proper* wave, i.e., a particular solution of the dynamic equations.

In the language of geometry, the dispersion relation describes in the space $(\omega, \boldsymbol{k})$ a certain three-dimensional hypersurface, dispersion surface, that may be multisheeted as well (e.g., in an anisotropic medium). At all points external to this hypersurface, the $\omega\boldsymbol{k}$-amplitudes of fields are zero, and so are the spectral densities in (1.90) and (1.101). In other words, $\zeta^{(i)}(\omega, \boldsymbol{k})$ and $G_{ij}^{(\zeta)}(\omega, \boldsymbol{k})$ contain a delta-function $\delta[\Delta(\omega, \boldsymbol{k})]$ as a factor, with the result that the multiplicity of the Fourier integrals is reduced: these in fact cover only the dispersion surface (all or parts of it), i.e., for ω and $\boldsymbol{k}$ obeying the dispersion relation. We will be looking at this in the next chapter.

In cases where the stationary field $\zeta(t, \boldsymbol{r})$ is *ergodic* as a function of t (Sect. I.4.7), the angular brackets in the earlier expressions may be treated (in the sense of the probabilistic convergence) as time averaging, i.e., we may think, for example, that an arbitrary deterministic function *for all practical purposes* obeys the following relation:

$$\langle f[\zeta(t, \boldsymbol{r})]\rangle = \lim_{T\to\infty} \widetilde{f[\zeta(t, \boldsymbol{r})]} \equiv \lim_{T\to\infty} \frac{1}{2T} \int_{-T}^{T} f[\zeta(t, \boldsymbol{r})]\, dt \ . \tag{1.104}$$

To obtain ensemble averages by time averaging we can, to a sufficient accuracy, limit ourselves to finite T that are markedly larger than the correlation time τ_{c}

$$T \gg \tau_{\mathrm{c}} \ . \tag{1.105}$$

Apart from temporal ergodicity, we can introduce the notion of *space* and *space-time* ergodicity. So, for homogeneous space-ergodic fields, the ensemble averages "coincide" with the space averages in the sense that they converge in probability. The implication is that an arbitrary deterministic function f may be thought of as obeying the equality

$$\langle f[\zeta(t, \boldsymbol{r})]\rangle = \lim_{V\to\infty} \widetilde{f[\zeta(t, \boldsymbol{r})]} \equiv \lim_{V\to\infty} \frac{1}{V} \int_{V} f[\zeta(t, \boldsymbol{r})]\, d^3r \ , \tag{1.106}$$

where V is the space region of averaging. The limiting process $V \to \infty$ can again be discontinued in the regions V whose transverse dimension L is large as compared with the correlation radius l_ζ, i.e., the following inequality, similar to (1.105), holds:

$$L \sim V^{1/3} \gg l_\zeta \ . \tag{1.107}$$

The space-time ergodicity of stationary and homogeneous fields occurs in the case of convergence in probability both as $T \to \infty$ and as $V \to \infty$, which amounts to (1.104) and (1.106) being satisfied simultaneously. Cases are also possible where the field $\zeta(t, \boldsymbol{r})$ is ergodic only in some of the space variables, e.g., in the plane (x, y) or on a spheric surface.

Lastly, it is sometimes convenient to make use of the notion of quasi-ergodic fields which, in relation to ergodic fields, are in about the same position as quasi-homogeneous fields in relation to homogeneous ones. In other words, these fields are only ergodic in volumes small as compared with the characteristic space scales L_ζ for the field statistics (mean, variance, etc.). Unlike (1.106), the domain of space averaging for quasi-ergodic fields should be bounded from above by L_ζ, but (1.107) should still hold. Accordingly, the quasi-ergodicity of fields only makes sense if the averaging domain V meets the inequalities

$$L_\zeta \gg V^{1/3} \gg l_\zeta \ . \tag{1.108}$$

In phenomenological physics, V is generally called the "physically infinitesimal volume". This volume should, on the one hand, be sufficiently small for fields under consideration to be statistically homogeneous within its confines (in mechanics and the electrodynamics of continuous media it is normally required that in V the mean fields be constant only); on the other hand, the volume should be so large that the field ζ within V should undergo sufficient spatial fluctuations.

In the physics of continuous media, l_ζ should be taken to be the mean separation between the perturbation sources, which may be individual molecules, dislocations, impurities, and so on. If n is the concentration of molecules (or other perturbing objects), then $l_\zeta \sim n^{-1/3}$ and the inequality $V^{1/3} \gg l_\zeta$ takes the form $nV \gg 1$, which corresponds to a large number of molecules in the physically infinitesimal volume V. If the inequality holds, the averaging over the ensemble of molecules may be replaced by the averaging over the small domain V. Macroscopic physics thus makes substantial use of the assumption that fields describing the state of the continuous media are spatially quasi-ergodic.

1.7 Functional Treatment of Random Fields

We conclude this chapter by discussing a method of describing random fields using a *characteristic functional*. The method uses more complex mathematical concepts, some relevant information about which will be given here (in addition, Exercises 1.8.13–21 should be worked through). Note that it pays here to employ more sophisticated mathematical techniques since this gives us a more complete and consistent description of random processes and fields. Specifically, special attention should be paid to Exercise 1.8.22, containing a *derivation* of the Einstein–Fokker–Planck equation for the distribution of probabilities for the solution of a set of ordinary differential equations with Gaussian delta-correlated coefficients. In Sect. I.6.4 the result was given without proof.

The functional treatment of random *fields* will for the most part be used in [1.1]. For the benefit of the readers, we will provide it here so that our treatment of purely mathematical issues is consistent.

As we know, a complete description of a real random variable ξ is given by its characteristic function $\varphi_\xi(u) = \langle \exp(\mathrm{i}u\xi) \rangle$, which allows us to find the probability densities for ξ, the moments of different orders, and the cumulants.

For an N-variate random variable $\boldsymbol{\xi} = \{\xi_1, \ldots, \xi_N\}$ the characteristic function has the form $\varphi_\xi(\boldsymbol{u}) = \langle \exp(\mathrm{i}\boldsymbol{u}\boldsymbol{\xi}) \rangle$, where $\boldsymbol{u}\boldsymbol{\xi} = \sum_{j=1}^{N} u_j \xi_j$. The function contains all the information about the collection of random variables ξ_j (Sect. I.3.2).

In this section it will be advantageous to lift for a moment the constraint on the number of variables of the functions of interest, and return to the n-dimensional space of variables $\boldsymbol{x} = \{x_1, \ldots, x_n\}$. Let $\xi(x_1, \ldots, x_n) = \xi(\boldsymbol{x})$ be a real random field in this space (in the particular case of a single variable, $n = 1$, we will, as earlier, consider that x is the time t).

Let us now consider a further way of defining a random function. The method is based on the so-called characteristic functionals and is more convenient and compact in many problems. First of all, we will make sure that the complete probabilistic description of the field $\xi(\boldsymbol{x})$ is given by the *characteristic functional*

$$\Phi[u] \equiv \langle \exp\{\mathrm{i} \textstyle\int \xi(\boldsymbol{x}) u(\boldsymbol{x}) d\boldsymbol{x}\} \rangle \;, \quad d\boldsymbol{x} = dx_1 \ldots dx_n \;. \tag{1.109}$$

Here $u(\boldsymbol{x})$ is an arbitrary function for which the integral in (1.109) converges for all the possible realizations of $\xi(\boldsymbol{x})$. The function $u(\boldsymbol{x})$ replaces the discrete collection of the variables, $\boldsymbol{u} = \{u_1, \ldots, u_N\}$, of the characteristic function for the N-variate random variable $\boldsymbol{\xi} = \{\xi_1, \ldots, \xi_N\}$. We would stress that the functional $\Phi[u]$ should be known for *any* $u(\boldsymbol{x})$.

Knowing $\Phi[u]$, we can derive a characteristic function for any collection of the random variables $\xi_1 = \xi(\boldsymbol{x}_1), \ldots, \xi_N = \xi(\boldsymbol{x}_N)$. To this end, it is sufficient to take the following argument $u(\boldsymbol{x})$ of the functional $\Phi[u]$:

$$u_N(\boldsymbol{x}) = u_1 \delta(\boldsymbol{x} - \boldsymbol{x}_1) + \ldots + u_N \delta(\boldsymbol{x} - \boldsymbol{x}_N) \;.$$

Then

$$\int \xi(\boldsymbol{x}) u_N(\boldsymbol{x}) d\boldsymbol{x} = \sum_1^N u_j \xi(\boldsymbol{x}_j)$$

and the functional $\Phi[u]$ will become the characteristic function of the N-variate random variable $\{\xi_1, \ldots, \xi_N\}$, i.e., of the values of the field $\xi(\boldsymbol{x})$ at N points $\boldsymbol{x}_1, \ldots, \boldsymbol{x}_N$

$$\Phi[u_N] = \left\langle \exp\left\{\mathrm{i}\sum_1^N u_j \xi(x_j)\right\}\right\rangle = \varphi(u_1, \ldots, u_N) \;.$$

But the characteristic function $\varphi(u_1, \ldots, u_N)$, as we know, allows us to find the N-dimensional probability density $w_N(\xi_1, \ldots, \xi_N)$. It is thus clear that the characteristic functional $\Phi[u]$ *uniquely* defines the densities $w_N(\xi_1, \ldots, \xi_N)$ at *any* N and *any selection* of the points $\boldsymbol{x}_1, \ldots, \boldsymbol{x}_N$, i.e., it completely defines the random field $\xi(\boldsymbol{x})$.

Given the functional $\Phi[u]$, we can also find the moments (in particular, covariances) and other statistics of $\xi(\boldsymbol{x})$. This, however, requires a knowledge of how $\Phi[u]$ is differentiated with respect to the *functional* argument $u(\boldsymbol{x})$. Therefore, part of this section will be devoted to the mathematics of functionals.

To begin with, we will give the *general definition* of a functional. A functional is said to be defined in some domain of functions, if a rule is given whereby *each function belonging to the domain is referred to a number, called the value of the functional of this function.* Let us take some examples of functionals:

$$F_1[u] = \int A(\boldsymbol{x})u(\boldsymbol{x})d\boldsymbol{x} \ ,$$

$$F_N[u] = \int A_N(\boldsymbol{x}_1, \ldots, \boldsymbol{x}_N)u(\boldsymbol{x}_1)\ldots u(\boldsymbol{x}_N)d\boldsymbol{x}_1 \ldots d\boldsymbol{x}_N \ ,$$

$$\Phi_{\mathrm{G}}[u] = \exp\{-\tfrac{1}{2}\int\int \psi(\boldsymbol{x}_1, \boldsymbol{x}_2)u(\boldsymbol{x}_1)u(\boldsymbol{x}_2)d\boldsymbol{x}_1\, d\boldsymbol{x}_2\} \ .$$

Here $F_1[u]$ is a linear functional, $F_N[u]$ a functional of the Nth degree and $\Phi_{\mathrm{G}}[u]$ a Gaussian functional. It will be shown below that $\Phi_{\mathrm{G}}[u]$ is a characteristic functional for Gaussian random fields. Note that the functional is dependent on the *form of the function* $u(\boldsymbol{x})$ *as a whole* (i.e., on its behavior in the *entire* space $\boldsymbol{x}$), whereby the notation $F[u]$ contains no explicit indication as to the variation of u with $\boldsymbol{x}$. In cases where it is necessary to indicate this dependence, we will mark the argument of u with a tilde: $F[u(\tilde{\boldsymbol{x}})]$. When using this representation, we should take into account that the functional F itself is *independent* of $\tilde{\boldsymbol{x}}$ (just as in tensor algebra the quantity $x_k u_k$ is independent of the "dummy" index).

We now consider the values of a functional F for two various "values" of its argument, viz., for the functions $u(\boldsymbol{x})$ and $u(\boldsymbol{x}) + \delta u(\boldsymbol{x})$. In so doing, we will assume that $\delta u(\boldsymbol{x})$ is only nonzero in a certain region $\Delta\boldsymbol{x}$ at the point $\boldsymbol{x}_0$. The *functional* (or *variational*) *derivative* of $F[u]$ at $\boldsymbol{x}_0$ is the limit

$$\frac{\delta F[u]}{\delta u[\boldsymbol{x}_0]} = \lim_{\substack{|\Delta \boldsymbol{x}|\to 0\\ \max|\delta u|\to 0}} \frac{F[u+\delta u] - F[u]}{\int_{\Delta \boldsymbol{x}} \delta u(\boldsymbol{x})d\boldsymbol{x}} \ , \tag{1.110}$$

if the limit exists and depends neither on the form of $\delta u(\boldsymbol{x})$, nor on the way in which the volume $|\Delta\boldsymbol{x}|$ contracts to zero, nor on the law whereby the maximum of the modulus of $\delta u(\boldsymbol{x})$ tends to zero. Several examples can be quoted.

We will find the functional derivative of the linear functional $F_1[u] = \int A(\boldsymbol{x}')u(\boldsymbol{x}')d\boldsymbol{x}'$ at $\boldsymbol{x}_0$. We have

$$F_1[u+\delta u] = \int A(\boldsymbol{x}')[u(\boldsymbol{x}') + \delta u(\boldsymbol{x}')]d\boldsymbol{x}' \ ,$$

$$F_1[u+\delta u] - F_1[u] = \int_{\Delta \boldsymbol{x}} A(\boldsymbol{x}')\delta u(\boldsymbol{x}')d\boldsymbol{x}' \ ,$$

so that

$$\frac{F_1[u+\delta u] - F_1[u]}{\int_{\Delta \boldsymbol{x}} \delta u(\boldsymbol{x}')d\boldsymbol{x}'} = \frac{\int_{\Delta \boldsymbol{x}} A(\boldsymbol{x}')\delta u(\boldsymbol{x}')d\boldsymbol{x}'}{\int_{\Delta \boldsymbol{x}} \delta u(\boldsymbol{x}')d\boldsymbol{x}'} \ .$$

Applying the mean-value theorem to the integral in the numerator, we obtain

$$\int_{\Delta x} A(x')\delta u(x')dx' = A(x_1)\int_{\Delta x} \delta u(x')dx' ,$$

where x_1 lies in the Δ-neighborhood of x_0. Taking the limit as $|\Delta x| \to 0$, we get

$$\frac{\delta}{\delta u(x_0)}\int A(x')u(x')dx' = A(x_0) . \tag{1.111}$$

More complicated functionals are differentiated in the same way (see Exercise 1.8.13).

Let $\delta u(x)$ be the function $\delta u(x) = \lambda\delta(x - x_0)$ [as before, $\delta(x - x_0) = \delta(x_1 - x_{01}) \ldots \delta(x_n - x_{0n})$]. Then $\int \delta u(x')dx' = \lambda$ and, by (1.110),

$$\frac{\delta F[u]}{\delta u(x_0)} = \lim_{\lambda\to 0}\frac{F[u(\tilde{x}) + \lambda\delta(\tilde{x} - x_0)] - F[u(\tilde{x})]}{\lambda}$$

$$= \left\{\frac{\partial}{\partial\lambda}F[u(\tilde{x}) + \lambda\delta(\tilde{x} - x_0)]\right\}_{\lambda=0} . \tag{1.112}$$

Formula (1.112) reduces the computing of the functional derivative to conventional differentiation.

We will now present formal rules of functional differentiation, which are similar to conventional ones.

If A_1 and A_2 are independent of u, then

$$\frac{\delta}{\delta u(x)}\{A_1\Phi_1[u] + A_2\Phi_2[u]\} = A_1\frac{\delta\Phi_1[u]}{\delta u(x)} + A_2\frac{\delta\Phi_2[u]}{\delta u(x)} . \tag{1.113}$$

Further,

$$\frac{\delta}{\delta u(x)}\{\Phi_1[u]\Phi_2[u]\} = \frac{\delta\Phi_1[u]}{\delta u(x)}\Phi_2[u] + \Phi_1[u]\frac{\delta\Phi_2[u]}{\delta u(x)} . \tag{1.114}$$

Let $f(z)$ be a function of z, and the functional $\varphi[u] = f(F[u])$ be a function of a functional. Then,

$$\frac{\delta}{\delta u(x)}f(F[u]) = f'(F[u])\frac{\delta F[u]}{\delta u(x)} , \tag{1.115}$$

where $f' = df(z)/dz$ at $z = F(u)$ is a conventional derivative of f with respect to its argument. Equations (1.113–115) follow directly from (1.112). Lastly, putting in (1.111) $A(x') = \delta(x' - x_1)$, we obtain

$$\frac{\delta u(x_1)}{\delta u(x_0)} = \delta(x_1 - x_0) . \tag{1.116}$$

Equations (1.113–116) enable most of the functionals occurring in specific problems to be differentiated without recourse to the formal definition (1.110) of functional differentiation. Specifically, (1.113) suggests that functionals can be differentiated under the sign of the integral or derivative.

As a further example, we will find the functional derivative of the Gaussian functional

$$\frac{\delta \Phi_G[u]}{\delta u(x)} = \frac{\delta}{\delta u(x)} \exp\left\{-\frac{1}{2}\int\int \psi(x_1, x_2) u(x_1) u(x_2) dx_1 dx_2\right\} .$$

From (1.115), we obtain

$$\frac{\delta \Phi_G[u]}{\delta u(x)} = \Phi_G[u]\left\{-\frac{1}{2}\frac{\delta}{\delta u(x)}\int\int \psi(x_1, x_2) u(x_1) u(x_2) dx_1 dx_2\right\} .$$

Using (1.113), we differentiate under the integral sign with the help of (1.114) and (1.116)

$$\begin{aligned}\frac{\delta \Phi_G[u]}{\delta u(x)} &= -\tfrac{1}{2}\Phi_G[u]\int\int \psi(x_1, x_2)[\delta(x_1 - x) u(x_2) \\ &\quad + u(x_1)\delta(x_2 - x)]\, dx_1 dx_2 \\ &= -\tfrac{1}{2}\Phi_G[u]\int[\psi(x, x') + \psi(x', x)] u(x') dx' .\end{aligned}$$

In this example we can take it that $\psi(x_1, x_2) = \psi(x_2, x_1)$, as the nonsymmetrical part of $\psi(x_1, x_2)$ does not contribute to the functional since the integral of the product of a nonsymmetrical and a symmetrical function, $u(x_1)u(x_2)$, is zero. Therefore, the resultant derivative may be written in a shorter way

$$\frac{\delta \Phi_G[u]}{\delta u(x)} = -\Phi_G[u]\int \psi(x, x') u(x') dx' .$$

The functional $F[u + v]$ can be expanded into a functional Taylor series in $v(\tilde{x})$ (see, e.g., [Ref. 1.6. Appendix I])

$$\begin{aligned}F[u + v] &= F[u] + \int dx\, v(x)\frac{\delta F[u]}{\delta u(x)} \\ &\quad + \frac{1}{2!}\int\int dx_1 dx_2\, v(x_1) v(x_2)\frac{\delta^2 F[u]}{\delta u(x_1)\delta u(x_2)} + \dots .\end{aligned} \tag{1.117}$$

Just as the operator representation of $f(x+\xi) = \exp\left\{\xi\frac{\partial}{\partial x}\right\} f(x)$ for a conventional Taylor series, (1.117) can conveniently be written in operator form

$$\begin{aligned}F[u + v] &= \left\{1 + \int dx\, v(x)\frac{\delta}{\delta u(x)}\right. \\ &\quad \left. + \frac{1}{2!}\int\int dx_1 dx_2\, v(x_1) v(x_2)\frac{\delta}{\delta u(x_1)}\frac{\delta}{\delta u(x_2)} + \dots\right\} F[u] \\ &= \left\{1 + \int dx\, v(x)\frac{\delta}{\delta u(x)} + \frac{1}{2!}\left[\int dx\, v(x)\frac{\delta}{\delta u(x)}\right]^2 + \dots\right\} F[u] .\end{aligned}$$

The expression in braces is a power expansion of an exponential function, so that

$$F[u+v] = \exp\left\{\int dx\, v(x)\frac{\delta}{\delta u(x)}\right\} F[u] \ . \tag{1.118}$$

We return now to the characteristic functional (1.109) and operate on it with $[\delta/\delta u(x)]/\mathrm{i}$. Differentiating under the averaging sign gives

$$\frac{1}{\mathrm{i}}\frac{\delta\Phi[u]}{\delta u(x)} = \langle \xi(x)\exp\{\mathrm{i}\int u(x')\xi(x')dx'\}\rangle \ . \tag{1.119}$$

Thus, operating with $[\delta/\delta u(x)]/\mathrm{i}$ on $\Phi[u]$ introduces under the averaging sign the factor $\xi(x')$. Repeating the operation, we arrive at

$$\frac{1}{\mathrm{i}}\frac{\delta}{\delta u(x_1)}\cdots\frac{1}{\mathrm{i}}\frac{\delta}{\delta u(x_N)}\Phi[u] = \langle \xi(x_1)\ldots\xi(x_N)\exp\{\mathrm{i}\int u(x')\xi(x')dx'\}\rangle \ . \tag{1.120}$$

If we now set $u=0$, we will have

$$\langle \xi(x_1)\ldots\xi(x_N)\rangle = \frac{1}{\mathrm{i}^N}\frac{\delta^N\Phi[u]}{\delta u(x_1)\ldots\delta u(x_N)}\bigg|_{u(\tilde{x})=0} \ . \tag{1.121}$$

Consequently, the moments (1.121) of the random field $\xi(x)$ can be found as values of functional derivatives of the characteristic functional at zero argument.

If we expand the exponential function (1.109) into a series and write the kth stage of the integral as a k-fold multiple integral, we will get

$$\Phi[u] = 1 + \sum_1^\infty \frac{\mathrm{i}^k}{k!}\int\langle \xi(x_1)\ldots\xi(x_k)\rangle u(x_1)\ldots u(x_k)dx_1\ldots dx_k \ , \tag{1.122}$$

i.e., the expansion of the characteristic functional $\Phi[u]$ into a functional Taylor series.

Now let $\Theta[u] = \ln\Phi[u]$ be a functional, such that $\Phi[u] = \exp(\Theta[u])$. The Taylor series expansion of $\Theta[u]$ defines the *cumulant* functions (or covariances) of various orders

$$\Theta[u] = \sum_1^\infty \frac{\mathrm{i}^k}{k!}\int\psi_k(x_1,\ldots,x_k)u(x_1)\ldots u(x_k)dx_1\ldots dx_k \ . \tag{1.123}$$

The kth-order cumulant function ($k = 1, 2, \ldots$) is given by the function

$$\psi_k(x_1,\ldots,x_k) = \frac{1}{\mathrm{i}^k}\frac{\delta^k\ln\Phi[u]}{\delta u(x_1)\ldots\delta u(x_k)}\bigg|_{u=0} \ , \tag{1.124}$$

which is similar to the expression for cumulants of a random variable. The relationship between the moment and the cumulant function follows from (1.121) and (1.124) (see Exercise 1.8.16). Specifically,

$$\psi_1(\boldsymbol{x}) = \overline{\xi(\boldsymbol{x})} \, ,$$
$$\psi_2(\boldsymbol{x}_1, \boldsymbol{x}_2) = \psi(\boldsymbol{x}_1, \boldsymbol{x}_2) = \langle \xi(\boldsymbol{x}_1)\xi(\boldsymbol{x}_2)\rangle - \overline{\xi(\boldsymbol{x}_1)}\,\overline{\xi(\boldsymbol{x}_2)} \, . \tag{1.125}$$

Up to this point we have considered several examples of functionals of the general form. Let us now turn to examples of *characteristic* functionals.

1) Let $\xi(\boldsymbol{x})$ be a *normal random field*, and $\mathscr{L}$ be the integral

$$\mathscr{L} = \int \xi(\boldsymbol{x}) u(\boldsymbol{x}) d\boldsymbol{x} \, .$$

The characteristic functional $\Phi_\mathrm{G}[u]$ will then take the form

$$\Phi_\mathrm{G}[u] = \langle \exp(\mathrm{i}\mathscr{L})\rangle \, .$$

The integral of a Gaussian random function is a Gaussian random variable, which allows the probability density for $\mathscr{L}$ to be written as

$$w(\mathscr{L}) = \frac{1}{\sqrt{2\pi}\sigma} \exp\left(-\frac{(\mathscr{L} - \bar{\mathscr{L}})^2}{2\sigma^2}\right) .$$

We see from the expression for $\mathscr{L}$ that $\bar{\mathscr{L}}$ and $\sigma^2 = \langle (\mathscr{L} - \bar{\mathscr{L}})^2 \rangle = \langle (\tilde{\mathscr{L}})^2 \rangle$ are given by

$$\bar{\mathscr{L}} = \int \overline{\xi(\boldsymbol{x})} u(\boldsymbol{x}) d\boldsymbol{x} \, ,$$
$$\sigma^2 = \int\int \psi_\xi(\boldsymbol{x}_1, \boldsymbol{x}_2) u(\boldsymbol{x}_1) u(\boldsymbol{x}_2) d\boldsymbol{x}_1 \, d\boldsymbol{x}_2 \, .$$

Next, we compute $\langle \exp(\mathrm{i}\mathscr{L})\rangle$ and arrive at the known result

$$\begin{aligned}\Phi_\mathrm{G}[u] &= \int e^{\mathrm{i}\mathscr{L}} w(\mathscr{L}) d\mathscr{L} \\ &= \frac{1}{\sqrt{2\pi}\sigma} \int \exp\left(\mathrm{i}\mathscr{L} - \frac{1}{2\sigma^2}(\mathscr{L} - \bar{\mathscr{L}})^2\right) d\mathscr{L} = \exp\left(\mathrm{i}\bar{\mathscr{L}} - \frac{1}{2}\sigma^2\right) .\end{aligned}$$

Substituting for $\bar{\mathscr{L}}$ and σ^2 gives

$$\Phi_\mathrm{G}[u] = \exp\{\mathrm{i}\int \overline{\xi(\boldsymbol{x})} u(\boldsymbol{x}) d\boldsymbol{x} - \tfrac{1}{2}\int\int \psi_\xi(\boldsymbol{x}_1, \boldsymbol{x}_2) u(\boldsymbol{x}_1) u(\boldsymbol{x}_2) d\boldsymbol{x}_1 \, d\boldsymbol{x}_2\} \, . \tag{1.126}$$

Hence

$$\begin{aligned}\Theta_\mathrm{G}[u] &= \ln \Phi_\mathrm{G}[u] \\ &= \mathrm{i}\int \overline{\xi(\boldsymbol{x})} u(\boldsymbol{x}) d\boldsymbol{x} - \tfrac{1}{2}\int\int \psi_\xi(\boldsymbol{x}_1, \boldsymbol{x}_2) u(\boldsymbol{x}_1) u(\boldsymbol{x}_2) d\boldsymbol{x}_1 \, d\boldsymbol{x}_2 \, .\end{aligned} \tag{1.127}$$

A comparison of (1.127) with (1.123) indicates that for Gaussian random fields all the cumulant functions of higher than second order are zero.

The central moments of a normal random field possess the following properties (Exercise I.3.7.6):

$$\begin{aligned}&\langle \tilde{\xi}_1 \ldots \tilde{\xi}_{2k+1} \rangle = 0, \quad \tilde{\xi}_i \equiv \tilde{\xi}(x_i) \ , \\ &\langle \tilde{\xi}_1 \ldots \tilde{\xi}_{2k} \rangle = \sum_{\text{d.p.}} \langle \tilde{\xi}_{\alpha_1} \tilde{\xi}_{\alpha_2} \rangle \ldots \langle \tilde{\xi}_{\alpha_{2k-1}} \tilde{\xi}_{\alpha_{2k}} \rangle \ ,\end{aligned} \tag{1.128}$$

where $\sum_{\text{d.p.}}$ is the sum over different permutations of pairs from $1, 2, \ldots, 2k$. We can readily find that the number of terms in the sum in (1.128) is $(2k-1)!! = 1.3.5 \ldots (2k-1)$. For example, at $k = 2$

$$\langle \tilde{\xi}_1 \tilde{\xi}_2 \tilde{\xi}_3 \tilde{\xi}_4 \rangle = \langle \tilde{\xi}_1 \tilde{\xi}_2 \rangle \langle \tilde{\xi}_3 \tilde{\xi}_4 \rangle + \langle \tilde{\xi}_1 \tilde{\xi}_3 \rangle \langle \tilde{\xi}_2 \tilde{\xi}_4 \rangle + \langle \tilde{\xi}_1 \tilde{\xi}_4 \rangle \langle \tilde{\xi}_2 \tilde{\xi}_3 \rangle \ .$$

A similar formula for the six-order moment contains 15 terms. The definitions of the Gaussian field (1.126) and (1.128) are equivalent (see Exercise 1.8.17).

2) The *Poisson* random field is constructed as follows. To begin with, consider a finite domain V of the space of variables $\boldsymbol{x} = \{x_1, \ldots, x_n\}$. Let m be a random variable distributed according to the Poisson law $P(m) = (\bar{m})^m \exp(-\bar{m})/m!$, where $\bar{m}$ is the mean of m. In V we select m points $\boldsymbol{x}_1, \ldots, \boldsymbol{x}_m$ at random, assuming that the probability density for the coordinates of each of these points is V^{-1} for $x_k \in V$ and zero elsewhere, and that the coordinates of all m points are statistically independent. Further, we take m mutually independent random variables that have the same probability densities $w(A)$, and hence the same characteristic function $\chi(\lambda)$. Let $g(x)$ be a fixed *deterministic* function. We will now define the Poisson random field as

$$\xi(\boldsymbol{x}) = \sum_{k=0}^{m} A_k g(\boldsymbol{x} - \boldsymbol{x}_k) \ . \tag{1.129}$$

As constructed, (1.129) contains the following random parameters: the "amplitude" A_k, the positions of the "centers", $\boldsymbol{x}_k$, and the number of terms, m. We generalize what in [1.2] was done for univariate random functions, i.e., for random processes (see, e.g., Chap. I.3), to the case of the random field in an n-variate space of $\boldsymbol{x}$.

The characteristic functional of the random field (1.129) has the form (see Exercise 1.8.18)

$$\Phi[u] = \exp\{\nu \textstyle\int [\chi(\int g(\boldsymbol{x} - \boldsymbol{x}')u(\boldsymbol{x})d\boldsymbol{x}) - 1] d\boldsymbol{x}'\} \ , \tag{1.130}$$

where $\nu = \bar{m}/V$ is the *mean density* of random points. In (1.130) V may also be taken to be unbounded ($V \to \infty$, $\bar{m} \to \infty$ at $\nu = \bar{m}/V = \text{const}$).

We can readily deduce from (1.130) the cumulant functions of the Poisson random field (see Exercise 1.8.10). These have the form

$$\psi_k(\boldsymbol{x}_1, \ldots, \boldsymbol{x}_k) = \nu \langle A^k \rangle \textstyle\int g(\boldsymbol{x}_1 - \boldsymbol{x}) g(\boldsymbol{x}_2 - \boldsymbol{x}) \ldots g(\boldsymbol{x}_k - \boldsymbol{x}) d\boldsymbol{x} \ . \tag{1.131}$$

The Gaussian random field is obtainable from the Poisson one if we somehow make ν tend to infinity, while decreasing the "amplitudes" A (see Exercise 1.8.20).

Many physics problems involve computing quantities of the form $\langle \xi(\boldsymbol{x}) R[\xi] \rangle$, where $R[\xi]$ is a certain functional of the random field $\xi(\boldsymbol{x})$. Let us derive, following [1.8], a useful formula that allows this quantity to be found. We now consider the mean of the expression

$$\langle \xi(\boldsymbol{x}) R[\xi + \eta] \rangle \equiv f[\boldsymbol{x}, \eta] ,$$

where $\eta = \eta(\tilde{\boldsymbol{x}})$ is a *deterministic* function. Using (1.118), we represent $R[\xi + \eta]$ in the form

$$R[\eta + \xi] = \exp\left\{\int d\boldsymbol{x}' \xi(\boldsymbol{x}') \frac{\delta}{\delta\eta(\boldsymbol{x}')}\right\} R[\eta] .$$

Then

$$f[\boldsymbol{x}, \eta] = \left\langle \xi(\boldsymbol{x}) \exp\left\{\int d\boldsymbol{x}' \xi(\boldsymbol{x}') \frac{\delta}{\delta\eta(\boldsymbol{x}')}\right\}\right\rangle R[\eta] . \tag{1.132}$$

Dividing and multiplying by the operator

$$\left\langle \exp\left\{\int d\boldsymbol{x}' \xi(\boldsymbol{x}') \frac{\delta}{\delta\eta(\boldsymbol{x}')}\right\}\right\rangle$$

[it commutes with the operator in (1.132)], we arrive at

$$f[\boldsymbol{x}, \eta] = \frac{\left\langle \exp\left[\int d\boldsymbol{x}' \xi(\boldsymbol{x}') \frac{\delta}{\delta\eta(\boldsymbol{x}')}\right] \xi(\boldsymbol{x}) \right\rangle}{\left\langle \exp\left[\int d\boldsymbol{x}' \xi(\boldsymbol{x}') \frac{\delta}{\delta\eta(\boldsymbol{x}')}\right]\right\rangle} \times \left\langle \exp\left\{\int d\boldsymbol{x}' \xi(\boldsymbol{x}') \frac{\delta}{\delta\eta(\boldsymbol{x}')}\right\}\right\rangle R[\eta] . \tag{1.133}$$

We will next introduce a functional Ω of the function $u(\tilde{\boldsymbol{x}})$ that, in addition, is a convenient function of $\boldsymbol{x}$

$$\Omega[\boldsymbol{x}, u] = \frac{\left\langle \xi(\boldsymbol{x}) \exp\left[\mathrm{i} \int d\boldsymbol{x}' \xi(\boldsymbol{x}') u(\boldsymbol{x}')\right]\right\rangle}{\left\langle \exp\left[\mathrm{i} \int d\boldsymbol{x}' \xi(\boldsymbol{x}') u(\boldsymbol{x}')\right]\right\rangle} . \tag{1.134}$$

This functional allows us to write (1.134) in the form

$$f[\boldsymbol{x}, u] = \Omega\left[\boldsymbol{x}, \frac{1}{\mathrm{i}} \frac{\delta}{\delta\eta}\right] \left\langle \exp\left[\int d\boldsymbol{x}' \xi(\boldsymbol{x}') \frac{\delta}{\delta\eta(\boldsymbol{x}')}\right] R[\eta] \right\rangle ,$$

where in the last factor a nonrandom quantity $R[\eta]$ is introduced under the averaging sign. Again making use of (1.118), we obtain

$$f[\boldsymbol{x}, \eta] = \Omega\left[\boldsymbol{x}, \frac{1}{\mathrm{i}} \frac{\delta}{\delta\eta}\right] \langle R[\eta + \xi] \rangle ,$$

or, introducing the nonrandom operator under the averaging sign,

$$f[x,\eta] = \left\langle \Omega\left[x, \frac{1}{\mathrm{i}}\frac{\delta}{\delta\eta}\right] R[\eta+\xi] \right\rangle .$$

The functional R being dependent on the sum $\eta+\xi$ only, we can substitute the operator of differentiation with respect to η for the operator of differentiation with respect to ξ. Thus,

$$f[x,\eta] = \langle \xi(x) R[\eta+\xi] \rangle = \left\langle \Omega\left[x, \frac{1}{\mathrm{i}}\frac{\delta}{\delta\xi}\right] R[\eta+\xi] \right\rangle .$$

It is worth noting that here we can already take the auxilliary function η to be zero. We thus arrive at the final result due to *Klyatskin* [1.8]

$$\langle \xi(x) R[\xi] \rangle = \left\langle \Omega\left[x, \frac{1}{\mathrm{i}}\frac{\delta}{\delta\xi}\right] R[\xi] \right\rangle . \tag{1.135}$$

According to (1.134), (1.119) and (1.109), we can express $\Omega[x, u]$ through the logarithm of the characteristic functional of the field $\xi(x)$ as follows:

$$\Omega[x,u] = \frac{1}{\Phi[u]}\frac{1}{\mathrm{i}}\frac{\delta\Phi[u]}{\delta u(x)} = \frac{1}{\mathrm{i}}\frac{\delta \ln\Phi[u]}{\delta u(x)} = \frac{1}{\mathrm{i}}\frac{\delta\Theta[u]}{\delta u(x)} . \tag{1.136}$$

Let us now return to the expansion (1.123). Differentiating and taking account of the result in Exercise 1.8.13, we have

$$\begin{aligned}\Omega[x,u] &= \frac{1}{\mathrm{i}}\frac{\delta\Theta[u]}{\delta u(x)} \\ &= \sum_{k=1}^{\infty} \frac{i^{k-1}}{(k-1)!} \int \psi_k(x, x_1, \dots, x_{k-1}) u(x_1) \dots u(x_{k-1}) \, dx_1 \dots dx_{k-1} .\end{aligned}$$

Substituting here $[\delta/\delta\xi(x_l)]/\mathrm{i}$ for $u(x_l)$ and inserting the result into (1.135), we arrive at

$$\begin{aligned}\langle \xi(x) R[\xi] \rangle = \sum_{k=1}^{\infty} \frac{1}{(k-1)!} &\int \psi_k(x, x_1, \dots, x_{k-1}) \\ &\times \left\langle \frac{\delta^{(k-1)} R[\xi]}{\delta\xi(x_1) \dots \delta\xi(x_{k-1})} \right\rangle dx_1 \dots dx_{k-1} .\end{aligned} \tag{1.137}$$

Next, let us take the special case of the Gaussian field $\xi(x)$, such that $\psi_3 = \psi_4 = \dots = 0$. Hence, in (1.137) we have only the first two terms with $\psi_1(x) = \overline{\xi(x)}$ and $\psi_2(x, x') = \psi_\xi(x, x')$

$$\langle \xi(x) R[\xi] \rangle = \overline{\xi(x)} \langle R[\xi] \rangle + \int \psi_\xi(x, x') \left\langle \frac{\delta R[\xi]}{\delta\xi(x')} \right\rangle dx' . \tag{1.138}$$

Equation (1.138) was derived by *Furutsu* [1.9], *Novikov* [1.10], and *Donsker* [1.11].

We will now see how some of the above relationships will change if we consider the collection of N random fields (N-component field) $\xi_1(x), \ldots, \xi_N(x)$. The characteristic functional is then given by

$$\Phi[u_1, \ldots, u_N] = \left\langle \exp\left\{ \mathrm{i} \sum_1^N \int u_k(\boldsymbol{x}) \xi_k(\boldsymbol{x})\, d\boldsymbol{x} \right\} \right\rangle . \tag{1.139}$$

Equations (1.113–115) retain their earlier form, but with $\delta/\delta u(\boldsymbol{x})$ replaced by $\delta/\delta u_j(\boldsymbol{x})$. Instead of (1.116), we now have

$$\frac{\delta u_i(\boldsymbol{x})}{\delta u_k(\boldsymbol{x}')} = \delta_{ik} \delta(\boldsymbol{x} - \boldsymbol{x}') . \tag{1.140}$$

The Furutsu–Novikov formula (1.138) becomes

$$\langle \xi_i(\boldsymbol{x}) R[\xi_1, \ldots, \xi_N] \rangle = \overline{\xi_i(\boldsymbol{x})} \langle R \rangle + \sum_{k=1}^N \int \langle \tilde{\xi}_i(\boldsymbol{x}) \tilde{\xi}_k(\boldsymbol{x}') \rangle \left\langle \frac{\delta R}{\delta \xi_k(\boldsymbol{x}')} \right\rangle d\boldsymbol{x}' . \tag{1.141}$$

To conclude this section, we will formulate the properties of homogeneity and isotropy of random fields using the formalism of characteristic functionals. Statistical homogeneity (in the strict sense) of the field $\xi(\boldsymbol{x})$ means that $\langle \xi(\boldsymbol{x}_1 + \boldsymbol{x}_0) \ldots \xi(\boldsymbol{x}_k + \boldsymbol{x}_0) \rangle = \langle \xi(\boldsymbol{x}_1) \ldots \xi(\boldsymbol{x}_k) \rangle$. Using this equality, we may write (1.122) in the form

$$\Phi[u(\tilde{\boldsymbol{x}})] = 1 + \sum_{k=1}^{\infty} \frac{\mathrm{i}^k}{k!} \int \langle \xi(\boldsymbol{x}_1 + \boldsymbol{x}_0) \ldots \xi(\boldsymbol{x}_k + \boldsymbol{x}_0) \rangle \times u(\boldsymbol{x}_1) \ldots u(\boldsymbol{x}_k)\, d\boldsymbol{x}_1 \ldots d\boldsymbol{x}_k ,$$

or, after the change of the variables $\boldsymbol{x}_l + \boldsymbol{x}_0 = \boldsymbol{x}_l'$,

$$\Phi[u(\tilde{\boldsymbol{x}})] = 1 + \sum_{k=1}^{\infty} \frac{\mathrm{i}^k}{k!} \int \langle \xi(\boldsymbol{x}_1') \ldots \xi(\boldsymbol{x}_k') \rangle \times u(\boldsymbol{x}_1' - \boldsymbol{x}_0) \ldots u(\boldsymbol{x}_k' - \boldsymbol{x}_0)\, d\boldsymbol{x}_1' \ldots d\boldsymbol{x}_k' .$$

But, as is easily seen, the right-hand side will contain the functional $\Phi[u(\tilde{\boldsymbol{x}} - \boldsymbol{x}_0)]$. Thus, for statistically homogeneous fields

$$\Phi[u(\tilde{\boldsymbol{x}})] = \Phi[u(\tilde{\boldsymbol{x}} - \boldsymbol{x}_0)] , \tag{1.142}$$

i.e., the characteristic functional does not change, if its argument u is taken at a displaced point.

In a similar way, if $\hat{R}\boldsymbol{x} = \boldsymbol{x}'$ implies the transformation of rotation in the space of variables $\boldsymbol{x}$, then the statistical isotropy of the field $\xi(\boldsymbol{x})$ will be given by

$$\Phi[u(\tilde{\boldsymbol{x}})] = \Phi[u(\hat{R}^{-1}\tilde{\boldsymbol{x}})] . \tag{1.143}$$

Cases are possible where the field is statistically homogeneous or isotropic only in some but not in all the coordinates $x_1, \ldots, x_n$, i.e., in a certain subspace of the variables x. Then $\Phi[u]$ must be invariant under translation [x_0 in (1.142)] or rotation [$\hat{R}$ in (1.143)] carried out in this subspace only.

An example is the characteristic functional of a statistically homogeneous Gaussian field with $\bar{\xi} = 0$:

$$\Phi[u] = \exp\{-\tfrac{1}{2}\textstyle\iint \psi_\xi(x_1 - x_2)u(x_1)u(x_2)\,dx_1\,dx_2\} \ . \tag{1.144}$$

It is clear that (1.142) holds here. If the field in question is statistically isotropic, then in place of (1.144) we will have

$$\Phi[u] = \exp\{-\tfrac{1}{2}\textstyle\iint \psi_\xi(|x_1 - x_2|)u(x_1)u(x_2)dx_1\,dx_2\} \ . \tag{1.145}$$

As stated, this section is mainly concerned with the rules governing the use of functionals and the derivation of a number of important expressions and equations. Applications of the functional technique to specific problems are illustrated in some of the later chapters. In Exercise 1.8.22 the technique is used to derive the Einstein–Fokker-Planck equation (I.6.45) for the transition probabilities of the random response of a discrete dynamic system to Gaussian actions delta-correlated in time.

1.8 Exercises

1.8.1 Find the spatial spectral density for an anisotropic homogeneous random field with the Gaussian covariance of the general form

$$\psi_\zeta(r) = \sigma_\zeta^2 \exp\left\{-\tfrac{1}{2}\sum_{i,j} \alpha_{ij} x_i x_j\right\} ,$$

where $x_1 = x$, $x_2 = y$, $x_3 = z$ and $\mathrm{Det}\,\|\alpha_{ii}\| \neq 0$.

Solution.

$$\Phi_\zeta(k) = \frac{\sigma_\zeta^2}{\sqrt{(2\pi)^3 \mathrm{Det}\,\|\alpha_{ij}\|}} \exp\left\{-\tfrac{1}{2}\sum_{i,j} A_{ij} k_i k_j\right\} ,$$

where $\|A_{ij}\| = \|\alpha_{ij}\|^{-1}$ is the inverse matrix to $\|\alpha_{ij}\|$. In the particular case where the covariance has the form (1.40), the spatial spectrum is given by (1.59).

1.8.2 Prove the following inequality for the $(n+n)$th order statistical moments of the complex field $\zeta(r)$

$$|B_{n,n}(1, \ldots, 2n)|^2$$
$$\leqslant B_{n,n}(1, \ldots, n; n, \ldots, 1) B_{n,n}(n+1, \ldots, 2n; 2n, \ldots, n+1) \ . \tag{1.146}$$

Solution. The inequality follows from the fact that for any complex λ, we have

$$\langle|\lambda\zeta(1)\ldots\zeta(n)+\zeta(n+1)\ldots\zeta(2n)|^2\rangle\geqslant 0\ .$$

If we set

$$\lambda=\alpha[B_{n,n}(1,\ldots,2n)]^*$$

where α is real, then (1.146) will be the condition that the trinomial which is quadratic in α be nonnegative. A special case of (1.146) at $n=1$ is

$$|B(1,2)|^2\leqslant B(1,1)B(2,2)\ ,$$

which is similar to (1.31).

1.8.3 If we draw a straight line in space and consider the values of a homogeneous and isotropic field $\zeta(\boldsymbol{r})$ on this line only, we will obtain a random function of a single spatial variable z, which is the coordinate reckoned along this line. We can represent the covariance $\psi_\zeta(z)$ as a one-dimensional Fourier integral

$$\psi_\zeta(z)=\int_{-\infty}^{\infty}\gamma_\zeta(k)\cos kz\,dk\ ,$$

where $\gamma_\zeta(k)$ is the one-dimensional spectral density expressed in terms of $\psi_\zeta(z)$ using the inverse Fourier transform

$$\gamma_\zeta(k)=\frac{1}{2\pi}\int_{-\infty}^{\infty}\psi_\zeta(z)\cos kz\,dz\ . \qquad (1.147)$$

Show that the three-dimensional spatial spectrum $\Phi_\zeta(k)$ of the homogeneous isotropic field $\zeta(r)$ is related to the one-dimensional spectrum $\gamma_\zeta(k)$ by

$$\Phi_\zeta(k)=-\frac{1}{2\pi k}\frac{\partial\gamma_\zeta(k)}{\partial k}\ . \qquad (1.148)$$

Hint: Differentiate (1.147) with respect to k and compare the result with (1.53).
Solution. Equation (1.148) can conveniently be used to calculate the three-dimensional spectrum in those frequent cases where the integral (1.147) for $\gamma_\zeta(k)$ is derived more simply than the integral (1.53) for $\Phi_\zeta(k)$. The simple covariances (1.35, 36) serve as examples.

Equation (1.148) is also used to find the three-dimensional spectrum of the isotropic random field from the one-dimensional spectrum $\gamma_\zeta(k)$ obtained from the experimentally found one-dimensional covariance $\psi_\zeta(z)$.

1.8.4 As in the preceding problem, let z be the coordinate along some preselected direction in space, and the random field $\zeta(\boldsymbol{r})$ be homogeneous in the plane $z=\text{const}$. Introducing in this plane the two-dimensional vector $\boldsymbol{\varrho}=(x,y)$ such that

$\boldsymbol{r} = (\boldsymbol{\varrho}, z)$, we can write the covariance of $\zeta(\boldsymbol{r})$ in the form

$$\psi_\zeta(\boldsymbol{\varrho}, z_1 - z_2) = \langle \tilde{\zeta}(\boldsymbol{\varrho}_1, z_1)\tilde{\zeta}^*(\boldsymbol{\varrho}_1 - \boldsymbol{\varrho}, z_2)\rangle \ .$$

This function can be represented as the two-dimensional Fourier integral

$$\psi_\zeta(\boldsymbol{\varrho}, z_1 - z_2) = \int_{-\infty}^{\infty} F_\zeta(\boldsymbol{k}_\perp, z_1 - z_2)\exp(\mathrm{i}\boldsymbol{k}_\perp \cdot \boldsymbol{\varrho})\, d^2k_\perp \ , \tag{1.149}$$

where $\boldsymbol{k}_\perp = (k_x, k_y)$. The function $F_\zeta(\boldsymbol{k}_\perp, z_1 - z_2)$ is called the *two-dimensional spectral density* of the field, or simply the two-dimensional (spatial) spectrum. Show that for a homogeneous, *isotropic* random field the two-dimensional spectrum $F_\zeta(\boldsymbol{k}_\perp, z_1 - z_2)$ and the three-dimensional spectrum $\Phi_\zeta(k)$ are related by

$$F_\zeta(\boldsymbol{k}_\perp, z_1 - z_2) = \int_{-\infty}^{\infty} \Phi_\zeta(\sqrt{k_z^2 + k_\perp^2})\cos k_z(z_1 - z_2)\, dk_z \ , \tag{1.150}$$

$$\Phi_\zeta(k) = \frac{1}{2\pi}\int_{-\infty}^{\infty} F_\zeta(k, z_1 - z_2)d(z_1 - z_2) \ . \tag{1.151}$$

Solution. Equation (1.150) follows from a comparison of (1.149) with the three-dimensional expansion (1.148) if we take account of the fact that the covariance $\psi_\zeta(r)$ is real for an isotropic field. Inverting (1.150) gives

$$\Phi_\zeta(\sqrt{k_z^2 + k_\perp^2}) = \frac{1}{2\pi}\int_{-\infty}^{\infty} F_\zeta(\boldsymbol{k}_\perp, z_1 - z_2)\cos k_z z\, d(z_1 - z_2) \ .$$

Equation (1.151) follows for $k_z = 0$. Note that $\Phi_\zeta(\sqrt{k_z^2 + k_\perp^2})$ as a function of k_z has the characteristic scale $k_\perp$. Accordingly, (1.150) suggests that $F_\zeta(k_\perp, z_1 - z_2)$ is concentrated within the confines given by

$$k_\perp |z_1 - z_2| < 2\pi \ . \tag{1.152}$$

1.8.5 Express the spectral expansion of the covariance of a field that is isotropic in the plane $z = \text{const}$ as a single integral.
Solution. In the general case, the covariance $\psi_\zeta(x, y) = \psi_\zeta(\boldsymbol{\varrho})$ and its two-dimensional spectrum $F_\zeta(\kappa_x, \kappa_y) = F_\zeta(\boldsymbol{\kappa})$ are connected by the direct and inverse Fourier transforms

$$\psi_\zeta(\boldsymbol{\varrho}) = \int F_\zeta(\boldsymbol{\kappa})\exp(\mathrm{i}\boldsymbol{\kappa}\cdot\boldsymbol{\varrho})\, d^2\kappa, \quad F_\zeta(\boldsymbol{\kappa}) = \frac{1}{(2\pi)^2}\int \psi_\zeta(\boldsymbol{\varrho})\exp(-\mathrm{i}\boldsymbol{\kappa}\cdot\boldsymbol{\varrho})\, d^2\varrho \ . \tag{1.153}$$

For an isotropic field we have $\psi_\zeta(\boldsymbol{\varrho}) = \psi_\zeta(\varrho)$ and $F_\zeta(\boldsymbol{\kappa}) = F_\zeta(\kappa)$. Passing in (1.153) to the polar coordinates (in $\boldsymbol{\varrho}$- and $\boldsymbol{\kappa}$-spaces, respectively) and integrating with respect to angular variables, we obtain the direct and inverse Hankel transforms

$$\psi_\zeta(\varrho) = 2\pi\int_0^{\infty} F_\zeta(\kappa)J_0(\kappa\varrho)\kappa\, d\kappa, \quad F_\zeta(\kappa) = \frac{1}{2\pi}\int_0^{\infty} \psi_\zeta(\varrho)J_0(\kappa\varrho)\varrho\, d\varrho \ , \tag{1.154}$$

where $\varrho = \sqrt{x^2 + y^2}$, $\kappa = \sqrt{\kappa_x^2 + \kappa_y^2}$, and J_0 is the zeroth-order Bessel function.

1.8.6 Find the frequency spectrum $g_\zeta(\omega)$ of the frozen field $\zeta(t, \boldsymbol{r})$, if the appropriate random field $v(\boldsymbol{r})$ at rest has the Gaussian spatial covariance $\psi_v(\boldsymbol{r}) = \sigma_v^2 \exp(-r^2/2l^2)$.
Solution:

$$g_\zeta(\omega) = \frac{\sigma_v^2 l}{\sqrt{2\pi}\, v} \exp\left\{-\frac{1}{2}\left(\frac{\omega l}{v}\right)^2\right\} .$$

1.8.7 Given the covariance $\psi_v(\boldsymbol{r})$ [or spatial spectrum $\Phi_v(\boldsymbol{k})$] of a field $v(\boldsymbol{r})$ at rest, find the covariance, space-time, spatial and frequency spectra of the frozen field $\zeta(t, \boldsymbol{r})$ for relativistic values of the velocity v.
Solution. If the velocity $\boldsymbol{v}$ is directed along the x-axis, then according to the Lorentz transformation

$$\Psi_\zeta(\tau, \boldsymbol{r}) = \psi_v\left(\frac{x - v\tau}{\sqrt{1 - v^2/c^2}}, y, z\right) ,$$

$$G_\zeta(\omega, \boldsymbol{k}) = \sqrt{1 - v^2/c^2}\, \Phi_v(k_x\sqrt{1 - v^2/c^2}, k_y, k_z)\delta(\omega - k_x v\sqrt{1 - v^2/c^2}) ,$$

$$\Phi_\zeta(\boldsymbol{k}) = \sqrt{1 - v^2/c^2}\, \Phi_v(k_x\sqrt{1 - v^2/c^2}, k_y, k_z) ,$$

$$g_\zeta(\omega) = \frac{1}{v} \iint_{-\infty}^{\infty} \Phi_v\left(\frac{\omega}{v}, k_y, k_z\right) dk_y\, dk_z .$$

Note that if the field $v(\boldsymbol{r})$ is isotropic, the frozen field $\zeta(t, \boldsymbol{r})$ is statistically anisotropic.

1.8.8 Let $\boldsymbol{a}(\boldsymbol{r})$ and $\boldsymbol{b}(\boldsymbol{r})$ be isotropic and isotropically connected vector random fields. Find the general form of the correlation tensor $\langle a_\alpha(\boldsymbol{r}_1) b_\beta^*(\boldsymbol{r}_2)\rangle$.
Solution. To construct the correlation tensor of homogeneous isotropic vector fields, we only have scalar functions of $\varrho \equiv |\boldsymbol{r}_1 - \boldsymbol{r}_2|$, the unit vector $\boldsymbol{\varrho}/\varrho$ and the unit tensor $\delta_{\alpha\beta}$. Therefore, the correlation tensor will in general have the form

$$\langle a_\alpha(\boldsymbol{r}_1) b_\beta^*(\boldsymbol{r}_2)\rangle = F(\varrho)\delta_{\alpha\beta} + [G(\varrho) - F(\varrho)]\frac{\varrho_\alpha \varrho_\beta}{\varrho^2} . \tag{1.155}$$

The selection of the scalar coefficients in (1.155) in the form of F and G–F is determined by the fact that $G(\varrho)$ and $F(\varrho)$ describe the *longitudinal* and *transverse* correlations. This becomes clear if we consider the correlation between the α-components a_α and b_α. If the points $\boldsymbol{r}_1$ and $\boldsymbol{r}_2$ lie on a line parallel to the x_α-axis, i.e., $\varrho = \varrho_\alpha$, then (Fig. 1.2a)

$$\langle a_\alpha(\boldsymbol{r}_1) b_\alpha^*(\boldsymbol{r}_2)\rangle = G(\varrho)$$

(the longitudinal correlation). If the vector $\boldsymbol{\varrho} = \boldsymbol{r}_1 - \boldsymbol{r}_2$ is orthogonal to the x_α axis, then $\varrho_\alpha = 0$ (Fig. 1.2b) and

$$\langle a_\alpha(\boldsymbol{r}_1) b_\alpha^*(\boldsymbol{r}_2)\rangle = F(\varrho)$$

(the transverse correlation).

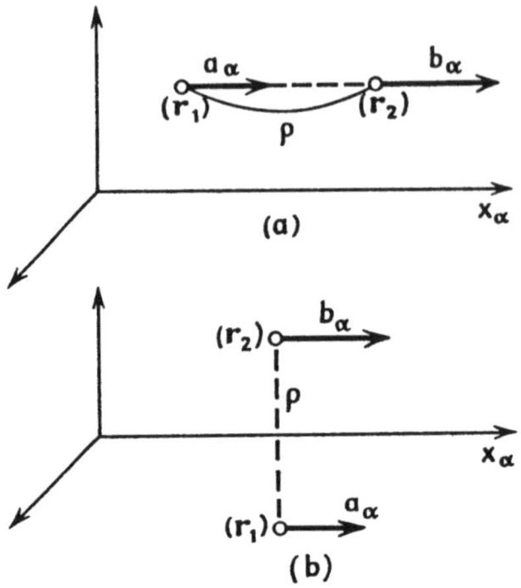

Fig. 1.2. Longitudinal (**a**) and transverse (**b**) separation between $\boldsymbol{a}$ and $\boldsymbol{b}$

Instead of $F(\varrho)$ and $G(\varrho)$, we can introduce two other scalar functions by setting

$$\langle a_\alpha(\boldsymbol{r}_1)b^*_\beta(\boldsymbol{r}_2)\rangle = U(\varrho)\delta_{\alpha\beta} + \frac{\partial^2 V(\varrho)}{\partial\varrho_\alpha\partial\varrho_\beta} \; . \tag{1.156}$$

Comparing (1.155) and (1.156), we find

$$F(\varrho) = U(\varrho) + \frac{1}{\varrho}V'(\varrho), \quad G(\varrho) = U(\varrho) + V''(\varrho) \; , \tag{1.157}$$

where the prime stands for the derivative with respect to ϱ.

1.8.9 Show that for the homogeneous, isotropic *potential* field $\boldsymbol{a}(\boldsymbol{r})$ (rot $\boldsymbol{a} = 0$), the longitudinal and transverse covariances $F(\varrho)$ and $G(\varrho)$, introduced in the preceding problem, are connected by the relationship

$$G(\varrho) = [\varrho F(\varrho)]' \tag{1.158}$$

due to A. M. Obukhov, but for the *solenoidal* field (div $\boldsymbol{a} = 0$), by the relationship

$$F(\varrho) = \frac{1}{2\varrho}[\varrho^2 G(\varrho)]' \tag{1.159}$$

due to Th. Karman.

Solution. Let $\boldsymbol{a} = -\nabla\varphi$ be a potential field, and $\psi_\varphi(\varrho)$ the covariance of the potential $\varphi(\boldsymbol{r})$. Then

$$\langle a_\alpha(\boldsymbol{r}_1)a^*_\beta(\boldsymbol{r}_2)\rangle = -\frac{\partial^2\psi_\varphi(\varrho)}{\partial\varrho_\alpha\partial\varrho_\beta} \; ,$$

i.e., in (1.156) $U = 0$, $V = -\psi_\varphi$. In consequence, from (1.157) $F(\varrho) = V'/\varrho$ and $G = V''$, and we thus obtain (1.158).

As for the solenoidal field, we multiply div $\boldsymbol{a}(\boldsymbol{r}_1) = 0$ by $\boldsymbol{a}(\boldsymbol{r}_2)$ and carry out the statistical averaging using (1.156). This leads to the condition $\nabla(U + \Delta V) = 0$, hence $U + \Delta V = 0$ up to a constant, which can always be included into U. We thus arrive at

$$U = -\Delta V = -\frac{1}{\varrho^2}\frac{d}{d\varrho}\left(\varrho^2 \frac{dV}{d\varrho}\right) = -\frac{2V'}{\varrho} - V'' \ .$$

According to (1.157)

$$F = -\frac{1}{\varrho} V' - V'', \qquad G = -\frac{2}{\varrho} V' \ ,$$

which is equivalent to (1.159).

1.8.10 Consider a vector solenoidal field $\boldsymbol{a}(\boldsymbol{r})$ that is homogeneous and isotropic. Show that it will not correlate with the scalar field $\varphi(\boldsymbol{r})$ (A. M. Obukhov) nor with the vector potential field $\boldsymbol{b}(\boldsymbol{r}) = -\nabla\varphi(\boldsymbol{r})$.
Solution. For a homogeneous isotropic field the covariance $\langle \varphi(\boldsymbol{r}_1)\boldsymbol{a}^*(\boldsymbol{r}_2)\rangle$ must be a vector parallel to $\boldsymbol{\varrho}$

$$\langle \varphi(\boldsymbol{r}_1)\boldsymbol{a}^*(\boldsymbol{r}_2)\rangle = H(\varrho)\frac{\boldsymbol{\varrho}}{\varrho} \equiv \boldsymbol{B}(\boldsymbol{\varrho}) \ ,$$

where $H(\varrho)$ is a scalar function. Since for a solenoidal field div $\boldsymbol{a} = 0$, we have

$$\nabla_2 \langle \varphi(\boldsymbol{r}_1)a^*(\boldsymbol{r}_2)\rangle = -\operatorname{div}_\varrho \boldsymbol{B}(\boldsymbol{\varrho}) = 0 \ ,$$

i.e., $\boldsymbol{B}$ must be solenoidal (∇_2 means differentiating with respect to $\boldsymbol{r}_2$). In the absence of sources, the only permissible solution is $\boldsymbol{B} = 0$, which suggests that the solenoidal and potential vector fields are uncorrelated

$$\langle b_\beta(\boldsymbol{r}_1)a_\alpha^*(\boldsymbol{r}_2)\rangle = -\frac{\partial}{\partial r_{1\beta}}\langle \varphi(\boldsymbol{r}_1)a_\alpha^*(\boldsymbol{r}_2)\rangle = -\frac{\partial B_\alpha(\boldsymbol{\varrho})}{\partial \varrho_\beta} = 0 \ .$$

1.8.11 Fluctuations with covariances of the type (1.35, 36) are referred to as *one-scale* fluctuational fields, which implies that the behavior of $\psi(\boldsymbol{r})$ is only characterized by a single scale l (clearly, the spatial spectrum of such fluctuations contains a multitude of scales concentrated within the region $k \lesssim 1/l$). Real covariances often have many characteristic scales (*multiscale* fluctuations). Sometimes such covariances can conveniently be approximated by a superposition of Gaussian covariances, i.e., the correlation coefficient is written as [1.12]

$$K(\varrho) = \int_0^\infty f(a)\mathrm{e}^{-\varrho^2/a^2} da \ .$$

The weight function $f(a)$ is subject, of course, to the normalization condition

$$K(0) = \int_0^\infty f(a)\, da = 1 \ .$$

The conventional Gaussian correlation coefficient $K(\varrho) = \exp(-\varrho^2/2l^2)$ is obtained at $f(a) = \delta(a - \sqrt{2}l)$.

Find the weight function $f(a)$ for:

a) the exponential correlation coefficient $\exp(-r/l)$
b) the locally homogeneous, isotropic model of turbulent fluctuations of the dielectric permittivity of air. The correlation coefficient for these fluctuations, by (1.81), has the asymptotic expressions

$$K(r) = \frac{\psi(r)}{\psi(0)} = 1 - \frac{D(r)}{D(\infty)} = \begin{cases} 1 - r^2 l_0^{-4/3} L_0^{-2/3}, & r \ll l_0 , \\ 1 - (r/L_0)^{2/3}, & l_0 \ll r \ll L_0 , \\ 0, & L_0 \ll r . \end{cases}$$

Solution.

$$\text{(a)}\ f(a) = \frac{1}{l\sqrt{\pi}} e^{-a^2/4l^2} .$$

$$\text{(b)}\ f(a) = \begin{cases} 0 & \text{at} \quad a < l_0 \quad \text{and} \quad a > L_0 , \\ \frac{2}{3} a^{-1/3} L_0^{-2/3} & \text{at} \quad l_0 \leqslant a \leqslant L_0 . \end{cases}$$

1.8.12 In Sect. II.2.6 an expression was derived for the covariance of intensities of the two normal complex oscillations $\zeta_1(t)$ and $\zeta_2(t)$, assuming that $\bar{\zeta}_1 = \bar{\zeta}_2 = 0$. Find a similar expression for the case where the mean of the random field $\zeta(t, r)$ is nonzero.

Solution. Using the notation $I_m = |\zeta_m|^2$, where $\zeta_m = \zeta(t_m, r_m)$, $m = 1, 2$, the covariance of the intensities is

$$\psi_I(1, 2) = \langle I_1 I_2 \rangle - \bar{I}_1 \bar{I}_2 = \langle \tilde{I}_1 \tilde{I}_2 \rangle , \quad \text{where}$$
$$\tilde{I} = I - \bar{I} = |\tilde{\zeta}|^2 - \langle |\tilde{\zeta}|^2 \rangle + (\tilde{\zeta}\bar{\zeta}^* + \text{complex conjugate}) .$$

Using the results of Exercise II.1.6.2, we obtain

$$\psi_I(1, 2) = |\psi_\zeta(1, 2)|^2 + |\tilde{\psi}_\zeta(1, 2)|^2 + [\psi_\zeta(1, 2)\bar{\zeta}_1^* \bar{\zeta}_2 + \tilde{\psi}_\zeta(1, 2)\bar{\zeta}_1^* \bar{\zeta}_2^*$$
$$+ \text{complex conjugate}] .$$

At $\bar{\zeta}_1 = \bar{\zeta}_2 = 0$ we arrive at the relationship

$$\psi_I(1,2) = |\psi_\zeta(1,2)|^2 + |\tilde{\psi}_\zeta(1,2)|^2 ,$$

which, at $\tilde{\psi}_\zeta = 0$, takes the form $\psi_I(1,2) = |\psi_\zeta(1,2)|^2$.

1.8.13 Let $F_N[u]$ be a functional of the Nth degree

$$F_N[u] = \int A(\boldsymbol{x}_1, \ldots, \boldsymbol{x}_N) u(\boldsymbol{x}_1) \cdot \ldots u(\boldsymbol{x}_N) d\boldsymbol{x}_1 \ldots d\boldsymbol{x}_N .$$

Show that:

a) the value of $F_N[u]$ is only dependent on the part of the function A that is symmetrical with respect to all the arguments

b) $$\frac{\delta F_N[u]}{\delta u(\boldsymbol{x})} = \int [A(\boldsymbol{x}, \boldsymbol{x}_1, \ldots, \boldsymbol{x}_{N-1}) + A(\boldsymbol{x}_1, \boldsymbol{x}, \boldsymbol{x}_2, \ldots, \boldsymbol{x}_{N-1})$$
$$\ldots + A(\boldsymbol{x}_1, \boldsymbol{x}_2, \ldots, \boldsymbol{x}_{N-1}, \boldsymbol{x})] u(\boldsymbol{x}_1) \ldots u(\boldsymbol{x}_{N-1}) d\boldsymbol{x}_1 \ldots d\boldsymbol{x}_{N-1}$$

c) in the case of symmetrical A

$$\frac{\delta F_N[u]}{\delta u(\boldsymbol{x})} = N \int A(\boldsymbol{x}, \boldsymbol{x}_1, \ldots, \boldsymbol{x}_{N-1}) u(\boldsymbol{x}_1) \ldots u(\boldsymbol{x}_{N-1}) d\boldsymbol{x}_1 \ldots d\boldsymbol{x}_{N-1}$$

(analog of $dx^N/dx = Nx^{N-1}$):

Hint: Three approaches are possible, based on (1.110), (1.112), or (1.113–116).

1.8.14 Find the variational derivative of the functional of the action arising from classical mechanics

$$S[x(\tilde{\tau})] = \int_{t_1}^{t_2} \left[\frac{m}{2} \left(\frac{dx(\tau)}{d\tau} \right)^2 - V(x(\tau)) \right] d\tau \ .$$

Solution. Differentiating the integrand gives

$$\frac{\delta S}{\delta x(t)} = \int_{t_1}^{t_2} \left[\frac{m}{2} \frac{\delta}{\delta x(t)} \left(\frac{dx(\tau)}{d\tau} \right)^2 - \frac{\delta V(x(\tau))}{\delta x(t)} \right] d\tau \ .$$

Using (1.115), we have

$$\frac{\delta}{\delta x(t)} \left(\frac{dx(\tau)}{d\tau} \right)^2 = 2 \frac{dx(\tau)}{d\tau} \frac{\delta}{\delta x(t)} \frac{dx(\tau)}{d\tau} \ .$$

Since the derivative $dx(\tau)/d\tau$ is the limit of the ratio $[x(\tau_2) - x(\tau_1)/(\tau_2 - \tau_1)]$, we can, from (1.113), introduce the operation of the functional differentiation under the operator $d/d\tau$, and then use (1.116):

$$\frac{\delta}{\delta x(t)} \frac{dx(\tau)}{d\tau} = \frac{\partial}{\partial \tau} \frac{\delta x(\tau)}{\delta x(t)} = \frac{\partial}{\partial \tau} \delta(\tau - t) = -\frac{\partial}{\partial t} \delta(\tau - t) \ .$$

For the second term in $S[x]$, we obtain, using (1.115, 116),

$$\frac{\delta V(x,(\tau))}{\delta x(t)} = V'(x(\tau)) \frac{\delta x(\tau)}{\delta x(t)} = \delta(\tau - t) V'(x(\tau)) \ .$$

We thus arrive at

$$\frac{\delta S[x]}{\delta x(t)} = \int_{t_1}^{t_2} \left[-m \frac{dx(\tau)}{d\tau} \frac{\partial}{\partial t} \delta(\tau - t) - V'(x(\tau))\delta(\tau - t) \right] d\tau$$

$$= -m \frac{\partial}{\partial t} \int_{t_1}^{t_2} \frac{dx(\tau)}{d\tau} \delta(\tau - t) d\tau - \int_{t_1}^{t_2} V'(x(\tau))\delta(\tau - t) d\tau \ ,$$

hence

$$\frac{\delta S[x]}{\delta x(t)} = -m \frac{d^2 x(t)}{dt^2} - V'(x(t)) \ .$$

From the principle of least action $\delta S[x]/\delta x(t) = 0$, the equation of motion will be $m\ddot{x} = -V'(x)$.

1.8.15 Find $\delta S[x]/\delta x(t)$ for the functional

$$S[x] = \int_{t_1}^{t_2} L(x(\tau), \dot{x}(\tau)) d\tau \ .$$

Solution.

$$\frac{\delta S[x]}{\delta x(t)} = \frac{\partial L}{\partial x} - \frac{d}{dt} \frac{\partial L}{\partial \dot{x}} \ ,$$

i.e., the variational derivative is the left-hand side of the Euler–Lagrange equation. This result is obtained using (1.113–116), as in the preceding problem.

1.8.16 Derive the equation for $\psi_3(\boldsymbol{x}_1, \boldsymbol{x}_2, \boldsymbol{x}_3)$ analogous to the expression (1.125) for $\psi_2(\boldsymbol{x}_1, \boldsymbol{x}_2)$.
Solution.

$$\begin{aligned}\psi_3(\boldsymbol{x}_1, \boldsymbol{x}_2, \boldsymbol{x}_3) = &\langle \xi(\boldsymbol{x}_1)\xi(\boldsymbol{x}_2)\xi(\boldsymbol{x}_3)\rangle - \bar{\xi}(\boldsymbol{x}_1)\langle \xi(\boldsymbol{x}_2)\xi(\boldsymbol{x}_3)\rangle \\ &- \bar{\xi}(\boldsymbol{x}_2)\langle \xi(\boldsymbol{x}_1)\xi(\boldsymbol{x}_3)\rangle - \bar{\xi}(\boldsymbol{x}_3)\langle \xi(\boldsymbol{x}_1)\xi(\boldsymbol{x}_2)\rangle \\ &+ 2\bar{\xi}(\boldsymbol{x}_1)\bar{\xi}(\boldsymbol{x}_2)\bar{\xi}(\boldsymbol{x}_3) \ .\end{aligned}$$

1.8.17 From the expressions (1.128) for the central moments of the Gaussian random field $\xi(\boldsymbol{x})$, find the characteristic functional for the fluctuations $\tilde{\xi}(\boldsymbol{x})$.
Solution. By (1.122), we have

$$\Phi_{\tilde{\xi}}[u] = 1 + \sum_{k=1}^{\infty} \frac{\mathrm{i}^k}{k!} \int \langle \tilde{\xi}(\boldsymbol{x}_1) \ldots \tilde{\xi}(\boldsymbol{x}_k)\rangle u(\boldsymbol{x}_1) \ldots u(\boldsymbol{x}_k) d\boldsymbol{x}_1 \ldots d\boldsymbol{x}_k$$

where, by virtue of (1.128), only the terms with $k = 2m$ remain

$$\Phi_{\tilde{\xi}}[u] = 1 + \sum_{m=1}^{\infty} \frac{(-1)^m}{(2m)!} \int \langle \tilde{\xi}(\boldsymbol{x}_1) \ldots \tilde{\xi}(\boldsymbol{x}_{2m})\rangle u(\boldsymbol{x}_1) \ldots u(\boldsymbol{x}_{2m}) d\boldsymbol{x}_1 \ldots d\boldsymbol{x}_{2m} \ .$$

Using now the second of the equations (1.128), we have

$$\Phi_{\tilde{\xi}}[u] = 1 + \sum_{m=1}^{\infty} \frac{(-1)^m}{(2m)!} \sum_{\text{d.p.}} \int \langle \tilde{\xi}(x_{\alpha_1}) \tilde{\xi}(x_{\alpha_2}) \rangle \ldots \langle \tilde{\xi}(x_{\alpha_{2m-1}}) \tilde{\xi}(x_{\alpha_{2m}}) \rangle$$
$$\times\, u(x_{\alpha_1}) u(x_{\alpha_2}) \ldots u(x_{\alpha_{2m-1}}) u(x_{\alpha_{2m}}) dx_1 \ldots dx_{2m} \;.$$

The integral of multiplicity $2m$ breaks down into the product of m double integrals that differ only in integration variables. Each of the double integrals has the form

$$I = \int\int \langle \tilde{\xi}(x') \tilde{\xi}(x'') \rangle u(x') u(x'')\, dx'\, dx'' \;.$$

Hence, each of the terms of the internal sum is I^m, and since the sum contains $(2m-1)!!$ terms, the functional $\Phi_{\tilde{\xi}}[u]$ will be

$$\Phi_{\tilde{\xi}}[u] = 1 + \sum_{m=1}^{\infty} \frac{(-1)^m}{(2m)!} (2m-1)!!\, I^m \;,$$

or, since $(2m)! = (2m)!!.\ (2m-1)!! = 2^m m! (2m-1)!!$

$$\Phi_{\tilde{\xi}}[u] = 1 + \sum_{m=1}^{\infty} \frac{(-1)^m}{m!} \left(\frac{I}{2}\right)^m = \exp\left(-\frac{I}{2}\right) .$$

Thus,

$$\Phi_{\tilde{\xi}}[u] = \exp\left\{ -\tfrac{1}{2} \int\int \psi_{\xi}(x_1, x_2) u(x_1) u(x_2)\, dx_1\, dx_2 \right\} .$$

Clearly, this expression is a special case of (1.126). It corresponds to the zero mean, as should be the case for $\tilde{\xi} = \xi - \bar{\xi}$ since $\langle \tilde{\xi} \rangle = 0$. This proves the equivalence of the definitions of the Gaussian random field in terms of (1.128) and the characteristic functional (1.126).

1.8.18 Find the characteristic functional of the Poisson random field (1.129).

Solution. By the definition of the Poisson field, we have

$$\Phi[u] = \left\langle \left\langle \left\langle \exp\left\{ \mathrm{i} \int dx\, u(x) \sum_{k=0}^{m} A_k g(x - x_k) \right\} \right\rangle_{A_k} \right\rangle_{x_k} \right\rangle_m ,$$

where each of the angle brackets with a subscript implies averaging over appropriate random variables. Clearly,

$$\Phi[u] = \left\langle \left\langle \left\langle \prod_{k=0}^{m} \exp\left\{ \mathrm{i} A_k \int g(x - x_k) u(x)\, dx \right\} \right\rangle_{A_k} \right\rangle_{x_k} \right\rangle_m .$$

Next, we average over A_k. Since $\langle \exp(\mathrm{i} A_k \lambda) \rangle_{A_k} = \chi(\lambda)$, where $\chi(\lambda)$ is the characteristic function for A_k, we have

$$\Phi[u] = \left\langle \left\langle \prod_{k=0}^{m} \chi\left(\int g(x - x_k) u(x) dx \right) \right\rangle_{x_k} \right\rangle_m .$$

Further, we average over $\boldsymbol{x}_k$. Essentially, this means integrating each multiplier with respect to $\boldsymbol{x}_k$, with the weight V^{-1} over the region V

$$\Phi[u] = \langle V^{-1} \int_V \chi(\int g(\boldsymbol{x}-\boldsymbol{x}_1)u(\boldsymbol{x})d\boldsymbol{x})d\boldsymbol{x}_1 \dots$$

$$\dots V^{-1} \int_V \chi(\int g(\boldsymbol{x}-\boldsymbol{x}_m)u(\boldsymbol{x})d\boldsymbol{x})d\boldsymbol{x}_m \rangle_m$$

$$= \langle [V^{-1} \int_V \chi(\int g(\boldsymbol{x}-\boldsymbol{x}')u(\boldsymbol{x})d\boldsymbol{x})d\boldsymbol{x}']^m \rangle_m \ .$$

If we denote the expression in brackets by a, the last averaging (using the Poisson distribution for m) gives

$$\langle a^m \rangle_m = \sum_{m=0}^{\infty} P(m) a^m = \mathrm{e}^{-\bar{m}} \sum_{m=0}^{\infty} \frac{(\bar{m})^m}{m!} a^m = \mathrm{e}^{-\bar{m}} \mathrm{e}^{a\bar{m}} = \exp[\bar{m}(a-1)] \ .$$

We thus obtain

$$\Phi[u] = \exp\{\bar{m}[V^{-1} \int_V \chi(\int g(\boldsymbol{x}-\boldsymbol{x}')u(\boldsymbol{x})d\boldsymbol{x})d\boldsymbol{x}' - 1]\}$$

which, by virtue of the identity

$$1 = V^{-1} \int_V 1 \cdot d\boldsymbol{x}' \ ,$$

can be written as

$$\Phi[u] = \exp\left\{\frac{\bar{m}}{V} \int_V [\chi(\int g(\boldsymbol{x}-\boldsymbol{x}')u(\boldsymbol{x})d\boldsymbol{x}) - 1]d\boldsymbol{x}'\right\} \ .$$

Introducing $\nu = \bar{m}/V$—the mean number of random points in a unit volume—we arrive at (1.130).

1.8.19 From (1.130), obtain cumulant functions of the Poisson random field.
Solution. We have

$$\Theta[u] = \ln \Phi[u] = \nu \int [\chi(\int g(\boldsymbol{x}-\boldsymbol{x}')u(\boldsymbol{x})d\boldsymbol{x}) - 1]d\boldsymbol{x}' \ .$$

Substituting the expansion of the characteristic function

$$\chi(\lambda) = \sum_{k=0}^{\infty} \frac{(\mathrm{i}\lambda)^k}{k!} \langle A^k \rangle$$

gives

$$\Theta[u] = \nu \sum_{k=1}^{\infty} \frac{\mathrm{i}^k}{k!} \langle A^k \rangle \int [\int g(\boldsymbol{x}-\boldsymbol{x}')u(\boldsymbol{x})d\boldsymbol{x}]^k d\boldsymbol{x}' \ .$$

We now write the kth power of the integral in the form of the k-fold multiple integral, and change the order of integration with respect to x' and x_l $(l = 1, 2, \ldots, k)$

$$\Theta[u] = \nu \sum_{k=1}^{\infty} \frac{\mathrm{i}^k}{k!} \langle A^k \rangle \int dx' \int dx_1 \ldots \int dx_k g(x_1 - x') \ldots$$
$$\ldots g(x_k - x') u(x_1) \ldots u(x_k)$$
$$= \sum_{k=1}^{\infty} \frac{\mathrm{i}^k}{k!} \int u(x_1) \ldots u(x_k)\, dx_1 \ldots dx_k \{ \nu \langle A^k \rangle$$
$$\times \int dx' g(x_1 - x') \ldots g(x_k - x') \} .$$

Comparing this with (1.123) gives

$$\psi_k(x_1, \ldots, x_k) = \nu \langle A^k \rangle \int g(x_1 - x') \ldots g(x_k - x')\, dx' .$$

1.8.20 Show that if A has the Gaussian probability distribution and $\bar{A} = 0$, then for $\nu \to \infty$, $\overline{A^2} \to 0$ and $\nu \overline{A^2} = \text{const}$, the Poisson random field tends to a Gaussian one. Show that a similar result can be obtained if we assume that A has the probability density $w(A) = (2A_0)^{-1} \exp(-|A|/A_0)$, i.e., the normality of the amplitude distribution is insignificant.
Hint: Use the formula (1.131) for the cumulants of the Poisson random field.

1.8.21 From (1.135), derive the equation to calculate the mean of the product of the Poisson random field $\xi(x)$ and the functional $R[\xi]$.
Solution. According to (1.130),

$$\Theta[u] = \nu \int [\chi(\int g(x'' - x') u(x'') dx'') - 1]\, dx' .$$

Operating on this with the operator $[\delta/\delta u(x)]/\mathrm{i}$ gives

$$\Omega[x, u] = \frac{\nu}{\mathrm{i}} \int dx' \chi'(\int g(x'' - x') u(x'') dx'') g(x - x') .$$

But

$$\chi(\lambda) = \int \mathrm{e}^{\mathrm{i}\lambda A} w(A)\, dA, \frac{1}{\mathrm{i}} \chi'(\lambda) = \int \mathrm{e}^{\mathrm{i}\lambda A} w(A) A\, dA .$$

Putting

$$\lambda = \frac{1}{\mathrm{i}} \int dx'' g(x'' - x') \frac{\delta}{\delta \xi(x'')} ,$$

we get

$$\Omega\left[x, \frac{1}{\mathrm{i}} \frac{\delta}{\delta \xi}\right] = \nu \int dx' g(x - x') \int w(A) A \exp\left[A \int dx'' g(x'' - x') \frac{\delta}{\delta \xi(x'')}\right] dA .$$

As a result, if we take into account the operator representation of the Taylor series (1.118), the equation (1.135) will take the form

$$\langle \xi(x) R[\xi] \rangle = v \int g(x - x') dx' \int w(A) A \langle R[\xi(\tilde{x}) + A g(\tilde{x} - x')] \rangle dA \ .$$

1.8.22 Let $X_i(x,t)$ be deterministic functions, and $f_i(x,t)$ *Gaussian* random functions of $(n+1)$ variables (x,t) such that

$$\langle f_i \rangle = 0 \ , \quad \langle f_i(x,t) f_k(x',t') \rangle = 2 F_{ik}(x,x',t) \delta(t-t') \ . \tag{1.160}$$

Let the functions $\xi_i(t)$ obey the set of dynamic equations

$$\frac{d\xi_i(t)}{dt} = X_i(\xi(t),t) + f_i(\xi(t),t) \tag{1.161}$$

with the initial condition $\xi_i(0) = \xi_i^0$.

Derive a differential equation for the probability density $w(x,t) \equiv \langle \delta[\xi(t) - x] \rangle$ of the solution to (1.161).

Solution. We differentiate $w(x,t)$ with respect to t

$$\frac{\partial w(x,t)}{\partial t} = \left\langle \frac{\partial}{\partial t} \delta(\xi(t) - x) \right\rangle = \left\langle -\frac{\partial \delta(\xi(t) - x)}{\partial x_i} \frac{d\xi_i(t)}{dt} \right\rangle$$

and substitute $d\xi_i(t)/dt$ from (1.161)

$$\begin{aligned} \frac{\partial w(x,t)}{\partial t} &= - \left\langle \frac{\partial \delta(\xi(t) - x)}{\partial x_i} [X_i(\xi(t), t) + f_i(\xi(t), t)] \right\rangle \\ &= - \frac{\partial}{\partial x_i} \langle \delta(\xi(t) - x) [X_i(\xi(t),t) + f_i(\xi(t),t)] \rangle \ . \end{aligned}$$

The last expression is the result of placing $\partial/\partial x_i$ outside the brackets and the averaging sign, since the only multiplier dependent on x is the delta-function. Using the equality

$$\delta(\xi(t) - x) F(\xi(t)) = \delta(\xi(t) - x) F(x) \ ,$$

we obtain

$$\begin{aligned} \frac{\partial w(x,t)}{\partial t} &= - \frac{\partial}{\partial x_i} \langle \delta(\xi(t) - x) [X_i(x,t) + f_i(x,t)] \rangle \\ &= - \frac{\partial}{\partial x_i} [X_i(x,t) \langle \delta(\xi(t) - x) \rangle + \langle f_i(x,t) \delta(\xi(t) - x) \rangle] \ , \end{aligned}$$

where we have placed the nonrandom multiplier $X_i(x,t)$ outside the averaging sign.

Since $\langle \delta[\boldsymbol{\xi}(t)-\boldsymbol{x}]\rangle = w(\boldsymbol{x},t)$, we get

$$\frac{\partial w(\boldsymbol{x},t)}{\partial t}+\frac{\partial}{\partial x_i}[X_i(\boldsymbol{x},t)w(\boldsymbol{x},t)] = -\frac{\partial Q_i}{\partial x_i}\,, \quad \text{where} \tag{1.162}$$

$$Q_i(\boldsymbol{x},t) \equiv \langle f_i(\boldsymbol{x},t)\delta(\boldsymbol{\xi}(t)-\boldsymbol{x})\rangle\ . \tag{1.163}$$

Equation (1.162) is not closed because, it contains the unknown functions Q_i, as well as the required function w. The next step is to express Q_i through w. For this purpose we use the Furutsu–Novikov formula (1.141) which, in this case, considering (1.160), has the form

$$\begin{aligned}\langle f_i(\boldsymbol{x},t)R[f]\rangle &= \langle f_i(\boldsymbol{x},t)\rangle\langle R\rangle \\ &\quad + \int d\boldsymbol{x}'\int dt'\langle \tilde{f}_i(\boldsymbol{x},t)\tilde{f}_k(\boldsymbol{x}',t')\rangle\left\langle\frac{\delta R[f]}{\delta f_k(\boldsymbol{x}',t')}\right\rangle \\ &= 2\int d\boldsymbol{x}'\int dt'\,F_{ik}(\boldsymbol{x},\boldsymbol{x}',t)\,\delta(t-t')\left\langle\frac{\delta R[f]}{\delta f_k(\boldsymbol{x}',t')}\right\rangle\ .\end{aligned} \tag{1.164}$$

To derive $Q_i(\boldsymbol{x},t)$, we should, according to (1.163), put $R[f]=\delta[\boldsymbol{\xi}(t)-\boldsymbol{x}]$. This is quite a legitimate equality since $\boldsymbol{\xi}(t)$, being a solution to the set of equations (1.161), is a functional of f_i.

To find $\langle \delta R/\delta f_k\rangle$, we return to (1.161). Integrating these equations with respect to t gives

$$\xi_i(t) = \xi_i^0 + \int_0^t [X_i(\boldsymbol{\xi}(\tau),\tau)+f_i(\boldsymbol{\xi}(\tau),\tau)]d\tau\ . \tag{1.165}$$

We see that $\xi_i(t)$ depend on the values of $f_k(\tau)$ within the interval $0\leqslant\tau\leqslant t$ only. Therefore,

$$\frac{\delta\xi_i(t)}{\delta f_k(t')} = 0, \quad \text{if} \quad t'<0 \quad \text{or} \quad t'>t\ . \tag{1.166}$$

Formula (1.166) shows that the solutions to (1.161) satisfy the dynamic causality principle. We write (1.165) in the form

$$\xi_i(t) = \xi_i^0 + \int_0^t d\tau\int d\boldsymbol{x}\,\delta(\boldsymbol{\xi}(\tau)-\boldsymbol{x})[X_i(\boldsymbol{x},\tau)+f_i(\boldsymbol{x},\tau)]$$

and apply to these expressions the operator $\delta/\delta f_j(\boldsymbol{x}',t')$, where $0<t'<t$

$$\begin{aligned}\frac{\delta\xi_i(t)}{\delta f_j(\boldsymbol{x}',t')} &= \int_0^t d\tau\int d\boldsymbol{x}\left\{-\frac{\partial\delta(\boldsymbol{\xi}(\tau)-\boldsymbol{x})}{\partial x_k}\frac{\delta\xi_k(\tau)}{\delta f_j(\boldsymbol{x}',t')}[X_i(\boldsymbol{x},\tau)+f_i(\boldsymbol{x},\tau)]\right. \\ &\quad \left. +\,\delta(\boldsymbol{\xi}(\tau)-\boldsymbol{x})\frac{\delta f_i(\boldsymbol{x},\tau)}{\delta f_j(\boldsymbol{x}',t')}\right\}\ .\end{aligned}$$

We will integrate by parts the first summand in braces. In the second summand, on integrating, we will use the property that, by (1.140),

$$\frac{\partial f_i(\boldsymbol{x},\tau)}{\delta f_j(\boldsymbol{x}',t')} = \delta_{ij}\delta(\boldsymbol{x}-\boldsymbol{x}')\,\delta(\tau-t') \ .$$

As a result,

$$\frac{\delta\xi_i(t)}{\delta f_j(\boldsymbol{x}',t')} = \delta_{ij}\delta(\boldsymbol{\xi}(t')-\boldsymbol{x}') + \int_0^t d\tau \int d\boldsymbol{x}\,\delta(\boldsymbol{\xi}(\tau)-\boldsymbol{x})\frac{\delta\xi_k(\tau)}{\delta f_j(\boldsymbol{x}',t')}\frac{\partial}{\partial x_k}[X_i(\boldsymbol{x},\tau)+f_i(\boldsymbol{x},\tau)] \ .$$

But, by virtue of the causality condition (1.166),

$$\frac{\delta\xi_k(\tau)}{\delta f_j(\boldsymbol{x}',t')} = 0 \quad \text{at} \quad \tau < t' \ ,$$

which enables the lower limit of the integral with respect to τ to be replaced by t'

$$\frac{\delta\xi_i(t)}{\delta f_j(\boldsymbol{x}',t')} = \delta_{ij}\delta(\boldsymbol{\xi}(t')-\boldsymbol{x}') + \int_{t'}^{t} d\tau \ldots .$$

Then, in the limit as $t' \to t$, the integral becomes zero and we obtain the exact relationship

$$\frac{\delta\xi_i(t)}{\delta f_j(\boldsymbol{x}',t)} = \delta_{ij}\delta(\boldsymbol{\xi}(t)-\boldsymbol{x}') \ , \tag{1.167}$$

which is a consequence of the dynamic equations (1.161). We return now to the expression (1.164) for Q_i. Since for $t' > t$ we have, by (1.166), $\langle \delta R/\delta f_k \rangle = 0$, in (1.164) the integration with respect to t' is within $(0,t)$. When integrating with respect to t', we use the property that $\int_0^t \delta(t-t')dt' = 1/2$, since the delta-function in the expression is taken to be the limit of an *even* covariance.
Therefore,

$$Q_i(\boldsymbol{x},t) = \int F_{ik}(\boldsymbol{x},\boldsymbol{x}',t)\left\langle \frac{\delta R[f]}{\delta f_k(\boldsymbol{x}',t)} \right\rangle d\boldsymbol{x}' \ . \tag{1.168}$$

Because $R[f] = \delta[\boldsymbol{\xi}(t)-\boldsymbol{x}]$, we have

$$\frac{\delta R[f]}{\delta f_k(\boldsymbol{x}',t)} = \frac{\delta}{\delta f_k(\boldsymbol{x}',t)}\delta(\boldsymbol{\xi}(t)-\boldsymbol{x}) = -\frac{\partial\delta(\boldsymbol{\xi}(t)-\boldsymbol{x})}{\partial x_l}\frac{\delta\xi_l(t)}{\delta f_k(\boldsymbol{x}',t)} \ ,$$

or, by (1.167),

$$\begin{aligned}\frac{\delta R[f]}{\delta f_k(\boldsymbol{x}',t)} &= -\delta_{lk}\delta(\boldsymbol{\xi}(t)-\boldsymbol{x}')\frac{\partial\delta(\boldsymbol{\xi}(t)-\boldsymbol{x})}{\partial x_l} \\ &= -\frac{\partial}{\partial x_k}[\delta(\boldsymbol{\xi}(t)-\boldsymbol{x}')\delta(\boldsymbol{\xi}(t)-\boldsymbol{x})] = -\frac{\partial}{\partial x_k}[\delta(\boldsymbol{x}-\boldsymbol{x}')\delta(\boldsymbol{\xi}(t)-\boldsymbol{x})] \ .\end{aligned}$$

Averaging this equality gives

$$\left\langle \frac{\delta R[f]}{\delta f_k(\boldsymbol{x}',t)} \right\rangle = -\frac{\partial}{\partial x_k}[\delta(\boldsymbol{x}-\boldsymbol{x}')w(\boldsymbol{x},t)] \ .$$

Taking the latter expression into account, (1.168) takes the form

$$Q_i(\boldsymbol{x},t) = -\int F_{ik}(\boldsymbol{x},\boldsymbol{x}',t)\frac{\partial}{\partial x_k}[w(\boldsymbol{x},t)\delta(\boldsymbol{x}-\boldsymbol{x}')]\,d\boldsymbol{x}' \ .$$

Writing the integrand in the form

$$-F_{ik}(\boldsymbol{x},\boldsymbol{x}',t)\frac{\partial}{\partial x_k}[w(\boldsymbol{x},t)\delta(\boldsymbol{x}-\boldsymbol{x}')]$$

$$= -\frac{\partial}{\partial x_k}[F_{ik}(\boldsymbol{x},\boldsymbol{x}',t)w(\boldsymbol{x},t)\delta(\boldsymbol{x}-\boldsymbol{x}')] + \frac{\partial F_{ik}(\boldsymbol{x},\boldsymbol{x}',t)}{\partial x_k}w(\boldsymbol{x},t)\delta(\boldsymbol{x}-\boldsymbol{x}') \ ,$$

we arrive at

$$Q_i(\boldsymbol{x},t) = -\frac{\partial}{\partial x_k}\int F_{ik}(\boldsymbol{x},\boldsymbol{x}',t)w(\boldsymbol{x},t)\delta(\boldsymbol{x}-\boldsymbol{x}')d\boldsymbol{x}'$$

$$+\int\frac{\partial F_{ik}(\boldsymbol{x},\boldsymbol{x}',t)}{\partial x_k}w(\boldsymbol{x},t)\delta(\boldsymbol{x}-\boldsymbol{x}')d\boldsymbol{x}'$$

$$= -\frac{\partial}{\partial x_k}[F_{ik}(\boldsymbol{x},\boldsymbol{x},t)w(\boldsymbol{x},t)] + \left[\frac{\partial F_{ik}(\boldsymbol{x},\boldsymbol{x}',t)}{\partial x_k}\right]_{\boldsymbol{x}'=\boldsymbol{x}} w(\boldsymbol{x},t) \ .$$

The function $Q_i(\boldsymbol{x},t)$ is thus represented via the derived probability density $w(\boldsymbol{x},t)$, whereby (1.162) becomes a *closed* equation

$$\frac{\partial w(\boldsymbol{x},t)}{\partial t} + \frac{\partial A_i(\boldsymbol{x},t)w}{\partial x_i} = \frac{\partial^2 F_{ik}(\boldsymbol{x},\boldsymbol{x},t)w}{\partial x_i\partial x_k} \ , \quad \text{where} \tag{1.169}$$

$$A_i(\boldsymbol{x},t) = X_i(\boldsymbol{x},t) + \left[\frac{\partial F_{ik}(\boldsymbol{x},\boldsymbol{x}',t)}{\partial x_k}\right]_{\boldsymbol{x}'=\boldsymbol{x}} \ .$$

Equation (1.169) is the Einstein–Fokker–Planck equation for the dynamic system (1.161). In Sect. I.6.4 it was given without derivation, the only distinction being that the factors at $\delta(t-t')$ in the correlation matrix of the actions $f_i(\boldsymbol{x},t)$ were denoted $C_{ik}(\boldsymbol{x},\boldsymbol{x}',t) = 2F_{ik}(\boldsymbol{x},\boldsymbol{x}',t)$.

2. Radiation and Diffraction of Random Wave Fields

Section 2.1 singles out basic types of statistical wave problems normally encountered in wave theory. These are problems with random sources, random medium variables, random boundaries, and random wave reception and registration conditions. This chapter is devoted to problems of the first type, namely the excitation of fields by random sources distributed over a volume or surface. In these cases, the statistical moments of a field are expressed directly in terms of the moments of sources (Sect. 2.2).

In Sects. 2.3–5, the general theory is applied to a variety of problems, among which are the passage of a wave through a chaotic screen, the focusing of random fields and the formation of optical images, and the statistics of fields excited by a large antenna or by a system of independent point sources.

2.1 Basic Types of Statistical Wave Problems

Among a variety of random fields dealt with in statistical radiophysics, *wave* fields occupy the central position. We will also concentrate on wave fields (primarily, electromagnetic) and will confine our discussion to *linear, nonquantum* problems[1]. A fairly broad class of such problems can be formulated as follows.

Let the propagation of waves of any physical nature (electromagnetic, elastic, surface, etc.) be described by a linear space-time operator $\hat{L}$ (generally differential and, less frequently, integro-differential), so that the wave field u obeys

$$\hat{L}u = q \ , \tag{2.1}$$

where the function $q(t, \boldsymbol{r})$ accounts for the wave sources. The fields u and q can also be multicomponent (in particular, vector), and then $\hat{L}$ is the operator matrix (in particular, a tensor). In many problems the space region, in which the field u, bounded by a surface S_0, is considered contains no sources ($q = 0$); instead there is

---

[1] We will indicate some of the reviews and monographs on nonlinear wave problems and the quantum statistics of radiation. The statistical phenomena in *nonlinear optics* are discussed in [2.1, 2]. Extensive research is under way on the theory of *weak turbulence*, a phenomenon which occurs in various situations including plasma waves [2.3, 4], waves on the surface of liquids [2.4, 5], and in active continuous systems, etc. The statistical issues of *nonlinear acoustics* are dealt with in [2.6] and [2.49]. The *quantum fluctuations* of electromagnetic fields, which have gained in importance in connection with the development of laser technology, are considered in [2.7–11].

a *primary* wave field u_0 that enters this region from the outside. Equation (2.1) will then be homogeneous

$$\hat{L}u = 0 \ . \tag{2.2}$$

But in S_0 values of the primary field (or its derivatives) are given, e.g.,

$$v \equiv u_0(t, \boldsymbol{r})|_{\boldsymbol{r} \in S_0} \ . \tag{2.3}$$

(In this case it is normally said that in S_0 "virtual" field sources are defined.) Here it is necessary to find the scattered, or diffracted field, i.e., these problems belong to diffraction theory.

If there are boundaries between various media or bodies (boundary surfaces S) inside S_0, the field u should, in addition, satisfy certain *boundary conditions.* If S_0 is not closed or embraces all the space, so that waves coming from real sources located within a finite region may propagate without limit, then known *radiation conditions* should be met (at sufficiently large distances from the sources only receding waves should exist).

In practice it is often of interest to *measure* a radiation or diffraction field in order to obtain information about field sources, scattering bodies or the medium through which the waves propagate. Then the scheme described above can also include a radiation *receiver*[2] and various types of noise, both of external (relative to the receiver) and internal origin. The response w of the receiver will depend both on the field u and on the noise ξ:

$$w = \hat{w}(u, \xi) \ , \tag{2.4}$$

where $\hat{w}$ is, in the general case, a nonlinear operator.

All that was said in Sect. I.6.1 regarding conventional stochastic differential equations in the case of random fields holds good for partial differential equations. Stochastic wave problems are formulated using the *same* equations and conditions as in dynamic problems, but here the equations and conditions will be *stochastic*, i.e., ones for *individual realizations of the random field u.* In other words, all or some of the parameters, functions and operators appearing in a problem are now *random* and, accordingly, specified by their respective probability distributions. Therefore, dynamic and statistical problems have a much wider variety of possibilities for space-time fields (as compared with processes occurring in time).

In accordance with the formulation of a dynamic wave problem, the following entities may be random:

1) *field sources* (real or virtual, so that it is possible to distinguish a given "source statistics" q from "primary field statistics" v);
2) *medium properties* ("medium statistics", and hence the operator $\hat{L}$);
3) *shape and location of boundaries S* ("boundary statistics");

---

[2] A receiver may be any measuring device, e.g., an antenna with an amplifying stage and a registering device, or a photodetector with measuring devices, etc.

4) *wave reception and registration conditions* (with "receiver statistics", i.e., the operator $\hat{w}$, and the "noise statistics" ξ).

The formulation of the overwhelming majority of problems of statistical wave theory reduces to these principal statistical schemes (which will be referred to as primary). It goes without saying that there are also problems of a mixed type, e.g., of thermal radiation in a randomly inhomogeneous medium [2.12], but not many of these problems have been treated so far (a brief overview is given in [2.13]).

If we had an *exact solution of the dynamic problem*, e.g., an integral representation of the desired field u in the form

$$u = \hat{G}q \ , \tag{2.5}$$

(where $\hat{G}q = \hat{L}^{-1}q$ is the solution of the nonhomogeneous equation (2.1) written, say, using Green's function subject to all the boundary and other conditions), then the calculation of the field moments would reduce to averaging the products of the form $u(1)u(2) \dots u(n)$ over the joint distribution of all the random variables and functions in the problem which characterize the statistics of sources, medium, boundaries, and so forth. But in actual practice this ideal situation does not occur frequently, e.g., in problems on the excitation of fields by random sources discussed in this chapter. More often than not we cannot find an exact solution for any deterministic functions, and hence for arbitrary realizations of all the random variables and functions. This makes it necessary, even at this stage of solving the dynamic problem, to turn to a variety of *approximate* methods. These numerous methods and procedures vary from problem to problem.

Fluctuations of random variables and functions (according to certain characteristic scales) may be classified as large and small, smooth and slow, or, on the contrary, abrupt and fast; correlation may be strong and "far", or weak and "short", and so on. These distinctions predetermine the use of various approximate treatments and secondary statistical schemes that are associated with corresponding approximate solution techniques.

It is worth noting that in many cases, especially when we are dealing with field moments, we can dispense with a dynamic solution (for the moments to be derived later) and we then derive equations and conditions for the *moments themselves* (from the equations and conditions for the field u). Even with this formulation, however, we in fact cannot dispense with the greatest possible simplification of the initial equations for u, a simplification that *a priori* takes into account the characteristic features of fluctuations and the character of deterministic functions.

The four primary statistical schemes listed above only reflect the frequently occurring division of variables and functions into deterministic and random. Now we will dwell a little upon these schemes and will furnish some relevant examples.

In scheme (1), if there are real sources, we deal with the *nonhomogeneous* equation (2.1), in which the *right-hand side* q is statistical in nature. The *homogeneous* boundary conditions are deterministic. Problems of this type abound in radiophysics, optics, and acoustics. These include, in particular, the statistical theory of antennas and the theory of thermal fluctuations in continuous systems.

Clearly, virtual random sources occur in all problems concerned with the diffraction of random (or rather, partly coherent) fields[3], where the homogeneous equation (2.2) and all the necessary conditions are deterministic, the only exception being the primary random field u_0. Such problems are encountered mainly in optics (the formation of optical and holographic images, interferometer operation, etc.), but they also occur in the radio frequency range (in radio astronomy specifically) and in the diffraction of X-rays.

In this chapter, we will be looking at problems of type (1). The thermal fluctuations in distributed systems, owing to their importance and specific nature, will be considered in a separate (next) chapter.

Scheme (2) includes the problem of the propagation and diffraction of waves through randomly inhomogeneous media (the random operator $\hat{L}$). These questions are of great interest for radio and laser communication, underwater acoustics, radio astronomy, plasma diagnostics, and so on. Chapter 4 is devoted to problems coming under the heading of scheme (2), but only in the special case where the inhomogeneities of the medium are so weak that we can confine ourselves to the single scattering approximation.

Scheme (3) includes wave problems concerned with bodies that have a *random* shape or a *random* position. These may, for example, be boundary surfaces with many random irregularities (so-called irregular or statistically rough surfaces). The various treatments of the scattering of waves from such surfaces are discussed in Chap. IV.5, see [2.14].

Problems concerned with randomly arranged bodies may involve scattering both from a few bodies, and from a host of discrete scatterers (rain, fog, aerosols, etc.). The latter, rather general, problem is strictly speaking, one of multiple scattering. We will confine our discussion of it to the single scattering approximation (Sect. 4.8).

Lastly, scheme (4) covers the many problems of the reception and processing of information about wave fields in noise. Knowing the statistical behavior of the field u, noise ξ and operator $\hat{w}$ that describes the receiver, we can, according to (2.4), in principle calculate the statistical characteristics of the receiver response w.

More important, and at the same time more complicated, is the question of selecting the optimal (in a definite sense) reception technique in noise, i.e., the optimal operator $\hat{w}$. An example of such a problem is that of image recognition, which is of paramount importance for optics, radio astronomy, radar, hydroacoustics, and so on. This problem of the optimal reception of random fields draws on the concepts and techniques of information theory—perhaps to a larger degree than the processing of random processes. But, as in Volumes I and II, see [2.15, 16], we shall not concern ourselves with these problems because they are more closely related to "radiomathematics" than to radiophysics. A number of issues concerning optimal reception and spatial filtration are considered in [2.17–24].

[3] As applied to wave fields, the terms "random" and "partly coherent" are equivalent.

2.2 Random Waves in an Unbounded Homogeneous Medium

The problem of excitation of fields by (real or virtual) random sources, which we referred to under scheme (1), is one of the few problems of statistical wave theory that permits of an essentially universal approach. This is due to the fact that the field u is related to the sources by a *linear deterministic* operator which, in principle, can be inverted.

According to (2.5), in exciting a field by real sources, any moments of the field can only be obtained by averaging the products of the form $u(1)u(2)\ldots u(n)$ over an ensemble of random sources q. So, for the first two moments we have

$$\langle u\rangle = \hat{G}\langle q\rangle, \quad \langle u_1 u_2^*\rangle = \hat{G}_1 \hat{G}_2^* \langle q_1 q_2^*\rangle \ . \tag{2.6}$$

In the case of *virtual* sources, the values v of the primary field (or its derivatives) are, by (2.3), defined on a certain surface S_0, and the desired field is expressed in terms of v using the linear deterministic operator $\hat{\mathscr{P}}$

$$u = \hat{\mathscr{P}} v \ . \tag{2.7}$$

The desired moments of u are related to the known moments of v by linear relationships similar to (2.6)

$$\langle u\rangle = \hat{\mathscr{P}}\langle v\rangle, \quad \langle u_1 u_2^*\rangle = \hat{\mathscr{P}}_1 \hat{\mathscr{P}}_2^* \langle v_1 v_2^*\rangle, \ldots \ . \tag{2.8}$$

The relationships of the form (2.6) or (2.8), in principle, give a complete solution to the stochastic problems of scheme (1), since the set of all the statistical moments uniquely defines all n-component probability densities of a random field. However, this formally simple procedure is rarely carried out in practice. Convenient expressions for the moments of the field u are obtainable only for the lowest moments, and only by introducing some approximation for the inverse operators $\hat{G}$ or $\hat{\mathscr{P}}$. But the probability density for the field is generally only obtainable when the Central Limit Theorem is applicable.

Even in the simplest case of a scalar wave field in a homogeneous unbounded medium, where we have relatively simple, exact expressions for $\hat{G}$ and $\hat{\mathscr{P}}$, we have to resort to various approximations. We will give the relevant dynamic relations necessary for a further analysis and, taking the scalar field as an example, will find some of the properties of random wave fields. The subsequent sections are devoted to specific radiophysical problems coming under scheme (1).

In a homogeneous stationary medium with no dispersion or absorption, a scalar field u satisfies the wave equation

$$\hat{L}u = \left(\Delta - \frac{1}{c^2}\frac{\partial^2}{\partial t^2}\right)u(t, \boldsymbol{r}) = q(t, \boldsymbol{r}) \ , \tag{2.9}$$

where c is the velocity of wave propagation, and q is a field of random sources. The solution to (2.9) meeting the conditions of radiation at infinity is known to have the form

$$u(t,\boldsymbol{r}) = -\frac{1}{4\pi}\int \frac{q(t-R/c,\boldsymbol{r}')}{R}d^3r' \ , \tag{2.10}$$

where $R = |\boldsymbol{r}-\boldsymbol{r}'|$. This is a special case of the linear expression (2.5) for a field with sources.

In many problems it is more convenient to deal not with the random field $u(t,\boldsymbol{r})$ itself, but with its spectral amplitude

$$u(\omega,\boldsymbol{r}) = \frac{1}{2\pi}\int_{-\infty}^{\infty} u(t,\boldsymbol{r})\mathrm{e}^{i\omega t}\,dt \ , \tag{2.11}$$

which, by virtue of (2.9), satisfies the equation

$$\Delta u(\omega,\boldsymbol{r}) + k^2 u(\omega,\boldsymbol{r}) = q(\omega,\boldsymbol{r}), \qquad k = \omega/c \ , \tag{2.12}$$

where $q(\omega,\boldsymbol{r})$ is the spectral amplitude of $q(t,\boldsymbol{r})$. For spectral amplitudes, (2.10) takes the form

$$u(\omega,\boldsymbol{r}) = -\frac{1}{4\pi}\int \frac{q(\omega,\boldsymbol{r}')\mathrm{e}^{ikR}}{R}d^3r' \equiv \int G(\boldsymbol{r},\boldsymbol{r}')q(\omega,\boldsymbol{r}')\,d^3r' \ . \tag{2.13}$$

Here we have introduced Green's function for the unbounded homogeneous medium

$$G(\boldsymbol{r},\boldsymbol{r}') = -\frac{1}{4\pi}\frac{\mathrm{e}^{ikR}}{R}, \qquad R = |\boldsymbol{r}-\boldsymbol{r}'| \tag{2.14}$$

that obeys the equation

$$\Delta G + k^2 G = \delta(\boldsymbol{r}-\boldsymbol{r}') \ .$$

Using (2.13), we can easily write expressions for all the moments of the field.

If the sources are concentrated in a bounded region, say within the confines of a sphere of radius a, and we wish to find the far (Fraunhofer) field u of the source distribution, then (2.10) and (2.13) are simplified, since we may now use the approximate expression for Green's function. We place the origin of the coordinate system at the center of the region occupied by the sources. Then, for $r \gg ka^2$ the distance $R = |\boldsymbol{r}-\boldsymbol{r}'|$ in (2.14) can be approximated by $r-(\boldsymbol{n}\cdot\boldsymbol{r}')$ in the exponent and by r in the denominator, which gives the Fraunhofer approximation. In this approximation, (2.13) takes the form

$$u(\omega,\boldsymbol{r}) \approx -\frac{\mathrm{e}^{ikr}}{4\pi r}\int q(\omega,\boldsymbol{r}')\mathrm{e}^{-ik\boldsymbol{n}\cdot\boldsymbol{r}'}\,d^3r', \qquad r \gg ka^2 \ , \tag{2.15}$$

where $\boldsymbol{n} = \boldsymbol{r}/r$ is the unit vector in the direction of the point of observation.

We will now give the dynamic relations for the case where *virtual* sources are available. We will start off with Green's formula, which relates the spectral amplitude of the field $u(\omega, \boldsymbol{r})$ inside the region bounded by the surface S_0 to the spectral amplitude of the boundary field $v(\omega, \boldsymbol{r}')$

$$u(\omega, \boldsymbol{r}) = -\frac{1}{4\pi}\oint\left[v(\omega, \boldsymbol{r}')\frac{\partial}{\partial N}\left(\frac{\mathrm{e}^{ikR}}{R}\right) - \frac{\mathrm{e}^{ikR}}{R}\frac{\partial v(\omega, \boldsymbol{r}')}{\partial N}\right]dS_0 \ . \tag{2.16}$$

Here $R = |\boldsymbol{r} - \boldsymbol{r}'|$ is the distance from the point of observation to a point $\boldsymbol{r}'$ lying on the surface S_0 over which integration is carried out, and $\partial/\partial N$ signifies differentiation along the external normal $\boldsymbol{N}$ to S_0. It is well known that the boundary values $v(\omega, \boldsymbol{r}')$ and $\partial v(\omega, \boldsymbol{r}')/\partial N$ should be specified in a mathematically consistent way. Therefore, the two-term Green's function (2.16) may only be used in cases where some other, additional considerations provide the relationship (precise or approximate) between the boundary field v and its normal derivative $\partial v/\partial N$.

We will confine our discussion to the special case where S_0 is a surface bounded by a hemisphere of infinite radius, so that the two-term formula (2.16) reduces to one-term.

If in the plane $z = 0$ the field $v(\omega, \boldsymbol{\varrho}) \equiv u(\omega, \boldsymbol{r})|_{z=0}$ itself is defined, where $\boldsymbol{\varrho} = (x, y)$ is a two-dimensional vector, then the integral over the infinite-radius sphere vanishes, and in the half-space $z > 0$

$$u(\omega, \boldsymbol{r}) = u(\omega, \boldsymbol{\varrho}, z) = -\frac{1}{2\pi}\int_{-\infty}^{\infty} v(\omega, \boldsymbol{\varrho}')\frac{\partial}{\partial z}\left(\frac{\mathrm{e}^{ikR}}{R}\right)d^2\boldsymbol{\varrho}' \ ,$$
$$R = \sqrt{(\boldsymbol{\varrho} - \boldsymbol{\varrho}')^2 + z^2} \ . \tag{2.17}$$

If $\partial v(\omega, \boldsymbol{\varrho})/\partial z = \partial u(\omega, \boldsymbol{r})/\partial z|_{z=0}$ is the normal derivative in the plane $z = 0$, then

$$u(\omega, \boldsymbol{\varrho}, z) = \frac{1}{2\pi}\int_{-\infty}^{\infty}\frac{\mathrm{e}^{ikR}}{R}\frac{\partial v(\omega, \boldsymbol{\varrho}')}{\partial z}d^2\varrho' \ . \tag{2.18}$$

In the case of the plane boundary $z = 0$, it is thus sufficient to specify on S_0 either the field u itself, or its normal derivative $\partial u/\partial z$.

Another technique for calculating the fields in the half-space $z > 0$ dates back to Rayleigh and is based on expanding the fields in plane waves. We will represent the boundary field $v(\omega, \boldsymbol{\varrho})$ in terms of the two-dimensional Fourier integral

$$v(\omega, \boldsymbol{\varrho}) = \int_{-\infty}^{\infty} v(\omega, \boldsymbol{\kappa})\exp(\mathrm{i}\boldsymbol{\kappa}\cdot\boldsymbol{\varrho})\,d^2\kappa \ , \tag{2.19}$$

where $\boldsymbol{\kappa} = (\kappa_x, \kappa_y)$ is a two-dimensional wave vector, and $v(\omega, \boldsymbol{\kappa})$ is the boundary field amplitude related to $v(\omega, \boldsymbol{\varrho})$ by the inverse Fourier transform

$$v(\omega, \boldsymbol{\kappa}) = \frac{1}{4\pi^2}\int_{-\infty}^{\infty} v(\omega, \boldsymbol{\varrho})\exp(-\mathrm{i}\boldsymbol{\kappa}\cdot\boldsymbol{\varrho})d^2\varrho \ .$$

In the half-space $z > 0$ each spatial harmonic of the boundary field $v(\omega, \boldsymbol{\kappa})\exp(\mathrm{i}\boldsymbol{\kappa}\boldsymbol{\varrho})$

gives rise to a plane wave eigen-solution $v(\omega, \boldsymbol{\kappa})\exp(\mathrm{i}\boldsymbol{\kappa}\cdot\boldsymbol{\varrho}+\mathrm{i}pz)$ [recall that the factor $\exp(-\mathrm{i}\omega t)$ is omitted], which obeys the wave equation subject to

$$\kappa^2 + p^2 = \omega^2/c^2 \equiv k^2 \ .$$

Thus, in the case of the wave equation for a homogeneous isotropic medium, the dispersion surface (see Sect. 1.6) in the four-dimensional space $(\omega, \boldsymbol{k})$ constitutes a three-dimensional cone with the axis in the ω-direction and the vertex at the origin. From this dispersion equation, in the source-free region $z > 0$ at a given κ, wave eigen-solutions of two types are possible—depending on whether the z-component p of the wave vector $\boldsymbol{k}$

$$p = \sqrt{k^2-\kappa^2} = \begin{cases} \sqrt{k^2-\kappa^2} & \kappa \leqslant k = \omega/c \\ \mathrm{i}\sqrt{|k^2-\kappa^2|} & \kappa > k \end{cases} \tag{2.20}$$

is real or imaginary.

For $\kappa < k$, the constant p is real and we have *travelling* waves. For $\kappa > k$, when the period of oscillations at the boundary $2\pi/\kappa$ is less than the wavelength $\lambda = 2\pi/k$ corresponding to a frequency ω, the component p is imaginary and thus the waves are *inhomogeneous*. They decay exponentially with distance from the boundary $z = 0$. In fact, even for $z \gtrsim \lambda$ only travelling waves remain.

The resulting wave field $u(\omega, \boldsymbol{r}) = u(\omega, \boldsymbol{\varrho}, z)$ satisfying the boundary condition $u(\omega, \boldsymbol{r})|_{z=0} = v(\omega, \boldsymbol{\varrho})$ is represented by a superposition of plane waves of both types

$$u(\omega, \boldsymbol{\varrho}, z) = \int_{-\infty}^{\infty} v(\omega, \boldsymbol{\kappa})\exp(\mathrm{i}\boldsymbol{\kappa}\cdot\boldsymbol{\varrho}+\mathrm{i}pz)\,d^2\kappa \ . \tag{2.21}$$

The field can also be represented in a similar way when the normal derivative $\partial u/\partial z$ is specified in the plane $z = 0$.

Clearly, the Green's function (2.17) and the expansion in plane waves (2.21) are equivalent. Either of them enables the moments $u(\omega, \boldsymbol{r})$ to be expressed through the moments $v(\omega, \boldsymbol{\varrho})$, and we will use the one that yields the final result the faster.

In calculating the moments of the field using the above equations, we often have to deal with special cases. Let us look at the most significant of them. For $kz \gg 1$, i.e., in the *wave zone*, the kernel of (2.17) will be

$$-\frac{1}{2\pi}\frac{\partial}{\partial z}\frac{\mathrm{e}^{\mathrm{i}kR}}{R} \approx \frac{kz}{2\pi\mathrm{i}R^2}\mathrm{e}^{\mathrm{i}kR} \qquad (kz \gg 1) \ .$$

If $|\boldsymbol{\varrho}-\boldsymbol{\varrho}'| \ll z$ and $k|\boldsymbol{\varrho}-\boldsymbol{\varrho}'|^4 \ll z^3$, the so-called *Fresnel approximation* holds, with $R = \sqrt{z^2+(\boldsymbol{\varrho}-\boldsymbol{\varrho}')^2}$ replaced by $z+(\boldsymbol{\varrho}-\boldsymbol{\varrho}')^2/2z$ in the exponent and by z in the denominator of the preceding relation. Green's function (2.17) then takes the form

$$u(\omega, \boldsymbol{\varrho}, z) = \frac{k\mathrm{e}^{\mathrm{i}kz}}{2\pi\mathrm{i}z}\int_{-\infty}^{\infty} v(\omega, \boldsymbol{\varrho}')\exp\left[\mathrm{i}k\frac{(\boldsymbol{\varrho}-\boldsymbol{\varrho}')^2}{2z}\right]d^2\varrho' \ . \tag{2.22}$$

Lastly, in the *Fraunhofer zone* $z \gg ka^2$, where a is the radius of the region where the boundary field $v(\omega, \boldsymbol{\varrho})$ is nonzero, we have

$$u(\omega, \boldsymbol{\varrho}, z) = \frac{k \exp[ik(z + \varrho^2/2z)]}{2\pi i z} \int_{-\infty}^{\infty} v(\omega, \boldsymbol{\varrho}') \exp\left[-ik\frac{(\boldsymbol{\varrho}\cdot\boldsymbol{\varrho}')}{z}\right] d^2\varrho' . \tag{2.23}$$

We now turn to some properties of the moments and spectra of wave fields. The statistical moments of random wave fields are often called *coherence functions*, since the appropriate correlation coefficients serve as a quantitative measure of coherence of these fields (Sect. II.25.6). As far as wave fields are concerned, the terms "coherence time", "coherence length", and "coherence degree" are synonyms for "correlation time", "correlation radius", and "correlation coefficient", respectively. In this volume we will adhere to our practice of referring to the theory of random waves, not to "coherence theory". The latter is simply considered to be one of the applications of the general theory of random fields. We will not, however, avoid the terms "coherence function" or "coherence degree", which have already been widely used in the literature.

The mixed moment of a wave field $u(t, \boldsymbol{r})$ of the $(m+n)$th order [the coherence function of the $(m+n)$th order] is defined by a relationship of the form (1.41) and is generally denoted by $\Gamma_{m,n}$ instead of by $B_{m,n}$. The correlation theory of random waves is only concerned with the moments of first and second order: the *mean field* $\langle u \rangle$ (the coherence function of the first order $\Gamma_1 \equiv \langle u \rangle$) and two mixed moments of the second order, termed the first and second coherence functions of the second order

$$\begin{aligned} \Gamma_{1,1}(1,2) &= \langle u(1)u^*(2)\rangle \equiv \Gamma(1,2) , \\ \Gamma_{2,0}(1,2) &= \langle u(1)u(2)\rangle \equiv \tilde{\Gamma}(1,2) . \end{aligned} \tag{2.24}$$

The latter functions are related to the first and second *correlation functions* of the field by

$$\begin{aligned} \psi(1,2) &= \langle \tilde{u}(1)\tilde{u}^*(2)\rangle = \Gamma(1,2) - \langle u(1)\rangle\langle u^*(2)\rangle , \\ \tilde{\psi}(1,2) &= \langle \tilde{u}(1)\,\tilde{u}(2)\rangle = \tilde{\Gamma}(1,2)\rangle - \langle u(1)\rangle\langle u(2)\rangle , \end{aligned} \tag{2.25}$$

(where, as usual, $\tilde{u} = u - \langle u \rangle$).

We will represent the time-dependence of wave fields by the analytic signal[4]

$$u(t, \boldsymbol{r}) = \int_0^{\infty} u(\omega, \boldsymbol{r}) e^{-i\omega t} d\omega , \tag{2.26}$$

where the spectral amplitude $u(\omega, \boldsymbol{r})$ is identically zero for negative frequencies. For $\omega \geqslant 0$ it is equal to the double spectral density of the initial real field $u_r(t, \boldsymbol{r}) = \mathrm{Re}\{u(t, \boldsymbol{r})\}$.

[4] It will be recalled that in the Fourier integral with respect to t, we take $\exp(-i\omega t)$, not $\exp(i\omega t)$ as in [2.15, 16]. Therefore, the field u as a function of t is now analytical in the lower half-plane of the complex time t, and not in the upper half.

If the field $u(t, \boldsymbol{r})$ as a function of t is an analytic signal and *stationary*, then the mean field $\langle u \rangle$ and its second mixed moment $\tilde{\Gamma}$ are zero (Sect. II.1.1).

If we do not assume an analytic signal, the mean of a stationary field may be distinct from zero and be a function of $\boldsymbol{r}$ alone, which satisfies the Laplace equation $\Delta \bar{u}(\boldsymbol{r}) = 0$. Since such static deterministic fields are of no interest to us, we may in that case, too, consider only the fluctuations $\tilde{u} = u - \bar{u}$, i.e., we may put $\bar{u} = 0$. As to nonstationary and, in particular, monochromatic fields $[\sim \exp(-\mathrm{i}\omega t)]$, their means and $\tilde{\Gamma}$ are, generally speaking, distinct from zero.

The energy-related quantities (intensity, energy density, energy flux density) are quadratic in the field, so that their means are representable in terms of statistical moments of first and second order (Sect. II.2.5 and Exercise 2.6.1).

Of special interest among the higher moments is the moment of fourth order

$$\Gamma_{2,2}(1, 2, 3, 4) = \langle u(1)u(2)u^*(3)u^*(4) \rangle \tag{2.27}$$

used, specifically, to express the covariance of *intensity fluctuations* of the complex field $I = |u|^2$

$$\psi_I(1, 2) = \langle \tilde{I}(1)\tilde{I}(2) \rangle = \Gamma_{2,2}(1, 2; 2, 1) - \bar{I}(1)\bar{I}(2) \ , \tag{2.28}$$

where $\bar{I}(1) = \langle |u(1)|^2 \rangle = \Gamma(1, 1)$. If the field u is normal, then all the higher moments are expressed in terms of the first ($\bar{u}$) and second (Γ and $\tilde{\Gamma}$) moments. The covariance for this case is given (in somewhat different notation) in Exercise 1.8.12.

Earlier we obtained expressions for the field moments using the dynamic solutions for u. As noted above, there is an alternative method for finding the moments, viz., from the equation governing the statistical moments themselves. This is a problem of the excitation of fields by real sources, which in the general case is described by (2.1) where the operator $\hat{L}$ and boundary conditions are deterministic. Equations for second moments are readily derived simply by multiplying together the left and right-hand sides of (2.1) taken at different space-time points. So, the second moment $\Gamma(1, 2) = \langle u(1)u^*(2) \rangle$ satisfies the equation

$$\hat{L}_1 \hat{L}_2^* \Gamma(1, 2) = \langle q(1)q^*(2) \rangle \ , \tag{2.29}$$

which becomes homogeneous in a source-free region.

Within the framework of scheme (1), when the dynamic solution of a problem for a field u is known, it is generally more convenient to find moments from equations of the type (2.29) than from (2.6) or (2.8). Perhaps here the wave equations for the moments are only of interest because they yield the wave properties of the moments themselves, and, accordingly, we can speak about the propagation and diffraction of these moments in almost the same sense as for u[5]. One important wave property of the statistical moments of u is that their values on

---

[5] In those cases, however, where we have no solution to a dynamic problem, the equations for the moments derived from the equations for u become valuable in their own right (Chaps. IV.2–4), see [2.14].

a certain surface S_0 control their behavior throughout the entire volume bounded by S_0, as was the case for the field itself.

As noted in Sect. 1.6, in the wave equation for a homogeneous stationary medium the dispersion hypersurface is a cone $\kappa^2 = \omega^2/c^2 \equiv k^2$ or $|\boldsymbol{\kappa}| = k$. Therefore, the spectral density $G(\omega, \boldsymbol{k})$ in the Fourier expansion of the covariance of the homogeneous stationary field (1.90) contains the factor $\delta(|\boldsymbol{\kappa}| - k)$

$$G(\omega, \boldsymbol{\kappa}) = f(\omega, \boldsymbol{n})\delta(|\boldsymbol{\kappa}| - k) \ , \tag{2.30}$$

where $f(\omega, \boldsymbol{n})$ is a function of the frequency ω and direction $\boldsymbol{n} = \boldsymbol{\kappa}/\kappa$.

In the coefficient of the delta-function we can separate the factor k^{-2}

$$G(\omega, \boldsymbol{\kappa}) = \mathscr{J}(\omega, \boldsymbol{n})k^{-2}\delta(\kappa - k) \ . \tag{2.31}$$

The quantity $\mathscr{J}(\omega, \boldsymbol{n})$ is called the *radiant intensity*, or *brightness*; it plays a major role in the theory of radiation transfer (Chap. IV.4, see [2.14]). We can establish the relationship between this quantity and the covariance by substituting (2.31) into (1.90) and putting $d^3\kappa = \kappa^2 d\kappa\, do(\boldsymbol{n})$ [here $do(\boldsymbol{n})$ is an element of the solid angle]. Integrating with respect to κ yields the desired relation ($\boldsymbol{r} = \boldsymbol{r}_1 - \boldsymbol{r}_2$, $t = t_1 - t_2$)

$$\begin{aligned}\Psi(t, \boldsymbol{r}) &= \iint \mathscr{J}(\omega, \boldsymbol{n})k^{-2}\delta(\kappa - k)\exp[i(\boldsymbol{\kappa}\cdot\boldsymbol{r} - \omega t)]\kappa^2 d\kappa\, do(\boldsymbol{n})d\omega \\ &= \int_0^\infty d\omega \oint \mathscr{J}(\omega, \boldsymbol{n})\exp[i(k\boldsymbol{n}\cdot\boldsymbol{r} - \omega t)]do(\boldsymbol{n}) \ .\end{aligned} \tag{2.32}$$

At $\boldsymbol{r} = 0$ and $t = 0$ the equation gives the variance

$$\sigma_u^2 = \langle|\tilde{u}|^2\rangle = \Psi(0, 0) = \int_0^\infty d\omega \oint \mathscr{J}(\omega, \boldsymbol{n})do(\boldsymbol{n}) \ . \tag{2.33}$$

It is seen from this expression that the radiant intensity $\mathscr{J}(\omega, \boldsymbol{n})$ describes the energy distribution in (positive) frequencies and angles, i.e., it constitutes the *frequency-angular spectrum* of the field. In particular, an isotropic field has an angle-independent radiant intensity (the general expression for the covariance is in this case given in Exercise 2.6.2). It should be noted that the angular spectrum often refers to the two-dimensional representation of the covariance as a Fourier integral

$$F(\omega; \boldsymbol{\kappa}_\perp, z) = \frac{1}{(2\pi)^2}\int_{-\infty}^{\infty} \psi(\omega; \boldsymbol{\varrho}, z)\exp(-i\boldsymbol{\kappa}_\perp\cdot\boldsymbol{\varrho})d^2\varrho, \quad \boldsymbol{\kappa}_\perp = (\kappa_x, \kappa_y) \ . \tag{2.34}$$

As shown in Exercise 2.6.4, when the angular spectrum is narrow, $F(\omega; \boldsymbol{\kappa}_\perp, z)$ is proportional to the radiant intensity (at $z = 0$ these quantities differ only in the factor k^{-2}).

Using (2.32), we can estimate the minimum correlation radius l_u of a free wave field. Let ω_{max} be the maximum frequency of the waves still noticeable in the spectrum $\mathscr{J}(\omega, \boldsymbol{n})$. By (2.32), the correlation radius of the field, l_u, may be estimated from the condition $k_{max}(\boldsymbol{n}, \boldsymbol{r}) \gtrsim 1$:

$$l_u \gtrsim \frac{1}{k_{max}} \sim \lambda_{min} \ .$$

In other words, the correlation radius of the random wave field cannot be smaller than the minimum wavelength in the spectrum of oscillations. Strictly speaking, this estimate applies to isotropic fields, where the radiant intensity is independent of $\boldsymbol{n}$. Exercise 2.6.3 contains estimates of the longitudinal and transverse correlation radii of a field where the radiant intensity is concentrated within a narrow cone.

The properties of statistic moments and spectral densities just described are characteristic not only of scalar, but also of vector random fields. We will not give here vector analogs of the above expressions for fields and their moments as they essentially lie beyond the scope of the general relationship (2.6) and (2.8). Note only that we have already considered the statistical behavior of polarization in plane quasi-monochromatic waves in Sect. II.2.7, see [2.16]. The polarization matrix introduced there characterizes the statistical relationship between different field components at the same point and at the same instant in time (see also [2.17, 18, 25] and [2.26]). The general form of the correlation matrix for a stationary, homogeneous, isotropic field is given in Exercise 2.6.6; the connection between the energy-related field characteristics and the correlation matrix and radiant intensity matrix is studied in Exercise 2.6.5.

2.3 Diffraction of Plane Wave on an Infinite Chaotic Screen

A primary field at the boundary of a region may fluctuate both due to fluctuations in the sources (lying outside the region), and to random perturbations caused in an initially deterministic primary wave as it propagates. An example is the propagation of such a wave through a randomly inhomogeneous medium, or a layer of such a medium.

If this layer is thin enough, it can be thought of as an infinitesimally thin screen. The applicability of the general techniques of calculating the field in a region at whose boundary S_0 the random field is specified (Sect. 2.2), is in no way associated with the cause of field fluctuations on the boundary. What is significant here is only the fact that the statistics of these fluctuations are known.

We will begin with the simplest problem, viz., the propagation of a plane monochromatic wave through a plane infinite chaotic screen. We will follow mainly the treatment given in [2.27][6].

1) Basic Equations. Let a monochromatic wave $u_{\mathrm{inc}}(\boldsymbol{\varrho}, z)\exp(-\mathrm{i}\omega t)$ fall on an infinite screen located in the plane $z=0$. The screen spatially modulates the incident wave in accordance with its complex transmittance function $f(\boldsymbol{\varrho})$. Its modulus describes the amplitude modulation and its argument the phase modulation. The boundary field, i.e., the one immediately beyond the screen $z=0$, is

$$v(\boldsymbol{\varrho}) = f(\boldsymbol{\varrho})u_{\mathrm{inc}}(\boldsymbol{\varrho}, 0) \; . \tag{2.35}$$

[6] This work also summarizes the results of earlier studies [2.28–30] referring largely to the propagation of radio waves through the randomly inhomogeneous ionosphere. A more complete bibliography on the latter question is contained in the review [2.31].

For simplicity, we omit the factor $\exp(-i\omega t)$ as well as the argument ω in the spectral amplitudes.

If the incident wave is plane and propagates normally to the screen, then $u_{\mathrm{inc}}(\varrho, z) = \exp(ikz)$, so that

$$v(\varrho) = f(\varrho) \ , \tag{2.36}$$

The field in the boundary plane $z = 0$ is simply equal to the transmittance function, and, accordingly, the statistics of $v(\varrho)$ for a *chaotic* screen are the same as for $f(\varrho)$. Equations (2.17) and (2.21) allow us to relate the statistical moments of the field $u(\boldsymbol{r}) = u(\varrho, z)$ beyond the screen (i.e., for $z > 0$) to the moments of the boundary field $v(\varrho)$. We will now proceed to find the first and second moments of $u(\boldsymbol{r})$ in the simplest case of a statistically homogeneous, chaotic screen, for which

$$\langle f(\varrho)\rangle \equiv f_0 = \mathrm{const}, \quad \sigma_f^2 = \langle|\tilde{f}|^2\rangle = \langle|f - f_0|^2\rangle = \mathrm{const} \ .$$

The covariance is only dependent on the coordinate difference

$$\psi_f(\varrho', \varrho'') = \langle f(\varrho') f^*(\varrho'')\rangle = \psi_f(\varrho' - \varrho'') \ .$$

By (2.36), we have

$$\langle v\rangle \equiv v_0 = f_0, \quad \sigma_v^2 = \sigma_f^2, \quad \psi_v(\varrho' - \varrho'') = \psi_f(\varrho' - \varrho'') \ .$$

From (2.17) we find the mean field beyond the screen

$$\langle u(\varrho, z)\rangle = -\frac{v_0}{2\pi} \int_{-\infty}^{\infty} \frac{\partial}{\partial z} \frac{\mathrm{e}^{ikR}}{R} d^3 \varrho' \ .$$

The integral is easily taken to give $-2\pi \exp(ikz)$. The mean field beyond the screen is thus the plane wave

$$\bar{u} = v_0 \mathrm{e}^{ikz} \ , \tag{2.37}$$

whose amplitude v_0 is equal to the mean transparency f_0.

Let us now find the covariance. It follows from (2.21) that

$$\begin{aligned}\psi_u(\varrho_1, z_1; \varrho_2, z_2) &= \langle \tilde{u}(\boldsymbol{r}_1)\tilde{u}^*(\boldsymbol{r}_2)\rangle \\ &= \iint_{-\infty}^{\infty} d^2\kappa' d^2\kappa'' \langle \tilde{v}(\boldsymbol{\kappa}')\tilde{v}^*(\boldsymbol{\kappa}'')\rangle \exp(\mathrm{i}\boldsymbol{\kappa}' \cdot \varrho_1 - \mathrm{i}\boldsymbol{\kappa}'' \cdot \varrho_2 + \mathrm{i}p'z_1 - \mathrm{i}p''z_2) \ ,\end{aligned} \tag{2.38}$$

where p' and p'' are given by (2.20) at $\kappa = \kappa'$ and $\kappa = \kappa''$, respectively. But for the homogeneous boundary field

$$\langle \tilde{v}(\boldsymbol{\kappa}')\tilde{v}^*(\boldsymbol{\kappa}'')\rangle = F_v(\boldsymbol{\kappa}')\delta(\boldsymbol{\kappa}' - \boldsymbol{\kappa}'') \ , \tag{2.39}$$

where $F_v(\boldsymbol{\kappa})$ is the two-dimensional spectral density in terms of which the covariance of the boundary field is expressed

$$\psi_v(\boldsymbol{\varrho}) = \int\limits_{-\infty}^{\infty} F_v(\boldsymbol{\kappa})\exp(\mathrm{i}\boldsymbol{\kappa}\cdot\boldsymbol{\varrho})d^2\kappa \ . \tag{2.40}$$

Substituting (2.39) into (2.38) gives

$$\psi_u = \psi_u(\boldsymbol{\varrho}_1 - \boldsymbol{\varrho}_2, z_1, z_2) = \int\limits_{-\infty}^{\infty} F_v(\boldsymbol{\kappa})\exp[\mathrm{i}\boldsymbol{\kappa}\cdot(\boldsymbol{\varrho}_1 - \boldsymbol{\varrho}_2) + \mathrm{i}(pz_1 - p^* z_2)]\, d^2\kappa \ . \tag{2.41}$$

The field $u(\boldsymbol{\varrho}, z)$ is thus statistically homogeneous in the plane $z = \text{const}$; this follows from the fact that $v(\boldsymbol{\varrho})$ is homogeneous at $z = 0$. Comparing (2.41) with the two-dimensional spectral representation

$$\psi_u(\boldsymbol{\varrho}_1 - \boldsymbol{\varrho}_2, z_1, z_2) = \int\limits_{-\infty}^{\infty} F_u(\boldsymbol{\kappa}, z_1, z_2)\exp[\mathrm{i}\boldsymbol{\kappa}\cdot(\boldsymbol{\varrho}_1 - \boldsymbol{\varrho}_2)]d^2\kappa \ , \tag{2.42}$$

we find the spatial spectral density

$$\begin{aligned} F_u(\boldsymbol{\kappa}; z_1, z_2) &= F_v(\boldsymbol{\kappa})\exp[\mathrm{i}(pz_1 - p^* z_2)] \\ &= F_v(\boldsymbol{\kappa}) \times \begin{cases} \exp[\mathrm{i}\sqrt{k^2 - \kappa^2}(z_1 - z_2)], & \kappa^2 \leqslant k^2 \ , \\ \exp[-\sqrt{|k^2 - \kappa^2|}(z_1 + z_2)], & \kappa^2 > k^2 \ . \end{cases} \end{aligned}$$

Spectral components with $\kappa > k$ [the inhomogeneities of $v(\boldsymbol{\varrho})$ are smaller than the wavelength] give rise in the half-space $z > 0$ to inhomogeneous waves that decay exponentially with the distance from the screen. For observation points z_1 and z_2 already separated from the screen by several wavelengths ($z_{1,2} \gg \lambda$), we approximately have

$$F_u(\boldsymbol{\kappa}; z_1, z_2) \approx \begin{cases} F_v(\boldsymbol{\kappa})\exp[\mathrm{i}\sqrt{k^2 - \kappa^2}(z_1 - z_2)], & \kappa^2 \leqslant k^2 \ , \\ 0, & \kappa^2 > k^2 \ . \end{cases} \tag{2.43}$$

Substituting (2.43) into (2.42), we obtain the covariance

$$\psi_u(\boldsymbol{\varrho}, \zeta) = \int\limits_{\kappa \leqslant k} F_v(\boldsymbol{\kappa})\exp[\mathrm{i}\boldsymbol{\kappa}\boldsymbol{\varrho} + \mathrm{i}\sqrt{k^2 - \kappa^2}\zeta]\, d^2\kappa \ , \tag{2.44}$$

where $\boldsymbol{\varrho} = \boldsymbol{\varrho}_1 - \boldsymbol{\varrho}_2$ and $\zeta = z_1 - z_2$. It is seen from (2.44) that for $z_{1,2} \gg \lambda$ the random field $u(\boldsymbol{\varrho}, z)$ becomes statistically homogeneous not only in the transverse planes $z = \text{const}$, but also in the longitudinal z-direction.

Using (2.44), we will look at the special cases of small- and large-scale fluctuations of the boundary field $v(\boldsymbol{\varrho})$.

Consider *small-scale* fluctuations. The correlation radius of the boundary field, l_v, is small as compared with the wavelength λ, i.e., $kl_v \ll 1$. We can assume that the spectral density $F_v(\boldsymbol{\kappa})$ is essentially constant within a circle $\kappa \leqslant k$ and is approximately equal to $F_v(0) \equiv F_0$. From (2.44) we then get

$$\psi_u(\boldsymbol{\varrho}, \zeta) = F_0 \int_{\kappa \leqslant k} \exp(\mathrm{i}\boldsymbol{\kappa} \cdot \boldsymbol{\varrho} + \mathrm{i}p\zeta) d^2\kappa, \quad kl_v \ll 1 \ . \tag{2.45}$$

The *transverse* covariance, i.e., the one in the plane $z = \text{const}$ normal to the direction of propagation, is obtained from (2.45) at $\zeta = z_1 - z_2 = 0$

$$\psi_\perp(\boldsymbol{\varrho}) \equiv \psi_u(\boldsymbol{\varrho}, 0) = F_0 \int_{\kappa \leqslant k} \exp(\mathrm{i}\boldsymbol{\kappa} \cdot \boldsymbol{\varrho}) d^2\kappa = 2\pi F_0 \frac{kJ_1(k\varrho)}{\varrho} \ .$$

Here J_1 is the first-order Bessel function. As ϱ increases, the ratio $J_1(k\varrho)/\varrho$ will for the first time be zero at $k\varrho = 3.83$, i.e., at $\varrho \approx 0.61\lambda$. Thus, even at a distance of several λ from the screen, the transverse correlation radius of the field, $l_\perp$, is of the order of the wavelength, i.e., substantially larger than the correlation radius l_v at $z = 0$: $l_\perp \sim \lambda \gg l_v$.

It follows from the expression for the *longitudinal* covariance

$$\psi_{||}(\zeta) \equiv \psi_u(0, \zeta) = F_0 \int_{\kappa \leqslant k} \mathrm{e}^{\mathrm{i}p\zeta} d^2\kappa = 2\pi F_0 \frac{d}{d\zeta}\left(\frac{1 - \mathrm{e}^{-\mathrm{i}k\zeta}}{\zeta}\right)$$

that the longitudinal correlation radius $l_{||}$ is also of the order of the wavelength: $l_{||} \sim \lambda \gg l_v$. Consequently, in the case of small scale fluctuations at the screen, both the transverse and the longitudinal correlation radii increase with distance and for $z \gg \lambda$ they are about the same as the wavelength.

At the same time, the fluctuation covariance decreases, i.e., the fluctuations smooth out. In fact, using (2.45), we obtain

$$\sigma_u^2 = \langle |\tilde{u}|^2 \rangle = \psi_u(0, 0) = F_0 \int_{\kappa \leqslant k} d^2\kappa = \pi k^2 F_0 \ .$$

But

$$F_0 \equiv F_v(0) = \frac{1}{4\pi^2} \int_{-\infty}^{\infty} \psi_v(\boldsymbol{\varrho}') d^2\varrho' \sim \frac{1}{4\pi^2} \sigma_v^2 l_v^2 \ ,$$

so that

$$\sigma_u^2 = \pi k^2 F_0 \sim (kl_v)^2 \sigma_v^2 \ll \sigma_v^2 \ .$$

In the $kl_v \to 0$ limit (very small inhomogeneities on the screen) the fluctuation field $\tilde{u}$ vanishes completely, because small inhomogeneities only produce inhomogeneous (exponentially decaying) waves beyond the screen.

We now turn to *large-scale* inhomogeneities of the boundary field ($kl_v \gg 1$). Such fluctuations produce travelling waves, and are thus especially important in applications. We will, therefore, concentrate our attention on them.

In the case of large-scale fluctuations of the boundary field, the two-dimensional spectrum $F_v(\boldsymbol{\kappa})$ is concentrated in the narrow interval $\kappa \lesssim \kappa_v \sim 1/l_v \ll k$.

This allows us, in (2.44), to expand the longitudinal wave number $p = \sqrt{k^2 - \kappa^2}$ into a Taylor series in κ^2 and only keep the first two terms

$$p = \sqrt{k^2 - \kappa^2} \approx k - \kappa^2/2k \ .$$

We can also extend the integration limits in (2.44) to $\pm\infty$, without thereby introducing a significant error. Then

$$\psi_u(\varrho, \zeta) \approx e^{ik\zeta} \int_{-\infty}^{\infty} F_v(\kappa) \exp(i\kappa \cdot \varrho - i\kappa^2\zeta/2k)\, d^2\kappa \quad (kl_v \gg 1) \ . \tag{2.46}$$

'It is a simple matter to verify that the assumptions made amount to using the Fresnel approximation (2.22) for $u(\varrho, z)$.

Putting into (2.46) $\zeta = 0$ gives the *transverse* covariance [see (2.40)]

$$\psi_\perp(\varrho) \equiv \psi_u(\varrho, 0) = \int_{-\infty}^{\infty} F_v(\kappa) \exp(i\kappa \cdot \varrho)\, d^2\kappa = \psi_v(\varrho) \ . \tag{2.47}$$

The transverse covariance of the wave field u is thus equal to the covariance of the boundary field and does not change with distance from the screen. Therefore, the transverse correlation radius is also the same as on the screen, i.e.,

$$l_\perp = l_v \ .$$

The conservation of transverse covariance, noted for the first time in [2.28], results in the conservation of the variance ($\sigma_u^2 = \sigma_v^2$) and transverse coherence function. Indeed, bearing in mind the behavior of the mean field (2.37), for the wave that is not bounded transversely we have

$$\begin{aligned} \Gamma_\perp(\varrho, z) &= \langle u(\varrho_1, z) u^*(\varrho_2, z) \rangle = \psi_\perp(\varrho) + \bar{u}(\varrho_1, z)\bar{u}^*(\varrho_2, z) \\ &= \psi_v(\varrho) + |v_0|^2 = \Gamma_v(\varrho) \ . \end{aligned} \tag{2.48}$$

It follows that at $\varrho = \varrho_1 - \varrho_2 = 0$, the mean intensity is constant beyond the screen

$$\bar{I}_u(z) = \langle |u|^2 \rangle = \Gamma_v(0) = \langle |v|^2 \rangle = \text{const} \ .$$

It should be stressed that, if the incident wave is not plane and the screen is statistically inhomogeneous, $\psi_\perp$, $\Gamma_\perp$, and $\bar{I}$ are not conserved beyond the plane $z = 0$, and higher moments are not conserved even for a plane wave and statistically homogeneous screen.

The *longitudinal* covariance $\psi_{\parallel}(\zeta)$ is derived from (2.46) at $\varrho = 0$

$$\psi_{\parallel}(\zeta) = e^{ik\zeta} \int_{-\infty}^{\infty} F_v(\kappa) e^{-i\kappa^2\zeta/2k}\, d^2\kappa \ .$$

The modulus of $\psi_{\parallel}(\zeta)$ decreases markedly as compared with the maximum value $\psi_{\parallel}(0) = \sigma_u^2$ at $\zeta \sim 2k/\kappa_v^2$, when the exponential function in the integrand begins to oscillate within the confines of the interval $\kappa \lesssim \kappa_v \sim 1/l_v$ where the spectrum $F_v(\kappa)$

is concentrated. From this we can estimate the longitudinal correlation radius $l_{\|}$

$$l_{\|} \sim \frac{2k}{\kappa_v^2} \sim kl_v^2 ,$$

i.e., $l_{\|}$ is larger than the transverse scale $l_{\perp} \sim l_v$ by a factor of $kl_v \gg 1$. The longitudinal covariance of the field may be said to vanish when the radius of the first Fresnel zone $\sqrt{\lambda|z_1 - z_2|} = \sqrt{\lambda|\zeta|}$ for $|\zeta| = |z_1 - z_2|$ becomes larger than the transverse correlation radius $\sqrt{\lambda l_{\|}} \sim l_v$.

What was said above regarding the relationship between the longitudinal and transverse scales of correlation will be illustrated by an example in which the covariance of the field is calculated accurately. Let the field at the screen have a Gaussian covariance, i.e.,

$$\psi_v(\varrho) = \sigma_v^2 \exp(-\varrho^2/2l_v^2) .$$

Then

$$F_v(\kappa) = \sigma_v^2 \exp(-\kappa^2 l_v^2/2) ,$$

and a straightforward calculation, using (2.46), gives

$$\psi_u(\varrho, \zeta) = \sigma_v^2 \gamma e^{ik\zeta} \exp(-\varrho^2\gamma/2l_v^2) = \gamma\psi_v(\varrho)\exp[(1-\gamma)\varrho^2/2l_v^2] ,$$

where $\gamma = [1 + i\zeta/kl_v^2]^{-1}$. The characteristic space scale for $|\psi_u|$ is of the order $l_{\perp} \sim l_v$ transversely, and $l_{\|} \sim kl_v^2 \gg l_v$ longitudinally. The surfaces of equal values of $|\psi_u|$ have here the shape of spheroids extended strongly along the z-axis.

2) The Passage of a Plane Wave Through a Phase Chaotic Screen. This is a transparent screen with the transmittance function $f(\varrho) = \exp[iS(\varrho)]$, where S is a real random function, i.e., the screen modulates the phase, but leaves the amplitude (and hence the intensity) of the wave unchanged. For $z > 0$ the phase screen produces both the phase and the amplitude modulation. The latter may be observed, for example, on a sheet of paper as light passes through optically inhomogeneous or irregular glass (conventional window-glass, say), if the paper is placed at a distance from the glass.

The phase screen is widely used as a model to describe a variety of phenomena both in optics and radiophysics. For example, a lens with optical inhomogeneities mostly modulates the phase of the transmitted light wave. The Earth's ionosphere has the same effect on ultrashort radio waves. The phase screen model is used (though not with such good reason) in treatments of the scintillation of radio waves sent out by extraterrestrial sources and transmitted through a statistically inhomogeneous interplanetary or interstellar medium.

If on an infinite plane phase screen a *plane wave* $u_{\text{inc}} = \exp(ikz)$ is incident, then the boundary value of the field in the screen plane ($z = 0$) will be

$$v(\varrho) = \exp[iS(\varrho)]u_{\text{inc}}|_{z=0} = \exp[iS(\varrho)] .$$

Let us now look at the way in which moments of the field of a transmitted wave are related to the phase covariance

$$\psi_S(\varrho) = \langle \tilde{S}(\varrho_1)\tilde{S}(\varrho_2)\rangle = \sigma_S^2 K_S(\varrho) ,$$

where σ_S^2 is the variance, and $K_S(\varrho)$ the correlation coefficient of the phase.

Suppose that the fluctuations of S are statistically homogeneous at $z = 0$, have a zero mean ($\bar{S} = 0$) and follow the normal law of probability distribution.

If we remember that for a normally distributed variable α with $\bar{\alpha} = 0$ we have $\langle \exp(\mathrm{i}\alpha)\rangle = \exp(-\overline{\alpha^2}/2)$, then

$$\langle v\rangle \equiv v_0 = \langle \mathrm{e}^{\mathrm{i}S}\rangle = \mathrm{e}^{-\sigma_S^2/2} ,$$

$$\Gamma_v(\varrho) = \langle v(\varrho_1)v^*(\varrho_2)\rangle = \langle \exp\{\mathrm{i}[S(\varrho_1) - S(\varrho_2)]\}\rangle$$
$$= \exp\{-\tfrac{1}{2}\langle [S(\varrho_1) - S(\varrho_2)]^2\rangle\} .$$

But, at $\bar{S} = 0$, the mean square of the phase difference is the structure function of the phase $D_S(\varrho)$ related to the covariance $\psi_S(\varrho)$ by (1.68). Therefore,

$$\Gamma_v(\varrho) = \exp[-\tfrac{1}{2}D_S(\varrho)] = \exp[\psi_S(\varrho) - \psi_S(0)] = \exp\{\sigma_S^2[K_S(\varrho) - 1]\} .$$

The mean and transverse coherence functions of the field beyond the screen may be found from (2.37) and (2.48), respectively,

$$\bar{u} = v_0\mathrm{e}^{\mathrm{i}kz} = \exp(-\sigma_S^2/2 + \mathrm{i}kz) ,$$

$$\Gamma_\perp(\varrho) = \Gamma_v(\varrho) = \exp[-\tfrac{1}{2}D_S(\varrho)] .$$

The transverse covariance is

$$\psi_\perp(\varrho) = \psi_v(\varrho) = \exp\{\sigma_S^2[K_S(\varrho) - 1]\} - \exp(-\sigma_S^2) . \tag{2.49}$$

If the phase fluctuations on the screen are weak ($\sigma_S^2 \ll 1$), then, by (2.49), we will have

$$\psi_\perp(\varrho) \approx \sigma_S^2 K_S(\varrho) = \psi_S(\varrho) \quad (\sigma_S^2 \ll 1) , \tag{2.50}$$

i.e., for small phase fluctuations the transverse covariance of the field for $z > 0$ coincides with $\psi_S(\varrho)$. If then the phase fluctuations are large ($\sigma_S^2 \gg 1$), the mean of the field $v_0 = \exp(-\sigma_S^2/2)$ is negligible as compared with unity, and the quantity $\exp\{\sigma_S^2[K_S(\varrho) - 1]\}$ is markedly different from zero only for small ϱ. Taking this into account, we will neglect the term $\exp(-\sigma_S^2)$ in (2.49) and expand the correlation coefficient $K_S(\varrho)$ into a Taylor series.

Assuming, for simplicity, the phase fluctuations at the screen to be isotropic, we have

$$K_S(\varrho) = 1 + K_S''(0)\frac{\varrho^2}{2} + \dots , \tag{2.51}$$

where the prime denotes differentiation with respect to ϱ [in (2.51) the term "linear in ϱ" is absent since the random field of $S(\boldsymbol{\varrho})$ is taken to be differentiable; $K_S'(0) = 0$]. As a result, we get

$$\psi_\perp(\varrho) \approx \mathrm{e}^{1/2\sigma_S^2 K_S''(0)\varrho^2} = \mathrm{e}^{-1/2|\psi_S''(0)|\varrho^2} \quad (\sigma_S^2 \gg 1) \;, \tag{2.52}$$

where allowance is made for the fact that at zero the values of the second derivatives $K_S''(0)$ and $\sigma_S''(0) = \sigma_S^2 K_S''(0)$ are negative.

It is seen from (2.52) that the correlation vanishes for $\varrho \gtrsim |\sigma_S^2 K_S''(0)|^{-1/2} \approx l_S/\sigma_S$, where $l_S \sim |K_S''(0)|^{-1/2}$ is the radius of phase correlation. Therefore, the transverse correlation radius is estimated to be

$$l_\perp \sim l_S/\sigma_S \quad (\sigma_S \gg 1) \;.$$

For $\sigma_S \gg 1$ the correlation radius is thus about σ_S times smaller than the correlation scale of the phase, l_S. It is easy to see the reason for this difference between $l_\perp$ and l_S.

Shifting along the screen by about l_S changes the phase S by $\Delta S \sim \sigma_S$. In the process S undergoes no more than *one* oscillation. At the same time, the boundary field $v(\boldsymbol{\varrho}) = \exp(\mathrm{i}S)$ undergoes, over the same distance $\varrho \sim l_S$, about $n \sim \sigma_S/\pi$ oscillations. Hence $l_\perp \sim l_S/n \sim l_S/\sigma_S \ll l_S$.

These relations are used, for instance, to interpret transmission data of ultrashort radiowaves from extraterrestrial sources through the Earth's ionosphere which contains inhomogeneities of electron concentration [2.27–30]. Under certain conditions, it may be considered that the field $v(\boldsymbol{\varrho})$ leaving the ionosphere undergoes only phase fluctuations.

By (2.50), if the phase variance σ_S^2 is small as compared with unity then the spatial covariance of the field at ground level coincides with the phase covariance $\psi_S(\boldsymbol{\varrho})$, which is related to the covariance of ionospheric inhomogeneities. In consequence, for $\sigma_S^2 \ll 1$ we can directly measure the covariance of the phase of the wave transmitted through the ionosphere, and are in a position to judge ionospheric inhomogeneities.

It is much more difficult to interpret observational evidence for $\sigma_S^2 \gg 1$. In this case the covariance of the isotropic field $\psi_\perp(\varrho)$ is related to $\psi_S(\varrho)$ by (2.52), this equation enabling only $|\psi_S''(0)| \sim \sigma_S^2/l_S^2$ to be evaluated. Additional information about σ_S and l_S can be derived from data on the relative fluctuations of wave intensity (to be discussed later), but even with these data, it is by no means easy to achieve an unambiguous interpretation.

The point is that ionospheric inhomogeneities are not located at a fixed altitude, but are distributed (nonuniformly at that) at altitudes from 100 to over 400 km. In addition, these inhomogeneities have a wide range of horizontal scales (from $l \lesssim 1$ to $l \gtrsim 500$ km). Their action is often masked by strongly focusing formations ("ionospheric lenses"), and so forth. Therefore, observations of field fluctuations on Earth allow only rough estimates of the behavior of ionospheric inhomogeneities: their horizontal scales, anisotropy degree (inhomogeneities, as a rule, are extended along the Earth's magnetic field lines), probable altitude, and also the mean velocity of travel (drift) of inhomogeneities [2.32, 33].

3) Amplitude and Phase Fluctuations Beyond the Infinite Phase Screen. In many applications, and in radio communication and radio navigation in particular, we are interested in the statistical characteristics of wave amplitudes and phases. To calculate these, we need the second covariance $\tilde{\psi}_\perp(\boldsymbol{\varrho}) = \langle \tilde{u}(\boldsymbol{\varrho}_1, z)\tilde{u}(\boldsymbol{\varrho}_2, z)\rangle$ as well as the first covariance of the complex field $\psi_\perp(\boldsymbol{\varrho}) = \langle \tilde{u}(\boldsymbol{\varrho}_1, z)\tilde{u}^*(\boldsymbol{\varrho}_2, z)\rangle$ (which for a plane wave and statistically homogeneous screen does not change with distance).

The variation of the second covariance with distance may be obtained using the Fresnel approximation (2.22), which is applicable to fields with large-scale inhomogeneities at the boundary. Using (2.22) and omitting ω for the sake of simplicity gives

$$\tilde{\psi}_\perp(\boldsymbol{\varrho}, z) = \left(\frac{ke^{ikz}}{2\pi iz}\right)^2 \iint_{-\infty}^{\infty} \tilde{\psi}_v(\boldsymbol{\varrho}' - \boldsymbol{\varrho}'')\exp\left\{i\frac{k}{2z}[(\boldsymbol{\varrho}_1 - \boldsymbol{\varrho}')^2 + (\boldsymbol{\varrho}_2 - \boldsymbol{\varrho}'')^2]\right\} d^2\varrho' d^2\varrho'' , \tag{2.53}$$

where $\tilde{\psi}_v(\boldsymbol{\varrho}) = \langle \tilde{v}(\boldsymbol{\varrho}_1)\tilde{v}(\boldsymbol{\varrho}_1 - \boldsymbol{\varrho})\rangle$ is the second covariance of the field at the screen. Let us introduce the new integration variables $\boldsymbol{\xi} = \boldsymbol{\varrho}' - \boldsymbol{\varrho}''$ and $\boldsymbol{\eta} = (\boldsymbol{\varrho}' - \boldsymbol{\varrho}'')/2$. The integral with respect to $\boldsymbol{\eta}$ is easily taken and this procedure eliminates from (2.53) the variation with $\boldsymbol{\varrho}_1 + \boldsymbol{\varrho}_2$, so that only the dependence on $\boldsymbol{\varrho}_1 - \boldsymbol{\varrho}_2$ remains. In the resultant integral we denote $\boldsymbol{\xi}$ by $\boldsymbol{\varrho}'$ and $\boldsymbol{\varrho}_1 - \boldsymbol{\varrho}_2$ by $\boldsymbol{\varrho}$ to obtain

$$\tilde{\psi}_\perp(\boldsymbol{\varrho}, z) = \frac{ke^{2ikz}}{4\pi iz} \int_{-\infty}^{\infty} \tilde{\psi}_v(\boldsymbol{\varrho}')\exp\left[\frac{ik}{4z}(\boldsymbol{\varrho} - \boldsymbol{\varrho}')^2\right] d^2\varrho' . \tag{2.54}$$

The second covariance is thus transformed in almost the same way as the field itself. The only distinction from (2.22) is that (2.54) includes the *double* distance from the screen, $2z$.

Subsequent treatment will be simplified if, instead of the fluctuational component of the field, $\tilde{u}$, we introduce an auxiliary random variable $a = \tilde{u}\exp(-ikz)$ ($\bar{a} = 0$), which is essentially the complex *amplitude of field fluctuations*. The first and second covariances of this variable are governed by equations similar to (2.47) and (2.54)

$$\psi_a(\boldsymbol{\varrho}, z) = \psi_v(\varrho), \tag{2.55}$$

$$\tilde{\psi}_a(\boldsymbol{\varrho}, z) = \frac{k}{4\pi iz} \int_{-\infty}^{\infty} \tilde{\psi}_v(\boldsymbol{\varrho}')\exp\left[\frac{ik}{4z}(\boldsymbol{\varrho} - \boldsymbol{\varrho}')^2\right] d^2\varrho' . \tag{2.56}$$

Unlike (2.54), (2.56) does not contain the factor $\exp(2ikz)$.

Furthermore, it is convenient to normalize the mean square of the field at the screen $v(\varrho)$ to unity

$$\langle |v|^2 \rangle = \sigma_{\tilde{v}}^2 + |v_0|^2 = 1 . \tag{2.57}$$

This normalization is related to nonabsorbing, nonreflecting screens, since an incident wave of unit intensity $I_{\text{inc}} = |\exp(ikz)|^2 = 1$ produces a wave of the same mean intensity beyond the screen

$$\bar{I}_u = \langle |u|^2 \rangle = \langle |v|^2 \rangle = 1 \ .$$

Clearly, this condition is met for a purely phase screen.

We will now find the relationship between a and the amplitude A and phase S of the wave $u = A\exp(\mathrm{i}S + \mathrm{i}kz)$ propagating beyond the screen. We have

$$\begin{aligned} u(\boldsymbol{\varrho}, z) &= A(\boldsymbol{\varrho}, z)\exp[\mathrm{i}kz + \mathrm{i}S(\boldsymbol{\varrho}, z)] \\ &= \bar{u} + \tilde{u} = (v_0 + a)\mathrm{e}^{\mathrm{i}kz} \ , \end{aligned}$$

where v_0 may, without loss of generality, be taken to be a real quantity. Separating a into real (a') and imaginary (a'') parts, we find

$$A\mathrm{e}^{\mathrm{i}S} = v_0 + a = (v_0 + a') + \mathrm{i}a'' \ ,$$

hence

$$A = \sqrt{(v_0 + a')^2 + a''^2}, \quad S = \arctan\frac{a''}{v_0 + a'} \ . \tag{2.58}$$

Figure 2.1 shows the appropriate vectors in the complex plane of the amplitude $U = A\exp(\mathrm{i}S)$.

The problem of finding the statistical behavior of amplitude A and phase S from (2.58) is solved in two partially overlapping cases: for *weak fluctuations* of the field at the screen, where $|a'|$ and $|a''|$ are small as compared with unity, and in the *Fraunhofer zone*, where a' and a'' are normally distributed. These cases will be discussed below.

To begin with, we examine the weak fluctuations of the phase and amplitude. Ignoring the small terms of order $|a'|^3$ and $|a''|^3$ and considering that, by (2.57), $(v_0 + a')^2 + (a'')^2 = 1$, using (2.58), we obtain the means of A and S

$$\bar{A} = 1 - \tfrac{1}{2}\langle a'^2 \rangle, \quad \bar{S} = -\langle a'a'' \rangle \ , \tag{2.59}$$

and also the expressions for their covariances ($\boldsymbol{\varrho} = \boldsymbol{\varrho}_1 - \boldsymbol{\varrho}_2$)

$$\begin{aligned} \psi_A(\boldsymbol{\varrho}, z) &\equiv \langle \tilde{A}(\boldsymbol{\varrho}_1, z)\tilde{A}(\boldsymbol{\varrho}_2, z)\rangle = \psi^a_{11}(\boldsymbol{\varrho}, z) \ , \\ \psi_S(\boldsymbol{\varrho}, z) &\equiv \langle \tilde{S}(\boldsymbol{\varrho}_1, z)\tilde{S}(\boldsymbol{\varrho}_2, z)\rangle = \psi^a_{22}(\boldsymbol{\varrho}, z) \ , \\ \psi_{AS}(\boldsymbol{\varrho}, z) &\equiv \langle \tilde{A}(\boldsymbol{\varrho}_1, z)\tilde{S}(\boldsymbol{\varrho}_2, z)\rangle = \psi^a_{12}(\boldsymbol{\varrho}\ z) \ , \end{aligned} \tag{2.60}$$

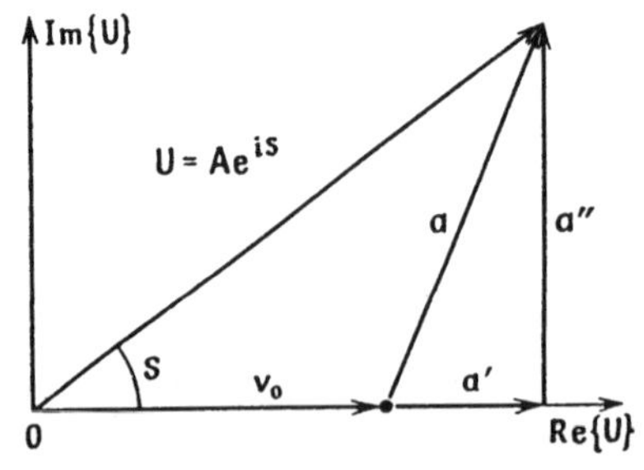

Fig. 2.1. Average (v_0) and fluctuating (a) parts of the complex amplitude $U = A\exp(\mathrm{i}S)$

where $\psi^a_{ij}(\boldsymbol{\varrho}, z)$ $(i, j = 1, 2)$ are the covariances of the real and imaginary parts of the complex random field a. In the general case, these are expressed through the first and second covariances $\psi_a(\boldsymbol{\varrho}, z)$ and $\tilde{\psi}_a(\boldsymbol{\varrho}, z)$ using (1.25). But if $\psi_a(\boldsymbol{\varrho}, z)$ is an even function of ϱ, we use the simpler relations (1.33)

$$\psi^a_{11} = \tfrac{1}{2}(\psi_a + \mathrm{Re}\{\tilde{\psi}_a\}) \ , \quad \psi^a_{22} = \tfrac{1}{2}(\psi_a - \mathrm{Re}\{\tilde{\psi}_a\}) \ , \qquad (2.61)$$
$$\psi^a_{12} = \psi^a_{21} = \tfrac{1}{2}\mathrm{Im}\{\tilde{\psi}_a\} \ .$$

With the help of (2.59, 60), the first two moments of the amplitude and phase are expressed in terms of the first and second covariances of the complex field a. The latter, in turn, vary with distance from the screen in accordance with (2.55, 56), whereby the problem is solved [2.27].

In the special case of a phase screen with an initial phase S having a normal distribution, we have

$$\psi_{A,S}(\boldsymbol{\varrho}, z) = \frac{1}{2}\left[\psi^0_S(\boldsymbol{\varrho}) \mp \frac{k}{4\pi z}\int\limits_{-\infty}^{\infty} \psi^0_S(\boldsymbol{\varrho}) \sin\frac{k(\boldsymbol{\varrho} - \boldsymbol{\varrho}')^2}{4z}\, d^2\varrho'\right] , \qquad (2.62)$$

$$\psi_{AS}(\boldsymbol{\varrho}, z) = \frac{k}{8\pi z}\int\limits_{-\infty}^{\infty} \psi^0_S(\boldsymbol{\varrho}) \cos\frac{k(\boldsymbol{\varrho} - \boldsymbol{\varrho}')^2}{4z}\, d^2\varrho' \ , \qquad (2.63)$$

where $\psi^0_S(\boldsymbol{\varrho})$ is the covariance of the phase in the plane $z = 0$.

For the Gaussian covariance of the phase

$$\psi^0_s(\varrho) = \sigma^2_{S_0} \mathrm{e}^{-\varrho^2/2l_S^2} \ , \qquad (2.64)$$

these integrals are easily taken and, at $\varrho = 0$ in particular,

$$\sigma^2_{A,S}(z) = \frac{\sigma^2_{S_0}}{2}\left(1 \mp \frac{1}{1+D^2}\right), \quad \langle \tilde{A}\tilde{S} \rangle = -\frac{\sigma^2_{S_0}}{2}\frac{D}{1+D^2} \ , \qquad (2.65)$$

where $D = D(z) = 2z/kl_S^2$. *Gorelik* [2.34] called the quantity D the *wave parameter*, and this is the term normally used. This parameter indicates by how many times the area of the first Fresnel zone $\pi(\sqrt{\lambda z})^2 = 2\pi^2 z/k$ is larger than the "area" of one inhomogeneity $\sim \pi l_S^2$, i.e., how many inhomogeneities there are in this zone. Depending on the value of the wave parameter, it is possible to single out three sections of the distance z (also known as zones): the near ($D \ll 1$), Fresnel ($D \sim 1$), and Fraunhofer ($D \gg 1$) zones (relative to an individual inhomogeneity). Each of them has its particular fluctuations.

The near zone ($D \ll 1$) is naturally dominated by phase fluctuations: $\sigma^2_A \ll \sigma^2_S \approx \sigma^2_{S_0}$. Farther away from the screen, the amplitude fluctuations build up whereas the phase ones go down, so that in the limit as $D \to \infty$ (Fraunhofer zone), the amplitude and phase variances level out

$$\sigma^2_{A,S}|_{z\to\infty} = \tfrac{1}{2}\sigma^2_{S_0} \ . \qquad (2.66)$$

The correlation between A and S is negligible in the near and far zones and maximal at $D \sim 1$.

We now turn to fluctuations in the *Fraunhofer zone* ($D \gg 1$) at arbitrary (not necessarily weak) fluctuations of the field at the screen. For $D \gg 1$, within the first Fresnel zone with a radius $\sqrt{\lambda z}$, which is the only one of significance for integration in (2.22), there are many inhomogeneities of the field at the screen. By virtue of the Central Limit Theorem in probability theory, the distribution law for a' and a'' therefore tends to normal.

These quantities are normalized by the "filtering" action of free space, much as time signals at the output of narrow-band filters (Sect. II.3.1). In fact, the variation of a random field following (2.22) is quite similar to the variation of a random process. An analog of the pulse function here is the difference kernel of (2.22), which is responsible for filtration with an effective band width of spatial frequency $\Delta\kappa \sim \sqrt{k/z} \sim (\lambda z)^{-\frac{1}{2}}$. This band narrows with distance z, and for $z \gg kl_v^2$ (i.e., in the far zone, $D \gg 1$) it becomes very much narrower than the initial width of the spatial spectrum $\kappa_v \sim 1/l_v$. Under these conditions, the field $\tilde{u}(\boldsymbol{\varrho}, z)$ is normalized.

The fact that a' and a'' in the far zone are normally distributed makes it possible to find the probability densities for the amplitude and phase and to work out the moments of these variables. This is essentially the well studied problem of the statistics of the envelope A and phase S of the signal $u(\boldsymbol{\varrho}, z) = A\,\exp(\mathrm{i}S + \mathrm{i}kz)$ $= (v_0 + a)\exp(\mathrm{i}kz)$, the latter being the sum of the harmonic oscillation $v_0 \exp(\mathrm{i}kz)$ and the Gaussian noise $\tilde{u} = a\exp(\mathrm{i}kz)$.

For Gaussian variables, the probability distribution is completely characterized by the lower moments alone—the means and covariances. In our case $\bar{a}' = \bar{a}'' = 0$, and the covariances ψ^a_{11}, ψ^a_{12} and ψ^a_{22} are expressed in terms of the first and second covariances of the complex field a by using (2.61). In the Fraunhofer zone, these equations are simplified, since for $D \gg 1$ the second covariance $\tilde{\psi}_a(\boldsymbol{\varrho}, z)$ is negligible as compared with the first. As a result, for $D \gg 1$, we have

$$\psi^a_{11}(\boldsymbol{\varrho}, z) = \psi^a_{22}(\boldsymbol{\varrho}, z) = \tfrac{1}{2}\psi_v(\boldsymbol{\varrho}), \quad \psi^a_{12}(\boldsymbol{\varrho}, z) = 0 \;, \tag{2.67}$$

i.e., in the Fraunhofer zone of the fields, a' and a'' are uncorrelated and their covariances are similar, differing only by a factor 1/2 from the first covariance of the field at the screen. In consequence, ψ^a_{11} and ψ^a_{22} and the related statistical characteristics of the amplitude and phase are independent of the distance from the screen and remain the same as at the "input" into the far zone $z \gtrsim kl_v^2$.

Knowing the (Gaussian) statistics of a' and a'', the statistical moments of the amplitude and phase are determined in the same way as in Sects. I.5.5 and II.2.2 [2.15, 16] (see also [2.35, 36]).

4. Intensity Fluctuations beyond the Infinite Phase Screen. When dealing with intensity fluctuations, we are generally interested in their covariance ($\boldsymbol{\varrho} = \boldsymbol{\varrho}_1 - \boldsymbol{\varrho}_2$)

$$\psi_I(\boldsymbol{\varrho}, z) = \langle \tilde{I}(\boldsymbol{\varrho}_1, z)\tilde{I}(\boldsymbol{\varrho}_2, z)\rangle = \langle I(\boldsymbol{\varrho}_1, z) I(\boldsymbol{\varrho}_2, z)\rangle - \bar{I}^2 \;, \tag{2.68}$$

and the so-called *scintillation index* β^2

$$\beta^2(z) = \frac{\langle I^2(z)\rangle - \bar{I}^2(z)}{\bar{I}^2} = \frac{\sigma_I^2(z)}{\bar{I}^2} \;, \tag{2.69}$$

which characterizes relative intensity fluctuations. If we follow our earlier practice and put $\bar{I} = 1$, then

$$\psi_I(\varrho, z) = \langle I(\varrho_1, z) I(\varrho_2, z) \rangle - 1 \ , \tag{2.68'}$$

$$\beta^2(z) = \langle I^2(z) \rangle - 1 = \sigma_I^2(z) \ . \tag{2.69'}$$

As in the treatment of amplitude and phase fluctuations, ψ_I and β^2 can only be worked out completely in the two partially overlapping cases—for weak fluctuations and for the Fraunhofer zone.

Using the results of item 3, we can easily show that in the case of *weak* fluctuations

$$\psi_I(\varrho, z) = 4\psi_A(\varrho, z) \ ,$$

$$\beta^2(z) = 4\sigma_A^2(z) \ .$$

The intensity fluctuations vary with distance from the screen, following the same law as the amplitude fluctuations. For $D \gg 1$ (the Fraunhofer zone), $\beta^2(z)$ tends to the limit

$$\beta_\infty^2 = 2\sigma_{S_0}^2, \quad \sigma_{S_0}^2 \ll 1 \tag{2.70}$$

regardless of the form of the covariance of the phase at the screen.

In calculating ψ_I in the *far zone* ($D \gg 1$), we will make use of the formula given in Exercise 1.8.12, which is valid for normally distributed fields. For $D \gg 1$, we have $\tilde{\psi}_u = \exp(2ikz)\tilde{\psi}_a \approx 0$, so that $\tilde{\Gamma}(1, 2) = \bar{u}(1)\bar{u}(2)$ and

$$\psi_I(\varrho, z) = |\Gamma_u(\varrho)|^2 - |\bar{u}|^4 = |\Gamma_v(\varrho)|^2 - |v_0|^4 \ . \tag{2.71}$$

As we put $\Gamma_v(0) = \bar{I} = 1$, we obtain the scintillation index in the far zone ($z \to \infty$)

$$\beta_{z\to\infty}^2 \equiv \beta_\infty^2 = \psi_I(0) = 1 - |v_0|^4 \ . \tag{2.72}$$

According to these equations, the intensity covariance ψ_I and the scintillation index β^2 for $D \gg 1$ are already independent of z.

In the special case of a phase screen with Gaussian phase fluctuations, where $v_0 = \exp(-\sigma_{S_0}^2/2)$ and $\psi_v(\varrho)$ is given by (2.49), we obtain

$$\psi_I(\varrho) = \mathrm{e}^{-2\sigma_{S_0}^2} \{\exp[2\psi_S^0(\varrho)] - 1\}, \quad \beta^2 = 1 - \mathrm{e}^{-2\sigma_{S_0}^2} \ . \tag{2.73}$$

For $\sigma_{S_0}^2 \ll 1$, the latter expression, as expected, goes over into (2.70). In another special case—a screen with zero mean field ($v_0 = 0$)—(2.71, 72) become especially simple

$$\psi_I(\varrho) = |\Gamma_v(\varrho)|^2 = |\psi_v(\varrho)|^2, \quad \beta_\infty^2 = 1 \ .$$

In the general case, the investigation of fluctuations is associated with considerable mathematical difficulties. If the fluctuations of the field at the screen are not small, and the observation point is not located in the Fraunhofer zone for an individual inhomogeneity, then the calculation of the scintillation index, within the Fresnel approximation (2.22), is reduced to evaluating the integral

$$\beta^2(z) = \left(\frac{k}{2\pi z}\right)^4 \iiiint\limits_{-\infty}^{\infty} \Gamma_v^{(2,2)}(1,2,3,4) \times \exp\left[\frac{ik}{2z}(\varrho_1^2+\varrho_2^2-\varrho_3^2-\varrho_4^2)\right] d^2\varrho_1 \dots d^2\varrho_4 - 1 , \quad (2.74)$$

where

$$\Gamma_v^{(2,2)}(1,2,3,4) = \langle v(\boldsymbol{\varrho}_1)v(\boldsymbol{\varrho}_2)v^*(\boldsymbol{\varrho}_3)v^*(\boldsymbol{\varrho}_4)\rangle$$

is the mixed moment of fourth order. For arbitrary $\Gamma_v^{(2,2)}(1,2,3,4)$, the integrals of the form (2.74) can only be taken numerically. Consider the results for a *phase* screen derived in this way [2.37].

If the phase fluctuations are normally distributed, then

$$\Gamma_v^{(2,2)}(1,2,3,4) = \langle \exp\{i[S(\boldsymbol{\varrho}_1)+S(\boldsymbol{\varrho}_2)-S(\boldsymbol{\varrho}_3)-S(\boldsymbol{\varrho}_4)]\}\rangle$$
$$= \exp\{-[2\sigma_{S_0}^2 - \psi_{13} - \psi_{14} - \psi_{23} - \psi_{24} + \psi_{12} + \psi_{34}]\} ,$$

where $\psi_{ik} = \psi_S(\boldsymbol{\varrho}_i - \boldsymbol{\varrho}_k)$ are the values of the covariance of the phase S at $\boldsymbol{\varrho} = \boldsymbol{\varrho}_i - \boldsymbol{\varrho}_k$, and $\sigma_{S_0}^2$ is the variance of the phase. In Fig. 2.2 (from [2.37]) the variations of the scintillation index β^2 with the wave parameter $D = 2z/kl_S^2$ for the Gaussian phase covariance (2.64) with $\sigma_{S_0}^2$ varying from 0.1 to 5 are represented.

We are acquianted with the behavior of these curves: in the Fraunhofer zone ($D \gg 1$) $\beta^2 = \beta_\infty^2 = 1 - \exp(-2\sigma_{S_0}^2)$ and for weak phase fluctuations ($\sigma_{S_0}^2 \ll 1$) $\beta^2 = 2\sigma_{S_0}^2 D^2/(1+D^2)$. A new element here is the presence, for $\sigma_{S_0}^2 > 1$, of maxima of β^2 at intermediate values of $D \lesssim 1$. In [2.37, 38], the maxima of β^2 for strong phase fluctuations ($\sigma_{S_0}^2 \gg 1$) were estimated to be $\beta_{max}^2 \sim \ln \sigma_{S_0}^2$. More accurate calculations, performed in [2.39, 40] for the phase covariance $\psi_S(\boldsymbol{\varrho})$ of an arbitrary form, lead to the following asymptotic value of β_{max}^2 for $\sigma_{S_0}^2 \gg 1$:

$$\beta_{max}^2 \approx \frac{1}{\pi e} \ln\left(\frac{2\sigma_{S_0}^2}{\gamma}\right) ,$$

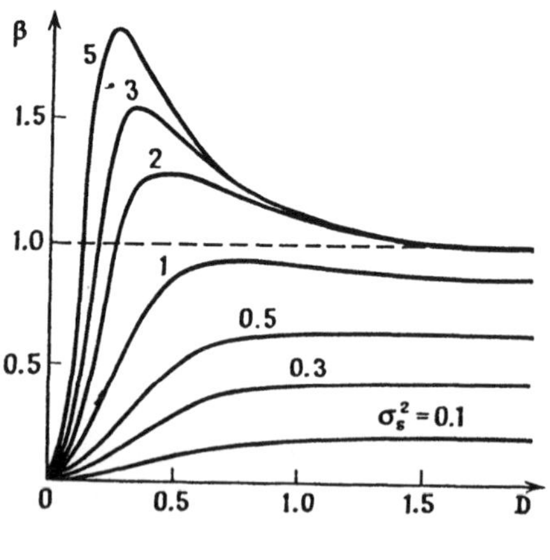

Fig. 2.2. Variation of β^2 beyond the phase screen with $D = 2z/kl_S^2$ [2.37]

where γ is a numerical coefficient equal to

$$\frac{l_1^4}{6}\frac{\partial^4 K_S(0)}{\partial \varrho^4} ,$$

and $l_1 = |K_S''(0)|^{-1/2}$ is a quantity of the order of the correlation radius for the phase. For the Gaussian covariance (2.64) $\gamma = 1/2$, $l_1 = l_S$. These estimates suggest a logarithmic "saturation" of the scintillation index $\beta^2_{\max}$ with increasing $\sigma^2_{S_0}$.

According to [2.37–40], the maximum of the scintillation index lies at a distance z_m

$$z_m = \frac{kl_S^2}{\sigma_{S_0}} . \tag{2.75}$$

Equation (2.71) permits of a simple geometrical interpretation, as does the appearance of maxima: the intensity fluctuations are maximal at the point where the waves beyond the phase screen are *focused*. Indeed, in the geometrical optics approximation, focusing occurs at a distance $z_m \sim 1/v$ from the phase screen, where v is the curvature of the phase front. The orders of magnitude are estimated to be

$$v \sim \frac{1}{k}\left(\frac{\partial^2 S}{\partial x^2}\right) ,$$

where x is one of the coordinates in the plane $z = 0$, and

$$\frac{\partial^2 S}{\partial x^2} \sim \frac{\sigma_{S_0}}{l_S^2} .$$

Accordingly,

$$z_m \sim 1/v \sim kl_S^2/\sigma_{S_0} ,$$

which coincides with (2.75). Clearly, the larger the phase variance $\sigma^2_{S_0}$, the closer the focusing zone lies to the screen.

It is quite obvious that the wave focusing effect and the scintillation index maxima only occur when the phase fluctuations are distinct at the screen. If only the amplitude fluctuates at the screen, then diffraction decreases the scintillation index. Examples of such smoothing effects are considered in Exercises 2.6.7 and 8. The smoothing of intensity fluctuations also occurs when not a plane wave, but a wave from a source of finite angular dimensions [2.37] is incident on a phase screen.

2.4 Random Field Diffraction in Simple Optical Systems

In the preceding section we considered one of the simplest statistical diffraction problems: the diffraction of a deterministic monochromatic wave at an infinite, statistically homogeneous, chaotic screen. We will now turn to the diffraction of random fields by deterministic objects. We will concentrate on weakly diverging wave beams which occur most often in optical and quasi-optical systems.

1) Passage of a Random Wave through a Hole in the Screen. Van Cittert–Zernike Theorem. Let $v(\omega, \boldsymbol{\varrho}) = u_{\text{inc}}(\omega, \boldsymbol{r})|_{z=0}$ be the spectral amplitude of a field produced by a source in the plane $z = 0$. If an opaque screen with a hole S, which produces a wave beam of finite diameter, is placed in this plane, then an "output" field $u_{\text{out}}(\omega, \boldsymbol{\varrho})$ will be formed directly outside the screen. In the Kirchhoff approximation it will be[7]

$$u_{\text{out}}(\omega, \boldsymbol{\varrho}) = \begin{cases} v(\omega, \boldsymbol{\varrho}) & \text{in} \quad S \ , \\ 0 & \text{elsewhere} \ . \end{cases}$$

This can be written in terms of the transmittance function for the hole

$$u_{\text{out}}(\omega, \boldsymbol{\varrho}) = M(\boldsymbol{\varrho}) v(\omega, \boldsymbol{\varrho}) \ , \tag{2.76}$$

where

$$M(\boldsymbol{\varrho}) = \begin{cases} 1 & \text{in} \quad S \ , \\ 0 & \text{elsewhere} \ . \end{cases} \tag{2.77}$$

In the case of smooth (on the scale of a wavelength) fluctuations of the boundary field $v(\omega, \boldsymbol{\varrho})$, we can compute the field beyond the hole using the Fresnel approximation (2.22) which, considering (2.76), gives,

$$u(\omega, \boldsymbol{r}) = \frac{k e^{ikz}}{2\pi i z} \int_{-\infty}^{\infty} M(\boldsymbol{\varrho}') v(\omega, \boldsymbol{\varrho}') \exp\left[\frac{ik}{2z}(\boldsymbol{\varrho} - \boldsymbol{\varrho}')^2\right] d^2\varrho' \ . \tag{2.78}$$

The average of $v(\omega, \boldsymbol{\varrho})$ over the source ensemble is zero. Consequently,

$$\langle u(\omega, \boldsymbol{r}) \rangle = 0 \ . \tag{2.79}$$

We now derive the transverse coherence function of the diffraction field, i.e., the coherence function at $z = \text{const}$. From (2.78) we have

$$\begin{aligned} \Gamma_{\perp}(\omega, \boldsymbol{\varrho}_1, \boldsymbol{\varrho}_2, z) &= \langle u(\omega, \boldsymbol{\varrho}_1, z) u^*(\omega, \boldsymbol{\varrho}_2, z) \rangle \\ &= \left(\frac{k}{2\pi z}\right)^2 \iint_{-\infty}^{\infty} M(\boldsymbol{\varrho}') M(\boldsymbol{\varrho}'') \Gamma_v(\omega, \boldsymbol{\varrho}' - \boldsymbol{\varrho}'') \\ &\quad \times \exp\left\{\frac{ik}{2z}[(\boldsymbol{\varrho}_1 - \boldsymbol{\varrho}')^2 - (\boldsymbol{\varrho}_2 - \boldsymbol{\varrho}'')^2]\right\} d^2\varrho' \, d^2\varrho'' \ , \end{aligned} \tag{2.80}$$

where $\Gamma_v(\omega, \boldsymbol{\varrho}' - \boldsymbol{\varrho}'')$ is the spatial coherence function of the boundary field

[7] The Kirchhoff approximation is known to be applicable if the size of a hole a is large as compared with the wavelength: $a \gg \lambda$.

$v(\omega, \boldsymbol{\varrho})$, which is supposed to be statistically homogeneous. The space-time coherence functions of the fields $u(t, \boldsymbol{r})$ and $v(t, \boldsymbol{\varrho})$ are related to $\Gamma_\perp(\omega, \boldsymbol{\varrho}_1, \boldsymbol{\varrho}_2, z)$ and $\Gamma_v(\omega, \boldsymbol{\varrho}' - \boldsymbol{\varrho}'')$ by the Fourier transform, e.g.,

$$\Gamma_\perp(\tau, \boldsymbol{\varrho}_1, \boldsymbol{\varrho}_2, z) = \int_0^\infty \Gamma_\perp(\omega, \boldsymbol{\varrho}_1, \boldsymbol{\varrho}_2, z)\, \mathrm{e}^{-\mathrm{i}\omega\tau}\, d\omega \ . \tag{2.81}$$

By virtue of (2.79), all these coherence functions coincide with the appropriate covariances.

For the general form of the coherence function (covariance) of the boundary field, the integral (2.80) can be taken in two limiting cases: for the large and small holes a as compared with the correlation radius l_v(but always for $a \gg \lambda$).

For $a \ll l_v$ (*small* hole), the coherence function of the boundary field Γ_v is essentially constant within the confines of the hole, and it can be taken outside the integral sign with the value $\Gamma_v(\omega, 0) \equiv \bar{I}_v(\omega)$. As a result[8]

$$\Gamma_\perp(\boldsymbol{\varrho}_1, \boldsymbol{\varrho}_2, z) = \bar{I}_v u_M(\boldsymbol{\varrho}_1, z)\, u_M^*(\boldsymbol{\varrho}_2, z) \ , \tag{2.82}$$

where the field $u_M(\boldsymbol{\varrho}, z)$ is given by

$$u_M(\boldsymbol{\varrho}, z) = \frac{k\mathrm{e}^{\mathrm{i}kz}}{2\pi\mathrm{i}z} \int M(\boldsymbol{\varrho}') \exp\left[\frac{\mathrm{i}k}{2z}(\boldsymbol{\varrho} - \boldsymbol{\varrho}')^2\right] d^2\varrho' \ , \tag{2.83}$$

which constitutes a diffraction field beyond the hole when a plane wave of unit amplitude strikes it normally.

The covariance for diffraction at a small hole is

$$K_\perp(\boldsymbol{\varrho}_1, \boldsymbol{\varrho}_2, z) = \frac{\Gamma_\perp(\boldsymbol{\varrho}_1, \boldsymbol{\varrho}_2, z)}{\sqrt{\Gamma_\perp(\boldsymbol{\varrho}_1, \boldsymbol{\varrho}_1, z)\, \Gamma_\perp(\boldsymbol{\varrho}_2, \boldsymbol{\varrho}_2, z)}} = \frac{u_M(\boldsymbol{\varrho}_1, z)\, u_M^*(\boldsymbol{\varrho}_2, z)}{|u_M(\boldsymbol{\varrho}_1, z)|\, |u_M(\boldsymbol{\varrho}_2, z)|} \ ,$$

so that $|K_\perp| = 1$. This means that a random wave that has passed through a small hole produces a spatially coherent field. Before the existence of lasers, the only way of producing spatially coherent light (apart from using small light sources) was essentially to pass light through a small hole. This technique is still used today where a field of high intensity is not required.

We now turn our attention to the other limiting case, that of $a \gg l_v$ (*large* hole). Note, first of all, that near the hole there is a region where the covariance is the same as in the absence of an opaque screen. The configuration and size of this region can be estimated using the spectral representation. The width of the two-dimensional spectrum $F_v(\boldsymbol{\kappa})$ of a random field v is about $\kappa_v \sim 1/l_v$ or, in terms of angles, $\delta \sim \kappa_v/k \sim \lambda/l_v$. Accordingly, if observation points $\boldsymbol{r}_1$ and $\boldsymbol{r}_2$ lie within a cone with base a and angle δ at the vertex (Fig. 2.3), then the field will not feel the influence of the diaphragm edges. For all points inside the cone, the transverse

[8] From (2.82) onwards, we omit ω whenever this does not lead to confusion.

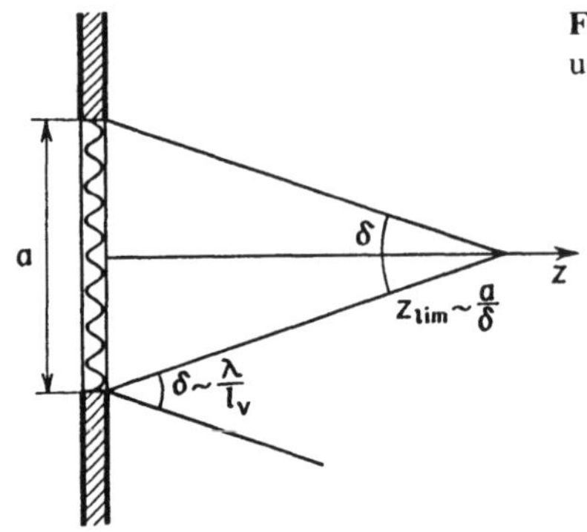

Fig. 2.3. The zone where the transverse covariance remains unchanged

covariance will be the same as for the boundary field

$$\Gamma_\perp(\boldsymbol{\varrho}_1, \boldsymbol{\varrho}_2, z) = \Gamma_v(\boldsymbol{\varrho}_1 - \boldsymbol{\varrho}_2) \ .$$

The limiting distance z_{lim} over which this relation still holds is estimated to be[9]

$$z \sim \frac{a}{\delta} \sim \frac{al_v}{\lambda} \sim kal_v \ . \tag{2.84}$$

For $a \gg l_v$, this distance corresponds to the *far* zone relative to the inhomogeneity size l_v, so that

$$\frac{z_{\text{lim}}}{kl_v^2} \sim \frac{a}{l_v} \gg 1 \ ,$$

and to the *near* zone relative to the hole size a

$$\frac{z_{\text{lim}}}{ka^2} \sim \frac{l_v}{a} \ll 1 \ .$$

Among other things, this leads to the practically important conclusion that the distribution of field fluctuations beyond the hole is already normalized in the near zone of the aperture a.

We will now proceed to carry out the integration of (2.80) for $z \gg z_{\text{lim}}$. To this end, we go over to the new variables $\boldsymbol{\xi} = \boldsymbol{\varrho}' - \boldsymbol{\varrho}''$ and $\boldsymbol{\eta} = (\boldsymbol{\varrho}' + \boldsymbol{\varrho}'')/2$ and denote $(\boldsymbol{\varrho}_1 + \boldsymbol{\varrho}_2)/2$ by $\boldsymbol{\varrho}_+$, and $\boldsymbol{\varrho}_1 - \boldsymbol{\varrho}_2$ by $\boldsymbol{\varrho}$. Then (2.80) will take the form

$$\Gamma_\perp(\boldsymbol{\varrho}, \boldsymbol{\varrho}_+, z) = \left(\frac{k}{2\pi z}\right)^2 \exp\left(\frac{ik\boldsymbol{\varrho}\cdot\boldsymbol{\varrho}_+}{z}\right) \iint_{-\infty}^{\infty} M\left(\boldsymbol{\eta}+\frac{\xi}{2}\right) M\left(\boldsymbol{\eta}-\frac{\xi}{2}\right) \Gamma_v(\xi)$$
$$\times \exp\left\{\frac{ik}{z}[\boldsymbol{\xi}\cdot\boldsymbol{\eta} - \boldsymbol{\eta}\cdot\boldsymbol{\varrho} - \boldsymbol{\xi}\cdot\boldsymbol{\varrho}_+]\right\} d^2\xi \, d^2\eta \ .$$

---

[9] If $l_v \lesssim a$, then one should put $z_{\text{lim}} \sim a$, since the width of the angular spectrum, δ, is then comparable to $\pi/2$.

The region of integration with respect to ξ is bounded here by the condition $\xi \lesssim l_v \ll a$, since for $\xi \gtrsim l_v$ the covariance $\Gamma_v(\xi)$ is close to zero. At the same time, $\eta \lesssim a$, since for $\eta > a$ the product $M(\boldsymbol{\eta}+\boldsymbol{\xi}/2)\ M(\boldsymbol{\eta}-\boldsymbol{\xi}/2) \approx M^2(\boldsymbol{\eta}) = M(\boldsymbol{\eta})$ vanishes. Therefore, the first term in the exponent is bounded from above by the inequality $k\,\boldsymbol{\xi}\cdot\boldsymbol{\eta}/z \lesssim kal_v/z \sim z_{\text{lim}}/z$. For $z \gg z_{\text{lim}}$ this term may be ignored, with the result that the integrals with respect to $\boldsymbol{\xi}$ and $\boldsymbol{\eta}$ are separated

$$\Gamma_\perp(\boldsymbol{\varrho}, \boldsymbol{\varrho}_+, z) = \left(\frac{k}{2\pi z}\right)^2 \exp\left(\frac{ik\boldsymbol{\varrho}\cdot\boldsymbol{\varrho}_+}{z}\right) \int\limits_{-\infty}^{\infty} d^2\eta\, M(\boldsymbol{\eta}) \exp\left(-\frac{ik\boldsymbol{\eta}\cdot\boldsymbol{\varrho}}{z}\right)$$
$$\times \int\limits_{-\infty}^{\infty} d^2\xi\, \Gamma_v(\boldsymbol{\xi}) \exp\left(-\frac{ik\boldsymbol{\xi}\cdot\boldsymbol{\varrho}_+}{z}\right). \tag{2.85}$$

Exactly the same expression follows directly from (2.23) for the field in the Fraunhofer zone

$$u(\boldsymbol{\varrho}, z) = \frac{k \exp\left[ik\left(z+\dfrac{\varrho^2}{2z}\right)\right]}{2\pi i z} \int\limits_{-\infty}^{\infty} M(\boldsymbol{\varrho}')v(\boldsymbol{\varrho}') \exp\left(-\frac{ik\boldsymbol{\varrho}\cdot\boldsymbol{\varrho}'}{z}\right) d^2\varrho'\,. \tag{2.86}$$

The approximation (2.86), however, is only valid for the far zone $z \gg ka^2$, whereas (2.85) is also applicable over much shorter distances $z \gg z_{\text{lim}} \sim kal_v$. In other words, in respect to the hole, the coherence function (2.85) is already formed in the *near* zone.

The variation of (2.83) with the transverse coordinates $\boldsymbol{\varrho} = \boldsymbol{\varrho}_1 - \boldsymbol{\varrho}_2$ and $\boldsymbol{\varrho}_+ = (\boldsymbol{\varrho}_1 + \boldsymbol{\varrho}_2)/2$ is determined by the product of three factors. The factor at the integral

$$\exp(ik\boldsymbol{\varrho}\cdot\boldsymbol{\varrho}_+/z) = \exp[ik(\varrho_1^2 - \varrho_2^2)/2z] \approx \exp[ik(r_1 - r_2)]$$

is simply due to the fact that beyond the hole, spherical waves are mismatched with the plane form of the surface $z = \text{const}$, where correlation of the field is considered. If we take the correlation on the sphere $r = \text{const}$, this factor becomes unity.

The integral with respect to $\boldsymbol{\eta}$ in (2.85), which depends only on the difference $\boldsymbol{\varrho} = \boldsymbol{\varrho}_1 - \boldsymbol{\varrho}_2$, characterizes the field of a coherent source with the amplitude distribution $M(\boldsymbol{\eta})$. If a is the size of the hole, the angular width of the directional pattern of such a radiator is of the order of λ/a, and the linear width of the central diffraction lobe in the plane $z=\text{const}$ is $\Delta\varrho \sim \lambda r/a$. It follows that the transverse correlation radius is

$$l_\perp \sim \frac{\lambda z}{a} \sim \frac{\lambda}{\gamma}\,, \tag{2.87}$$

where $\gamma \sim a/z$ is the angular size of the hole relative to the observation point. The transverse correlation radius, like the width of diffraction lobes, grows with increasing distance from the plane $z = 0$. The increase in the coherence radius can be accounted for geometrically by the fact that farther away from the hole, the phase fronts of all the elementary spherical waves (up to λ) can be thought of as coinciding over an ever increasing surface area.

The second integral in (2.85) governs the spatial distribution of the mean intensity of the diffraction field. Putting $\boldsymbol{\varrho}=0$ into (2.85) and considering that the integral with respect to $\boldsymbol{\eta}$ at $\boldsymbol{\varrho}=0$ equals the area of the hole S, we have

$$\bar{I}_u(\boldsymbol{\varrho}_+, z) = \frac{k^2 S}{(2\pi z)^2} \int_{-\infty}^{\infty} \Gamma_v(\xi) \exp\left(-\frac{ik\boldsymbol{\varrho}_+ \cdot \xi}{z}\right) d^2\xi \; . \tag{2.88}$$

The space scale l_v for Γ_v is assumed to be small as compared with a. Therefore, the width of the intensity distribution $\bar{I}_u$ in the plane $z=\text{const}$, which equals $\varrho_+ \sim \lambda z/l_v$ (the angular width has the order $\varrho_+/z \sim \lambda/l_v \sim 1/kl_v$), is large as compared with the transverse correlation radius $l_\perp$

$$\frac{\lambda z}{l_v} \gg l_\perp \sim \frac{\lambda z}{a} \quad (l_v \ll a) \; .$$

This is an example of a quasi-homogeneous field (in the plane $z=\text{const}$). In this case, all realizations of the diffraction field beyond the hole are fast spatial oscillations with the space scale $l_\perp$ and characteristic amplitude $[\bar{I}_u(\varrho)]^{1/2}$ (Fig. 2.4).

If we divide the coherence function (2.85) by the mean intensity of the diffraction field, we will obtain the coefficient of transverse spatial correlation

$$K_\perp(\boldsymbol{\varrho}) = \exp\left(\frac{ik\boldsymbol{\varrho}\cdot\boldsymbol{\varrho}_+}{z}\right) \mathscr{F}_M\left(\frac{k\boldsymbol{\varrho}}{z}\right) ,$$

where

$$\mathscr{F}_M(\boldsymbol{\kappa}) = \frac{1}{S} \int_{-\infty}^{\infty} M(\boldsymbol{\eta}) \exp(-i\boldsymbol{\kappa}\cdot\boldsymbol{\eta})\, d^2\eta \tag{2.89}$$

is the unity-normalized $[\mathscr{F}(0)=1]$ Fourier transform of $M(\boldsymbol{\eta})$ and, simultaneously, the radiation pattern of the hole when a monochromatic wave strikes it

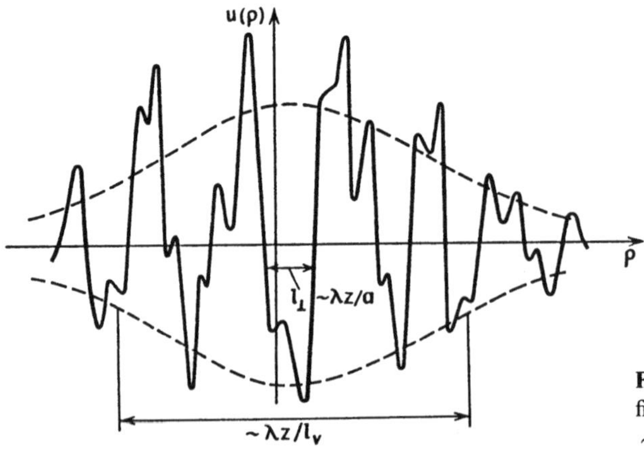

Fig. 2.4. Typical variation of the field beyond the hole for $z \gg z_{\text{lim}} \sim kal_v$

normally. In the particular case of a round hole where $M(\boldsymbol{\varrho}) = 1$ for $\varrho < a$ and vanishes elsewhere, we have

$$\mathscr{F}_M(\kappa) = 2J_1(\kappa a)/\kappa a \ ,$$

where $J_1(\kappa a)$ is the first-order Bessel function.

In the limiting case, when the correlation radius of the boundary field l_v is small compared with the wavelength, the integral with respect to ξ is no longer dependent on $\boldsymbol{\varrho}_+$

$$\int\limits_{-\infty}^{\infty} \Gamma_v(\xi)\exp\left(-\frac{ik\boldsymbol{\varrho}_+\cdot\xi}{z}\right)d^2\xi \approx \bar{I}_v S_v \quad (l_v \ll \lambda) \ ,$$

where

$$S_v = \frac{1}{\bar{I}_v}\int\limits_{-\infty}^{\infty} \Gamma_v(\xi)\,d^2\xi = \int\limits_{-\infty}^{\infty} K_v(\xi)\,d^2\xi \tag{2.90}$$

is the effective coherence area of the boundary field, which is about l_v^2. As a result, for $l_v \ll \lambda$ we obtain from (2.85) the so-called *Van Cittert–Zernike theorem*

$$\Gamma_\perp(\boldsymbol{\varrho}, \boldsymbol{\varrho}_+, z) = \frac{1}{z^2}\exp\left(\frac{ik\boldsymbol{\varrho}\cdot\boldsymbol{\varrho}_+}{z}\right)\int\limits_{-\infty}^{\infty} J(\boldsymbol{\varrho}')\exp\left(-\frac{ik\boldsymbol{\varrho}\cdot\boldsymbol{\varrho}'}{z}\right)d^2\varrho' \ , \tag{2.91}$$

The quantity

$$J(\boldsymbol{\varrho}') = \frac{k^2}{4\pi^2}\bar{I}_v S_v M(\boldsymbol{\varrho}') \tag{2.92}$$

in this relation is called the *reduced intensity*.

According to this theorem, the modulus of the transverse coherence function $\Gamma_\perp$ varies with $\boldsymbol{\varrho} = \boldsymbol{\varrho}_1 - \boldsymbol{\varrho}_2$ in exactly the same way as the field in the Fraunhofer zone produced by a fully coherent source with the amplitude distribution proportional to $J(\boldsymbol{\varrho}')$. Equation (2.91) differs from conventional diffraction equations in two respects. First, $\Gamma_\perp$, which is quadratic in field, is inversely proportional to the square (rather than to the first power) of the distance. Second, as already noted, this expression, together with the initial formula (2.85), applies over distances $z \gg z_{\mathrm{lim}}$. In the limit considered ($l_v \ll a$), z_{lim} is comparable with a, as the pattern of a coherent source of size a is formed at substantially larger distances $z \gg ka^2 \gg z_{\mathrm{lim}} \sim a$.

The Van Cittert–Zernike theorem (2.91) is generally deduced by formally introducing delta-correlated fluctuations of the boundary field, i.e., by introducing (2.80) or (2.85)

$$\Gamma_v(\boldsymbol{\varrho} - \boldsymbol{\varrho}') = \bar{I}_v(\boldsymbol{\varrho}')\,\delta(\boldsymbol{\varrho}' - \boldsymbol{\varrho}'')\,S_v \ . \tag{2.93}$$

Then, in the expression (2.91) for the reduced intensity, the variable intensity $\bar{I}_v(\boldsymbol{\varrho}')$ appears instead of the product $\bar{I}_v M(\boldsymbol{\varrho}')$. A source with independent values of the field at arbitrarily close points is said to be spatially *incoherent*.

The concept of a spatially incoherent source is an idealization with limited scope of applications. The point is that a field of *travelling* waves cannot, by its very nature, be delta-correlated in space, because (Sect. 2.2) the space scale l_v of the field cannot be less than the wavelength[10]. Therefore, in the problem under consideration (diffraction of a random field at a hole) the transition to the $l_v \to 0$ limit is, strictly speaking, illegitimate, thus making the validity of the derivation of the Van Cittert–Zernike theorem based on (2.92) questionable. We are going to deal with the theorem (2.91) in another important case, namely when discussing the field radiated by a system of independent point sources (Sect. 2.5). Clearly, the reduced intensity in that problem will have a different sense than in (2.92).

The results for the spatial correlation of spectral amplitudes obtained above hold good for fields which are quasi-monochromatic. In fact, according to (2.81) the spatial covariance is

$$\begin{aligned}\Gamma_\perp(\tau=0,\boldsymbol{\varrho}_1,\boldsymbol{\varrho}_2,z) &= \int_0^\infty \Gamma_\perp(\omega,\boldsymbol{\varrho}_1,\boldsymbol{\varrho}_2,z)\,\mathrm{e}^{\mathrm{i}\omega\tau}\,d\omega\,|_{\tau=0}\\ &= \int_0^\infty d\omega\left(\frac{k}{2\pi z}\right)^2 \exp\left(\frac{\mathrm{i}k\boldsymbol{\varrho}\cdot\boldsymbol{\varrho}_+}{z}\right)\int_{-\infty}^{\infty} d^2\xi\,\Gamma_v(\omega,\boldsymbol{\xi})\exp\left(-\frac{\mathrm{i}k\boldsymbol{\xi}\cdot\boldsymbol{\varrho}_+}{z}\right)\\ &\quad\times\int_{-\infty}^{\infty} d^2\eta\,M(\boldsymbol{\eta})\exp\left(-\frac{\mathrm{i}k\boldsymbol{\eta}\cdot\boldsymbol{\varrho}}{z}\right).\end{aligned}\tag{2.94}$$

For the quasi-monochromatic field, $\Gamma_v(\omega,\boldsymbol{\xi})$ is concentrated within a narrow frequency band $\Delta\omega$ at ω_0. Therefore, approximately ($k_0=\omega_0/c$) we have

$$\begin{aligned}\Gamma_\perp(\tau=0,\boldsymbol{\varrho}_1,\boldsymbol{\varrho}_2,z) &= \left(\frac{k_0}{2\pi z}\right)^2\exp\left(\frac{\mathrm{i}k_0\boldsymbol{\varrho}\cdot\boldsymbol{\varrho}_+}{z}\right)\int_{-\infty}^{\infty} d^2\xi\,\Gamma_v(\tau=0,\xi)\\ &\quad\times\exp\left(-\frac{\mathrm{i}k_0\boldsymbol{\xi}\cdot\boldsymbol{\varrho}_+}{z}\right)\int_{-\infty}^{\infty} d^2\eta\,M(\boldsymbol{\eta})\exp\left(-\frac{\mathrm{i}k_0\boldsymbol{\eta}\cdot\boldsymbol{\varrho}}{z}\right),\end{aligned}\tag{2.95}$$

where

$$\Gamma_v(\tau=0,\boldsymbol{\xi}) = \int_0^\infty \Gamma_v(\omega,\boldsymbol{\xi})\,\mathrm{e}^{\mathrm{i}\omega\tau}\,d\omega\,|_{\tau=0} = \Delta\omega_{\mathrm{ef}}\,\Gamma_v(\omega_0,\boldsymbol{\xi})$$

is the spatial covariance of the boundary field, and $\Delta\omega_{\mathrm{ef}}$ is the effective width of the frequency band.

The transition from (2.94) to (2.95) is possible if

$$\left|\frac{\Delta k}{z}(\boldsymbol{\varrho}\cdot\boldsymbol{\varrho}_+ - \boldsymbol{\xi}\cdot\boldsymbol{\varrho}_+ - \boldsymbol{\eta}\cdot\boldsymbol{\varrho})\right| = \frac{\Delta k}{2z}|(\boldsymbol{\varrho}_1-\boldsymbol{\varrho}')^2-(\boldsymbol{\varrho}_2-\boldsymbol{\varrho}'')^2| \ll 1\ .\tag{2.96}$$

[10] Note in passing that the identification of the reduced intensity (2.92) with the mean field intensity in the hole, $\bar{I}_v(\boldsymbol{\varrho})$, leads to the same conclusion. Such an identification is possible at $k^2S_v/4\pi^2=1$. This means that $S_v=\lambda^2$, which contradicts the assumption (2.93) of delta-correlation, according to which $S_v \sim l_v^2 \ll \lambda^2$.

The physical significance of the inequality is that the path difference of waves coming from different parts of the hole must be small as compared with the coherence length $l_c = c\tau_c \sim c/\Delta\omega$. We can readily estimate that for the inequality (2.96) to hold, it is sufficient that the quasi-monochromaticity condition ($\Delta\omega \ll \omega_0$), together with the condition $\varrho_+ \lesssim a$, be met. This requires that the "center of mass" of the observation points $\boldsymbol{\varrho}_+ = (\boldsymbol{\varrho}_1 + \boldsymbol{\varrho}_2)/2$ lie within the confines of a.

2) Focusing of Random Waves. The diffraction pattern at the focal point of the lens can be obtained using the above results and some simple transformations, since the field in the focal plane of the lens is similar to the field of a hole at infinity.

Suppose that there is a thin lens with a main focal length F in the hole of an opaque screen. The action of the lens can be described by introducing into (2.78) the phase factor $\exp(-ik\varrho'^2/2F)$ under the integration sign

$$u(\boldsymbol{r}) = \frac{ke^{ikz}}{2\pi iz} \int_{-\infty}^{\infty} M(\boldsymbol{\varrho}')v(\boldsymbol{\varrho}')\exp\left(-\frac{ik\varrho'^2}{2F}\right)\exp\left(\frac{ik(\boldsymbol{\varrho}-\boldsymbol{\varrho}')^2}{2z}\right)d^2\varrho' \;. \tag{2.97}$$

In the main focal plane $z = F$ the exponent is simplified and the field $u(\boldsymbol{\varrho}, F)$ is conveniently expressed in terms of the Fourier transform of $M(\boldsymbol{\varrho}')v(\boldsymbol{\varrho}')$

$$u(\boldsymbol{\varrho}, F) = \frac{k\exp\left[ik\left(F + \dfrac{\varrho^2}{2F}\right)\right]}{2\pi iF} \int_{-\infty}^{\infty} M(\boldsymbol{\varrho}')v(\boldsymbol{\varrho}')\exp\left(-\frac{ik\boldsymbol{\varrho}\cdot\boldsymbol{\varrho}'}{F}\right)d^2\boldsymbol{\varrho}' \;. \tag{2.98}$$

Equation (2.98) differs from the Fraunhofer approximation (2.86) for the field of a hole (without a lens) only in that F is substituted for z, so that equations for the covariances are counterparts of the appropriate expressions in Sect. 2.4.1. So, when $a \ll l_v$ (small hole), the intensity distribution in the focal plane is the same as for a spatially coherent plane wave. In the opposite case of $a \gg l_v$ (large hole), the intensity distribution can be obtained from (2.88)

$$\bar{I}_u(\boldsymbol{\varrho}_+, F) = \frac{k^2 S}{(2\pi F)^2} \int_{-\infty}^{\infty} \Gamma_v(\boldsymbol{\xi})\exp\left(-\frac{ik\boldsymbol{\varrho}_+\cdot\boldsymbol{\xi}}{F}\right)d^2\xi \quad (a \gg l_v) \;.$$

In this case, the size of the focal spot is given by the expression $\varrho_f \sim \lambda F/l_v$, i.e., it is a/l_v times larger than for the deterministic field of a plane wave ($\varrho_f \sim \lambda F/a$). However, the intensity at the center of the diffraction pattern is now

$$\bar{I}_u(\varrho_+ = 0, F) \approx \frac{k^2 S S_v \bar{I}_v}{(2\pi F)^2} \;,$$

which is smaller by a factor of $S/S_v \sim (a/l_v)^2$ than in coherent exposure (small hole). It is quite obvious that the total intensity derived by integrating $\bar{I}_u$ over all the plane $z = F$ is in both cases equal to the total "input" intensity $\bar{I}_v S$.

As with diffraction at the hole, the intensity distribution for the quasi-monochromatic field appears to be essentially the same as for strictly monochromatic waves. Moreover, in optics, even with *white light*, for which $\Delta\omega \sim \omega_0$, the intensity distribution in the plane $z = F$ differs only insignificantly from the appropriate pattern for the monochromatic field. That is why, in problems concerned with the formation of an optical image, it is legitimate to ignore effects caused by the temporal incoherence of the field [2.25, 41].

3) Role of Spatial Coherence of Light in the Formation of an Optical Image. Consider a simple optical system containing a single thin lens (Fig. 2.5). A semi-opaque object, e.g., a slide characterized by the complex transmission coefficient $f(\boldsymbol{\varrho}_1)$, is located in the object plane $z = -z_1$. When the object is exposed to a monochromatic wave $u_{1\text{inc}}(\boldsymbol{\varrho}, z)$ from the left, the field directly beyond the object will be

$$u_1(\boldsymbol{\varrho}_1) = f(\boldsymbol{\varrho}_1)v(\boldsymbol{\varrho}_1), \qquad v(\boldsymbol{\varrho}_1) \equiv u_{1\text{inc}}(\boldsymbol{\varrho}, z)|_{z=-z_1} , \tag{2.99}$$

where $\boldsymbol{\varrho}_1$ is the radius vector of a point in the object plane.

In the plane $z = 0$, where the lens is located, the field $u_{2\text{inc}}(\boldsymbol{\varrho}_2)$ ($\boldsymbol{\varrho}_2$ is a vector in the plane $z = 0$) is formed. In the Fresnel approximation (2.22) it is given by

$$u_{2\text{inc}}(\boldsymbol{\varrho}_2) = \frac{k e^{ikz_1}}{2\pi i z_1} \int_{-\infty}^{\infty} u_1(\boldsymbol{\varrho}_1) \exp\left[\frac{ik(\boldsymbol{\varrho}_2 - \boldsymbol{\varrho}_1)^2}{2z_1}\right] d^2\varrho_1 . \tag{2.100}$$

At the "output" of the lens the field is

$$u_2(\boldsymbol{\varrho}_2) = u_{2\text{inc}}(\boldsymbol{\varrho}_2) M(\boldsymbol{\varrho}_2) \exp\left(-\frac{ik\varrho_2^2}{2F}\right) . \tag{2.101}$$

The factor $M(\boldsymbol{\varrho}_2)$, which is unity at the lens and zero elsewhere, describes the action of the boundary diaphragm of the lens, and the factor $\exp(-ik\varrho_2^2/2F)$ describes the action of the lens with focal distance F.

Lastly, the field (2.101) from the lens can also be related to the image plane $z = z_2$ using the Fresnel approximation

$$u(\boldsymbol{\varrho}) = \frac{k e^{ikz_2}}{2\pi i z_2} \int_{-\infty}^{\infty} u_2(\boldsymbol{\varrho}_2) \exp\left(\frac{ik(\boldsymbol{\varrho} - \boldsymbol{\varrho}_2)^2}{2z_2}\right) d^2\varrho . \tag{2.102}$$

Expressions (2.99–102) relate the field $u(\boldsymbol{\varrho})$ in the image plane to the transmission coefficient $f(\boldsymbol{\varrho}_1)$ characterizing the object, and to the initial field $v(\boldsymbol{\varrho}_1)$ whose properties are determined by the light source. We will not concern ourselves with

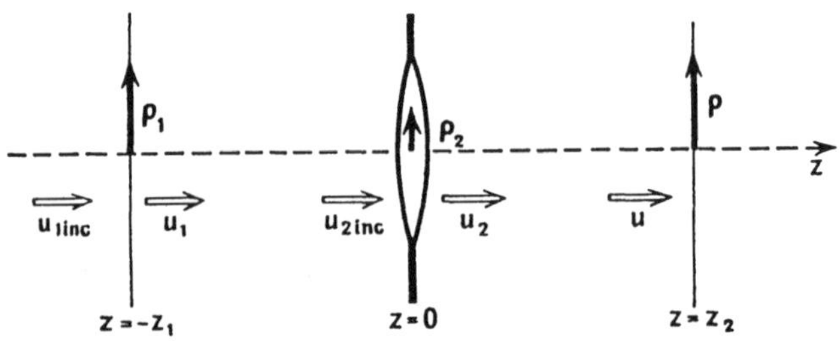

Fig. 2.5. Image formation in a simple optical system

the optical aspects of the problem, but will concentrate on the dependence of the image formation process on the degree of spatial coherence of the light incident on the semi-transparent object.

Bearing this in mind, we assume from the very beginning that the image and object planes are conjugate in the geometrical optics sense, i.e., $1/z_1 + 1/z_2 = 1/F$. Substituting in succession (2.101), (2.100), and (2.99) into (2.102), we have

$$u(\boldsymbol{\varrho}) = \frac{-k^2}{4\pi^2 z_1 z_2} \exp\left[ik\left(z_1 + z_2 + \frac{\varrho^2}{2z_2}\right)\right]$$

$$\times \iint_{-\infty}^{\infty} f(\boldsymbol{\varrho}_1) v(\boldsymbol{\varrho}_1) M(\boldsymbol{\varrho}_2) \exp\left[-ik\boldsymbol{\varrho}_2\left(\frac{\boldsymbol{\varrho}}{z_2} + \frac{\boldsymbol{\varrho}_1}{z_1}\right) + ik\frac{\varrho_1^2}{2z_1}\right] d^2\varrho_1\, d^2\varrho_2 \ .$$

Integrating with respect to $\boldsymbol{\varrho}_2$ gives

$$\int M(\boldsymbol{\varrho}_2) \exp\left[-ik\boldsymbol{\varrho}_2 \cdot \left(\frac{\boldsymbol{\varrho}}{z_2} + \frac{\boldsymbol{\varrho}_1}{z_1}\right)\right] d^2\varrho_2 = S\mathscr{F}_M\left[k \cdot \left(\frac{\boldsymbol{\varrho}}{z_2} + \frac{\boldsymbol{\varrho}_1}{z_1}\right)\right],$$

where S is the surface area of the hole, and $\mathscr{F}_M(\kappa)$ is the unity-normalized $[\mathscr{F}_M(0) = 1]$ Fourier transform of $M(\boldsymbol{\varrho})$ [see (2.89)]. As a result

$$u(\boldsymbol{\varrho}) = -\frac{k^2 S \exp[ik(z_1 + z_2 + \varrho^2/2z_2)]}{4\pi^2 z_1 z_2} \int_{-\infty}^{\infty} v(\boldsymbol{\varrho}_1) f(\boldsymbol{\varrho}_1)$$

$$\times \mathscr{F}_M\left[k \cdot \left(\frac{\boldsymbol{\varrho}}{z_2} + \frac{\boldsymbol{\varrho}_1}{z_1}\right)\right] \exp\left(\frac{ik\varrho_1^2}{2z_1}\right) d^2\varrho_1 \ . \tag{2.103}$$

Expression (2.103) is a special case of the more general formula

$$u(\boldsymbol{\varrho}) = \int v(\boldsymbol{\varrho}_1) f(\boldsymbol{\varrho}_1) \mathscr{P}(\boldsymbol{\varrho}, \boldsymbol{\varrho}_1) d^2\varrho_1 \ , \tag{2.104}$$

which is a relationship of the type (2.7), and in the problem at hand relates the field in the object plane of an arbitrary optical system to the incident field $v(\boldsymbol{\varrho}_1)$ and the transmittance function of the object, $f(\boldsymbol{\varrho}_1)$. The function $\mathscr{P}(\boldsymbol{\varrho}, \boldsymbol{\varrho}_1)$ is called the *spread function* of the system. The characteristic space scale of the spread function in the image plane (i.e., in $\boldsymbol{\varrho}$) is the radius of the diffraction spot corresponding to the point object, and the space scale of $\mathscr{P}$ in $\boldsymbol{\varrho}_1$ defines the size of the region in the object plane contributing noticeably to the field at a given point in the image plane. In other words, this is the limit of the resolution of the optical system.

In this case, i.e., a single-lens system with a round diaphragm, the spread function is

$$\mathscr{P}(\boldsymbol{\varrho}, \boldsymbol{\varrho}_1) = -\frac{k^2 S \exp[ik(z_1 + z_2 + \varrho^2/2z_2 + \varrho_1^2/2z_1)]}{4\pi^2 z_1 z_2} \frac{2J_1(\kappa a)}{\kappa a} \ , \tag{2.105}$$

where $\kappa = k|\boldsymbol{\varrho}/z_2 + \boldsymbol{\varrho}_1/z_1|$. According to (2.105), corresponding to a point object at $\boldsymbol{\varrho}_{10}$ there is a diffraction spot with center at $\boldsymbol{\varrho}_0' = -\boldsymbol{\varrho}_{10} z_2/z_1$. The ratio z_2/z_1 characterizes the magnification of a given system, and the minus sign implies an inverted image. Since the Bessel function $J_1(\kappa a)$ has its first zero at $\kappa a = 3.83$, the characteristic scale of the spread function in ϱ may be the radius $3.83 z_2/ka = 0.61\lambda z_2/a$ of the first dark Airy ring in the diffraction image of the point object. On the other hand, the space scale in ϱ_1 is $\delta \equiv 0.61\,\lambda z_1/a$, which up to a factor of order unity is the Rayleigh resolution limit.

In examining the image behavior we will begin with the general formula (2.104), and will use only its particular form (2.103) to illustrate general conclusions. According to (2.104), when objects are exposed to partially coherent light, the mean intensity of the field in the image plane is

$$\bar{I}(\varrho) = \int \Gamma_v(\varrho', \varrho'') f(\varrho') f^*(\varrho'') \mathscr{P}(\varrho, \varrho') \mathscr{P}^*(\varrho, \varrho'') d^2\varrho' d^2\varrho'' \ , \tag{2.106}$$

where

$$\Gamma_v(\varrho', \varrho'') = \langle v(\varrho') v^*(\varrho'') \rangle$$

is the spatial coherence function of the primary field.

For a given object (f is fixed) the behavior of the intensity distribution in the object plane depends on the relationship between the coherence radius of the primary field l_v (the characteristic space scale of the coherence function Γ_v), the size of the resolution zone δ (characteristic space scale of the spread function in ϱ_1) and the finest detail of the object, $\varrho_{\min}$. The relationship between $\varrho_{\min}$ and δ is governed by the characteristics of the object and the optical device, respectively. If these variables are known, the image can only be affected by changing l_v, i.e., the character of primary radiation.

Consider the limiting cases $l_v \gg \varrho_{\min}, \delta$ and $l_v \ll \varrho_{\min}, \delta$, which are generally called the coherent and incoherent illumination of the object. With *coherent* light ($l_v \gg \varrho_{\min}, \delta$) we can assume that in the region essential for integration in (2.106) (in this region both $\mathscr{P}(\varrho, \varrho')$ and $\mathscr{P}^*(\varrho, \varrho'')$ are non-zero and $|\varrho' - \varrho''| \lesssim \delta$), the coherence function $\Gamma_v(\varrho' - \varrho'')$ is practically constant (since $|\varrho' - \varrho''| \lesssim \delta \ll l_v$). Bringing it outside the integration sign as $\bar{I}_v$, where $\bar{I}_v = \Gamma_v(0)$ is the intensity of the incident field, we obtain

$$\bar{I}(\varrho)_{\text{coh}} = \bar{I}_v \left| \int_{-\infty}^{\infty} f(\varrho_1) \mathscr{P}(\varrho, \varrho_1) d^2\varrho_1 \right|^2 . \tag{2.107}$$

The equation can, clearly, be obtained directly from (2.106), if the primary field $v(\varrho_1)$ is taken to be coherent over the entire object plane. It is seen from this treatment that the requirement of coherence over the entire object is superfluous: for (2.107) to hold, it is sufficient that the primary field $v(\varrho)$ be coherent only within the confines of the resolution region δ.

In the other limiting case of *incoherent* light ($l_v \ll \delta, \varrho_{\min}$), the region essential for integration in (2.107) is bounded by the coherence area with linear dimensions $|\varrho' - \varrho''| \lesssim l_v$. Within this area the functions $f^*(\varrho'')$ and $\mathscr{P}^*(\varrho, \varrho'')$ can approximately be replaced by $f^*(\varrho')$ and $\mathscr{P}^*(\varrho, \varrho')$, respectively, with the result that

$$\bar{I}(\varrho)_{\text{incoh}} = \int_{-\infty}^{\infty} |f(\varrho') \mathscr{P}(\varrho, \varrho')|^2 d^2\varrho' \int_{-\infty}^{\infty} \Gamma_v(\varrho' - \varrho'') d^2\varrho'' \ .$$

The internal integral is $\bar{I}_v S_v$. Here, as in (2.90), we introduced the coherence area S_v that is equal to the integral of the correlation coefficient of the primary field $K_v = \Gamma_v / \bar{I}_v$.

The distribution of the mean intensity in the image plane for incoherent light is thus given by

$$\bar{I}(\boldsymbol{\varrho})_{\text{incoh}} = \bar{I}_v S_v \int |f(\boldsymbol{\varrho}_1)|^2 |\mathscr{P}(\boldsymbol{\varrho}, \boldsymbol{\varrho}_1)|^2 d^2 \varrho_1 \ . \tag{2.108}$$

Taking, as in (2.93), the primary field to be delta-correlated, the expression can formally be derived from (2.106).

Equation (2.107) differs markedly from (2.108). It can be said that with coherent (in the above sense, i.e., for $l_v \gg \delta$, $\varrho_{\min}$) light, the complex transmission coefficient $f(\boldsymbol{\varrho}_1)$ itself [with the transfer function $\sqrt{\bar{I}_v}\,\mathscr{P}(\boldsymbol{\varrho}, \boldsymbol{\varrho}_1)$] is transformed; and with incoherent light ($l_v \ll \delta$, $\varrho_{\min}$), $|f(\boldsymbol{\varrho}_1)|^2$ (with the transfer function $\bar{I}_v S_v |\mathscr{P}(\boldsymbol{\varrho}, \boldsymbol{\varrho}_1)|^2$) is transformed. It follows that with incoherent light the image is independent of the phase of the complex transmittance function f, whereas coherent light can bring about abrupt variations of the phase of f even at $|f| = \text{const}$. For example, if on a certain line in the object plane the phase changes abruptly by π, but $|f| = \text{const}$, then, by (2.107), on an appropriate (conjugate) line in the image plane the intensity $\bar{I}(\boldsymbol{\varrho})_{\text{coh}}$ will vanish and a dark fringe will be seen. It is true that if the modulus and argument of the transparency coefficient vary *smoothly* (in the scale of δ), i.e., for $\varrho_{\min} \gg \delta$, both kinds of light give an essentially identical image, because in both cases the intensity distribution in the image plane, $\bar{I}(\boldsymbol{\varrho})$, is proportional to $|f(\boldsymbol{\varrho}_1)|^2$.

Two further distinctions between coherent and incoherent light are worth mentioning. First, given the same light intensity I_v, the intensity in the image plane for coherent light will be higher than for incoherent light, since the width of the angular spectrum is larger for incoherent than for coherent light. Second, the limiting cases also differ in resolving power: with incoherent light the size of the smallest discernable (according to Rayleigh) detail is somewhat smaller (by about 30%) than with coherent light. This distinction is associated simply with the fact that $|\mathscr{P}(\boldsymbol{\varrho}, \boldsymbol{\varrho}_1)|^2$ in (2.108) has a stronger dependence on $\boldsymbol{\varrho}_1$ than the spread function $\mathscr{P}(\boldsymbol{\varrho}, \boldsymbol{\varrho}_1)$ in (2.107). As a result, two point objects with centers at points $\boldsymbol{\varrho}_1'$ and $\boldsymbol{\varrho}_1''$ are so resolved that the trough between the crests of the intensity distribution in the image plane appears to be deeper if the light oscillations at $\boldsymbol{\varrho}_1'$ and $\boldsymbol{\varrho}_1''$ are incoherent. The trough will be made still deeper if *negative correlation* of the incident field at $\boldsymbol{\varrho}_1'$ and $\boldsymbol{\varrho}_1''$ is attained. (Negative values may, for example, be assumed by the coherence function of the form $2J_1(x)/x$ that corresponds to a uniformly glowing disc or a pencil of light passed through a small round hole.) This elegant way of improving the resolving power gives, however, an insignificant gain (only several per cent) as compared with incoherent light. A more detailed treatment of this issue is given in [2.25, 41].

These examples by no means exhaust the stochastic problems which occur in optics. Among others we mention the problems of spatial filtration, statistical questions connected with optical systems being necessarily imperfect and photographic materials being granular, the use of optical systems for correlation analysis, and so forth. The reader can become acquainted with these and other applications in [2.17–26]. Many ideas that originated in optics have also been applied in radio engineering [2.42].

2.5 Excitation of Fields by Random Sources

Fluctuations of a field excited by random sources are of interest in many branches of physics. In this section we will deal with the radiation from radio antennas having random variations of the current in the aperture and consider the radiation of a large number of independent sources.

1) Statistics of the Field Produced by a Large Antenna. Fluctuations of the field emitted by an antenna may be caused by random deviations from specified values of amplitudes and phases of currents flowing through the aperture. These deviations may be due to various factors, the major ones being irregularities of the antenna reflector, variation in the parameters of the radiating oscillators, random deviations in excitation systems, and so forth. The deformation of antennas due to temperature variations or wind loads also plays a role. We consider one such model of an antenna with fluctuating currents.

If a flat antenna is large as compared with the wavelength, and currents through the aperture are monochromatic [they are proportional to $\exp(-\mathrm{i}\omega t)$] and have only one component (say j_x), then in directions not markedly deviating from the normal to the aperture plane ($z=0$), the electric field may be supposed to have only one component E_x, that satisfies the equation

$$\Delta E_x + k^2 E_x = -\frac{4\pi \mathrm{i} k}{c} j_x \, .$$

Written as ($u = E_x$, $q = -4\pi\mathrm{i}kj_x/c$), the problem thus reduces to the earlier one (2.12) with the solution (2.13).

For antennas with a flat aperture, the currents are concentrated at $z=0$, i.e., the volume integral (2.13) becomes a surface one. Retaining the notation q for the density of surface currents, we obtain

$$u(\boldsymbol{r}) = \frac{1}{4\pi}\int_S q(\boldsymbol{\varrho}')\frac{\mathrm{e}^{\mathrm{i}kR}}{R}\, d^2\varrho', \quad R = |\boldsymbol{r}-\boldsymbol{r}'| \, , \tag{2.109}$$

where ω is omitted for simplicity, and the integration covers the aperture area S.

We are normally interested in the field in the far (Fraunhofer) zone $z \gg ka^2$, where a is the aperture size. It is therefore possible to use the Fraunhofer diffraction approximation, i.e., (2.15), which in this case (surface sources; radiation concentrated within a narrow cone near the z-axis) takes the form

$$u(\boldsymbol{\varrho}, z) = \frac{1}{4\pi z}\exp\left[\mathrm{i}k\left(z+\frac{\varrho^2}{2z}\right)\right]\int_S q(\boldsymbol{\varrho}')\exp\left(-\frac{\mathrm{i}k\boldsymbol{\varrho}\cdot\boldsymbol{\varrho}'}{z}\right)d^2\varrho' \, , \tag{2.110}$$

where $\boldsymbol{\varrho} = (x, y)$ are the coordinates of an observation point in a distant plane $z = \text{const}$. We will confine ourselves to finding the mean and spatial covariance of the field radiated by a large antenna, concentrating on the basic effects caused by current fluctuations in antennas. A systematic exposition of the statistical theory of antennas is to be found in [2.43].

As in Sect. 2.4, it is convenient to introduce a function $M(\boldsymbol{\varrho})$, which is unity at the aperture S and zero elsewhere [see (2.77)]. We thus have

$$q(\boldsymbol{\varrho}) = M(\boldsymbol{\varrho})[\bar{q}(\boldsymbol{\varrho}) + \tilde{q}(\boldsymbol{\varrho})] \ .$$

This allows us to shift the integration limits in (2.110) to infinity

$$u(\boldsymbol{\varrho}, z) = \frac{1}{4\pi z}\exp\left[ik\left(z + \frac{\varrho^2}{2z}\right)\right]\int_{-\infty}^{\infty} M(\boldsymbol{\varrho}')[\bar{q}(\boldsymbol{\varrho}') + \tilde{q}(\boldsymbol{\varrho}')] \times \exp\left(-\frac{ik\boldsymbol{\varrho}\cdot\boldsymbol{\varrho}'}{z}\right)d^2\varrho' \ . \tag{2.111}$$

The ensemble average field for current realizations is given by

$$\bar{u}(\boldsymbol{\varrho}, z) = \frac{1}{4\pi z}\exp\left[ik\left(z + \frac{\varrho^2}{2z}\right)\right]\int_{-\infty}^{\infty} \bar{q}(\boldsymbol{\varrho}')M(\boldsymbol{\varrho}')\exp\left(-\frac{ik\boldsymbol{\varrho}\cdot\boldsymbol{\varrho}'}{z}\right)d^2\varrho' \ .$$

This can be represented as

$$\bar{u}(\boldsymbol{\varrho}, z) = \frac{1}{4\pi z}\exp\left[ik\left(z + \frac{\varrho^2}{2z}\right)\right]\bar{q}_{\mathrm{m}} S_A \mathscr{F}_A\left(\frac{k\boldsymbol{\varrho}}{x}\right) , \tag{2.112}$$

where $\bar{q}_{\mathrm{m}}$ is the maximum value of the mean current through the aperture, and $\mathscr{F}_A(\boldsymbol{\kappa})$ is the unity-normalized $[\mathscr{F}_A(0) = 1]$ antenna radiation pattern

$$\mathscr{F}_A(\boldsymbol{\kappa}) = \frac{1}{S_A}\int_{-\infty}^{\infty} M(\boldsymbol{\varrho}')\frac{\bar{q}(\varrho')}{\bar{q}_{\mathrm{m}}}\exp(\mathrm{i}\boldsymbol{\kappa}\cdot\boldsymbol{\varrho}')d^2\varrho' \ . \tag{2.113}$$

Here the direction at the observation point (x, y) in the plane $z = \text{const}$ is given by the vector $\boldsymbol{\kappa} = k\boldsymbol{\varrho}/z = (kx/z, ky/z)$. Finally,

$$S_A = \frac{1}{\bar{q}_{\mathrm{m}}}\int M(\boldsymbol{\varrho}')\bar{q}(\boldsymbol{\varrho}')d^2\varrho'$$

is the normalization factor whose value is close to the aperture area ($S_A = S$ at $\bar{q} = \text{const}$).

The covariance of the field in the plane $z = \text{const}$ (transverse covariance) is expressed by the double integral

$$\psi_\perp(\boldsymbol{\varrho}_1, \boldsymbol{\varrho}_2, z) = \langle\tilde{u}(\boldsymbol{\varrho}_1, z)\tilde{u}^*(\boldsymbol{\varrho}_2, z)\rangle = \frac{\exp[ik(\varrho_1^2 - \varrho_2^2)/2z]}{(4\pi z)^2}\iint_{-\infty}^{\infty} M(\boldsymbol{\varrho}')M(\boldsymbol{\varrho}'')\psi_q(\boldsymbol{\varrho}', \boldsymbol{\varrho}'') \times \exp\left[\frac{ik(\boldsymbol{\varrho}_1\cdot\boldsymbol{\varrho}' - \boldsymbol{\varrho}_2\cdot\boldsymbol{\varrho}'')}{z}\right]d^2\varrho'\, d^2\rho'' \ , \tag{2.114}$$

where $\psi_q(\boldsymbol{\varrho}', \boldsymbol{\varrho}'') = \langle \tilde{q}(\boldsymbol{\varrho}')\tilde{q}^*(\boldsymbol{\varrho}'')\rangle$ is the covariance of the current. Generally speaking, current fluctuations in the aperture are statistically inhomogeneous, due to boundary effects: owing to the mutual influence of elements of the radiating system, the current statistics at the aperture edges may differ from those in the middle of the antenna. But for large antennas, whose size a is large as compared with the correlation radius of the currents, l_q, we can neglect the boundary effect and consider that for the entire aperture,

$$\psi_q(\boldsymbol{\varrho}', \boldsymbol{\varrho}'') \approx \psi_q(\boldsymbol{\varrho}' - \boldsymbol{\varrho}''), \quad l_q \ll a \ .$$

Assuming the small-scale current fluctuation model ($l_q \ll a$) at the aperture, we can make use of the technique of taking integrals of the type (2.70), as presented in Sect. 2.4. If we denote

$$\boldsymbol{\varrho} = \boldsymbol{\varrho}_1 - \boldsymbol{\varrho}_2, \quad \boldsymbol{\varrho}_+ = (\boldsymbol{\varrho}_1 + \boldsymbol{\varrho}_2)/2 \ ,$$

then (2.114) will take the form

$$\psi_\perp(\boldsymbol{\varrho}, \boldsymbol{\varrho}_+, z) = \frac{\exp(ik\boldsymbol{\varrho}\cdot\boldsymbol{\varrho}_+/z)}{(4\pi z)^2}\sigma_q^2 \int_{-\infty}^{\infty} M(\boldsymbol{\eta})\exp\left(\frac{ik\boldsymbol{\varrho}\cdot\boldsymbol{\eta}}{z}\right)d^2\boldsymbol{\eta}$$
$$\times \int_{-\infty}^{\infty} K_q(\boldsymbol{\xi})\exp\left(\frac{ik\boldsymbol{\varrho}_+\cdot\boldsymbol{\xi}}{z}\right)d^2\boldsymbol{\xi} \ , \tag{2.115}$$

where $\sigma_q^2 = \langle|\tilde{q}|^2\rangle$ is the variance, and K_q the correlation coefficient of the current fluctuations.

The first integral in (2.115) is equal to $S\mathscr{F}_M(k\boldsymbol{\varrho}/z)$, where S is the aperture area, and $\mathscr{F}_M$ is given by (2.89). This quantity differs somewhat from the antenna pattern (2.199), but at $\bar{q} = \text{const}$ we have $\mathscr{F}_M = \mathscr{F}_A$. The second integral (with respect to ξ) can also be written as the product of two factors

$$\int_{-\infty}^{\infty} K_q(\boldsymbol{\xi})\exp\left(\frac{ik\boldsymbol{\varrho}_+\cdot\boldsymbol{\xi}}{z}\right)d^2\xi = S_q\mathscr{F}_q\left(\frac{k\boldsymbol{\varrho}_+}{z}\right), \quad \text{where} \tag{2.116}$$

$$S_q = \int_{-\infty}^{\infty} K_q(\boldsymbol{\xi})d^2\boldsymbol{\xi}$$

is the *effective area of coherence* of the currents, and $\mathscr{F}_q(\boldsymbol{\kappa})$ is the unity-normalized [$\mathscr{F}_q(0) = 1$] *radiation pattern of the fluctuational currents.*

In this notation the covariance $\psi_\perp$ takes the form

$$\psi_\perp(\boldsymbol{\varrho}, \boldsymbol{\varrho}_+, z) = \exp\left(ik\frac{\boldsymbol{\varrho}\cdot\boldsymbol{\varrho}_+}{z}\right)\frac{\sigma_q^2 SS_q}{(4\pi z)^2}\mathscr{F}_M\left(\frac{k\boldsymbol{\varrho}}{z}\right)\mathscr{F}_q\left(\frac{k\boldsymbol{\varrho}_+}{z}\right) . \tag{2.117}$$

This formula can be viewed as a generalization of the Van Cittert–Zernike theorem to the case of a finite correlation radius for sources. For delta-correlated fluctuations, $\mathscr{F}_q = 1$, and then (2.117) differs from (2.91) only in notation. It follows that the spatial distribution of the field variance is determined by a relatively wide (with angular flare $\sim \lambda/l_q$) pattern $\mathscr{F}_q$

$$\sigma_u^2(\boldsymbol{\varrho}_+, z) = \psi_\perp(0, \boldsymbol{\varrho}_+, z) = \frac{\sigma_z^2 S S_q}{(4\pi z)^2} \mathscr{F}_q\left(\frac{k\boldsymbol{\varrho}_+}{z}\right), \tag{2.118}$$

whereas the spatial covariance of the field depends on $\mathscr{F}_M$

$$K_\perp(\varrho, z) = \frac{\psi_\perp(\boldsymbol{\varrho}, \boldsymbol{\varrho}_+, z)}{\sigma_u^2(\boldsymbol{\varrho}_+, z)} = \exp\left(ik\frac{\boldsymbol{\varrho}\cdot\boldsymbol{\varrho}_+}{z}\right)\mathscr{F}_M\left(\frac{k\boldsymbol{\varrho}}{z}\right).$$

The pattern $\mathscr{F}_M$ has an angular width $\sim \lambda/a$, i.e., smaller than the width λ/l_q, since $a \gg l_q$. The transverse correlation radius $l_\perp$ appears to be the same as in (2.87), i.e., $l_\perp \sim \lambda z/a$.

The spatial distribution of the mean intensity $\bar{I}_u$ is obtainable from (2.112) and (2.118)

$$\begin{aligned}\bar{I}_u(\boldsymbol{\varrho}_+, z) &= |\bar{u}(\boldsymbol{\varrho}_+, z)|^2 + \sigma_u^2(\boldsymbol{\varrho}_+, z)\\ &= \frac{1}{(4\pi z)^2}\left[\overline{q_{\mathrm{m}}^2}\, S_A^2 \left|\mathscr{F}_A\left(\frac{k\boldsymbol{\varrho}_+}{z}\right)\right|^2 + \sigma_q^2 S_q S\, \mathscr{F}_q\left(\frac{k\boldsymbol{\varrho}_+}{z}\right)\right].\end{aligned} \tag{2.119}$$

The first term here is contributed by the mean field, the second by the fluctuational field. Figure 2.6 shows the $\bar{I}_u$ vs ϱ_+ behavior for various relationships between these terms, assuming that the total power supplied to the antenna is constant, i.e.,

$$|\bar{q}_{\mathrm{m}}|^2 + \sigma_q^2 = \mathrm{const} \equiv Q^2\,. \tag{2.120}$$

In the absence of fluctuations, when $\sigma_q = 0$ and $|\bar{q}_{\mathrm{m}}| = Q$, we have the unperturbed intensity

$$I_0(\boldsymbol{\varrho}_+, z) = \frac{S_A^2 Q^2}{(4\pi z)^2}\left|\mathscr{F}_A\left(\frac{k\boldsymbol{\varrho}_+}{z}\right)\right|^2, \tag{2.121}$$

which is proportional to the squared modulus of $\mathscr{F}_A$. The width of $\mathscr{F}_A$, say the distance between zeros, is $\lambda z/a$ (Curve 1 in Fig. 2.6). At small perturbations, a

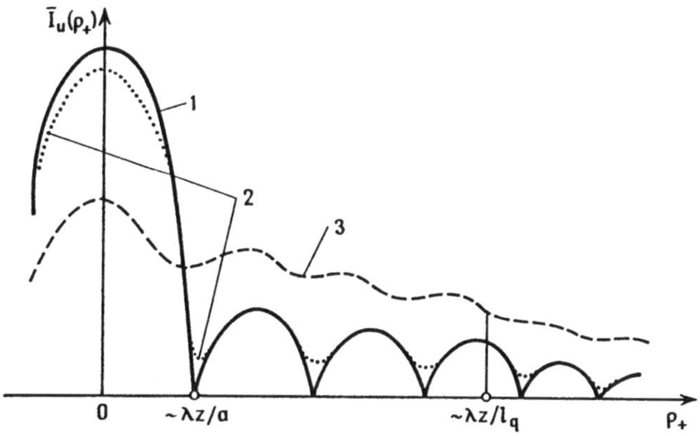

Fig. 2.6. The radiation pattern for an antenna with fluctuational currents: (*1*) unperturbed; (*2*) weak fluctuations; (*3*) strong fluctuations

fluctuational component emerges that is proportional to the angular spectrum of the fluctuations of $\mathscr{F}_q(k\varrho_+/z)$ and has the width $\Delta\varrho_+ \sim \lambda z/l_q \gg \lambda z/a$. This component smears out the zeros (or minima) of the regular pattern (Curve 2 in Fig. 2.6) and increases the side-lobe radiation, i.e., radiation outside the central maximum. At the same time, the contribution of the regular field component is somewhat reduced since, when the power fed to the antenna is unchanged, $|\bar{q}_{\mathrm{m}}|^2$ decreases as compared with Q^2: $|\bar{q}_{\mathrm{m}}|^2 = Q^2 - \sigma_q^2$.

If the regular and fluctuation components are similar at the center of the pattern, the intensity distribution at $z = \text{const}$ is represented by Curve 3 in Fig. 2.6. According to (2.119), the condition $|\bar{u}|^2 = \sigma_u^2$ is attained at

$$\sigma_q^2 = Q^2\left(1 - \frac{SS_q}{S_A^2}\right) .$$

Here $|\bar{q}_{\mathrm{m}}|^2 = Q^2 - \sigma_q^2$ is small as compared with Q^2, i.e.,

$$|\bar{q}_{\mathrm{m}}|^2 \approx \frac{SS_q}{S_A^2} Q^2 \ll Q^2 .$$

When the perturbations are large, one speaks in terms of the *destruction* of the antenna radiation pattern.

As the fluctuations increase further, the regular term in (2.119) becomes negligible and the distribution of the mean intensity is no longer similar to the unperturbed distribution (2.121)

$$\bar{I}_u(\varrho_+, z) \approx \sigma_u^2(\varrho_+, z) = \frac{SS_q}{(4\pi z)^2} \sigma_q^2 \mathscr{F}_q\left(\frac{k\varrho_+}{z}\right) .$$

The pattern width here is of order $\lambda z/l_q$, and the intensity maximum is S_q/S times smaller than in the absence of fluctuations.

The decline in the intensity in the direction of the radiation maximum is conveniently characterized by the *directive gain* that indicates by how many times the field intensity at the center of the main lobe of the pattern is larger than that of the nondirectional antenna beaming the same power. We will now proceed to calculate the mean directive gain of the antenna, $\langle G\rangle$, under the above assumptions. Clearly, if G_0 is the directive gain of the unperturbed antenna, then

$$\langle G\rangle = G_0 \frac{\bar{I}_{\max}}{I_{0\max}} .$$

Since

$$I_{0\max} = \frac{S_A^2 Q^2}{(4\pi z)^2} , \quad \bar{I}_{\max} = \frac{1}{(4\pi z)^2}[|\bar{q}_{\mathrm{m}}|^2 S_A^2 + SS_q\sigma_q^2] ,$$

we obtain, from (2.120),

$$\frac{\langle G\rangle}{G_0} = \frac{|\bar{q}_{\mathrm{m}}|^2 + (S_q S/S_A^2)\sigma_q^2}{Q^2} = 1 - \left(1 - \frac{SS_q}{S_A^2}\right)\frac{\sigma_q^2}{Q^2} . \tag{2.122}$$

For small perturbations of the currents in the aperture ($\sigma_q \ll Q$), the mean directive gain of the antenna differs only slightly from G_0: $\langle G \rangle \approx G_0(1 - \sigma_q^2/Q^2)$ (recall that $S_q \ll S$). The mean directive gain halves at noticeable fluctuations, when $\sigma_q^2 \sim Q^2/2$. With strong fluctuations, when $\sigma_q \to Q$ and the regular current component $|\bar{q}_m| = \sqrt{Q^2 - \sigma_q^2}$ tends to zero, we obtain

$$\frac{\langle G \rangle}{G_0} \approx \frac{S_q S}{S_A^2} \sim \frac{S_q}{S_A} . \tag{2.123}$$

Thus, with strong fluctuations, the directive gain is the smaller, the larger the number of inhomogeneities $N = S/S_q$ in the aperture: $\langle G \rangle = G_0/N$. If we take into account that $G_0 \sim S/\lambda^2$, (2.123) can be written as

$$\langle G \rangle \sim \frac{S_q}{\lambda^2} .$$

This suggests that if the correlation radius l_q is comparable with the wavelength λ, then $S_q \approx l_q^2 \sim \lambda^2$ and $\langle G \rangle \sim 1$. In other words, with strong ($\sigma_q \gg |\bar{q}_m|$) and small ($l_q \sim \lambda$) fluctuations, the antenna radiation becomes omnidirectional.

Many other problems arise in the statistical theory of antennas. We may be interested, for example, in the direction of the central lobe of the pattern, the level of side-lobe radiation, the amplitude and phase statistics, and so on. These issues are discussed in some detail in [2.43]. It should be noted that in treatments of the radiation statistics of an antenna, the main difficulty seems to reside not in the calculation of the integrals in the expressions for the moments of the field, but in the physically significant definition of the currents fluctuating over the antenna aperture. It is clearly impossible to predict these statistics from theoretical considerations alone. At the same time, direct measurements of statistical characteristics pose considerable difficulties, in particular because introducing measuring probes into the antenna changes the current distribution. Also, indirect determination of the fluctuations by measuring the fields beamed by the antenna does not provide a complete solution to the problem: such measurements are subject to additional external influences (nearby objects and structures, irregularities in the terrain, etc.), let alone the fact that field measurements as such are complicated in the near as well as in the far zones.

2) Statistics of a Field Radiated by a System of Independent Sources. For simplicity we will treat the wave field as scalar, and the sources as points. If the jth point source travels along the path $\boldsymbol{r} = \boldsymbol{r}_j(t)$, and its intensity varies according to the law (generally speaking, random) $Q_j(t)$, then for a system of N point sources the right-hand side of the wave equation (2.9) will be written in the form

$$q(t, \boldsymbol{r}) = \sum_{j=1}^{N} Q_j(t)\delta[\boldsymbol{r} - \boldsymbol{r}_j(t)] .$$

As a result, the radiation field (2.10) will be given by

$$u(t,\boldsymbol{r})=\frac{1}{4\pi}\sum_{j=1}^{N}\frac{Q_j\left(t-\dfrac{|\boldsymbol{r}-\boldsymbol{r}_j(t)|}{c}\right)}{|\boldsymbol{r}-\boldsymbol{r}_j(t)|}\,, \tag{2.124}$$

where $|\boldsymbol{r}-\boldsymbol{r}_j(t)|$ is the distance from the observation point to the jth source.

Since the mean of the field is zero in the overwhelming majority of cases, we proceed to calculate the space-time coherence function—at first for a system of *fixed* sources ($\boldsymbol{r}_j=\text{const}$). By (2.124),

$$\begin{aligned}\Gamma(t',\boldsymbol{r}';t'',\boldsymbol{r}'')&=\langle u(t',\boldsymbol{r}')u^*(t'',\boldsymbol{r}'')\rangle\\&=\frac{1}{(4\pi)^2}\sum_{j=1}^{N}\sum_{l=1}^{N}\left\langle\overline{\frac{Q_j\left(t'-\dfrac{1}{c}|\boldsymbol{r}'-\boldsymbol{r}_j|\right)Q_l^*\left(t''-\dfrac{1}{c}|\boldsymbol{r}''-\boldsymbol{r}_l|\right)}{|\boldsymbol{r}'-\boldsymbol{r}_j||\boldsymbol{r}''-\boldsymbol{r}_l|}}\right\rangle\,,\end{aligned} \tag{2.125}$$

where the angle brackets indicate averaging over the ensemble of source positions, and the bar averaging over the ensemble of realizations of the random processes $Q_j(t)$.

Equation (2.125) has an especially simple form for a system of *independent identical* sources. For such a system, at $j\neq l$,

$$\overline{Q_j(t')Q_l^*(t'')}=0\,,$$

and if the covariance, which is the same for all the processes $Q_j(t)$, is denoted by $\psi_Q(\tau)$ (the processes are supposed to be stationary), the double sum (2.125) becomes the sum of N identical terms

$$\Gamma_u(\tau;\boldsymbol{r}_1,\boldsymbol{r}_2)=\frac{N}{(4\pi)^2}\left\langle\frac{\psi_Q(\tau-s/c)}{|\boldsymbol{r}_1-\boldsymbol{R}||\boldsymbol{r}_2-\boldsymbol{R}|}\right\rangle\,,$$

where $\tau=t_1-t_2$, and $s=|\boldsymbol{r}_1-\boldsymbol{R}|-|\boldsymbol{r}_2-\boldsymbol{R}|$ is the path difference from the source at $\boldsymbol{R}$ to the observation points $\boldsymbol{r}_1$ and $\boldsymbol{r}_2$ (Fig. 2.7).

The averaging over the source positions can be represented as a volume integral with the weight function $w(\boldsymbol{R})$, which is the probability density of the

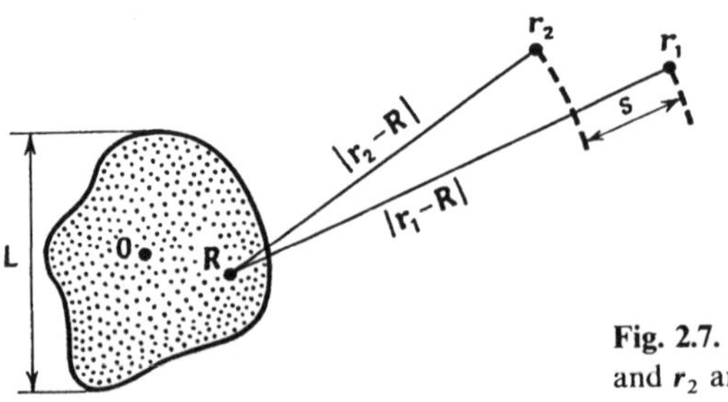

Fig. 2.7. The arrangement of the observation points $\boldsymbol{r}_1$ and $\boldsymbol{r}_2$ and the scattering point $\boldsymbol{R}$

radius vector $\boldsymbol{R}$

$$\Gamma_u(\tau; \boldsymbol{r}_1, \boldsymbol{r}_2) = \frac{N}{(4\pi)^2} \int \frac{\psi_Q(\tau - s/c)}{|\boldsymbol{r}_1 - \boldsymbol{R}||\boldsymbol{r}_2 - \boldsymbol{R}|} w(\boldsymbol{R})\, d^3R \ ,$$

or, if we introduce the mean concentration of sources $\bar{n}(\boldsymbol{R}) = Nw(\boldsymbol{R})$,

$$\Gamma_u(\tau; \boldsymbol{r}_1, \boldsymbol{r}_2) = \frac{1}{(4\pi)^2} \int \frac{\psi_Q(\tau - s/c)}{|\boldsymbol{r}_1 - \boldsymbol{R}||\boldsymbol{r}_2 - \boldsymbol{R}|} \bar{n}(\boldsymbol{R})\, d^3R \ . \tag{2.126}$$

In the particular case where the sources are uniformly distributed over the volume, $\bar{n} = N/V = \text{const.}$

The *temporal* coherence function is obtainable from (2.126), if we put $\boldsymbol{r}_1 = \boldsymbol{r}_2 = \boldsymbol{r}$. Then, $s = 0$, and

$$\Gamma_u(\tau; \boldsymbol{r}, \boldsymbol{r}) = \frac{\psi_Q(\tau)}{(4\pi)^2} \int \frac{\bar{n}(\boldsymbol{R})}{|\boldsymbol{r} - \boldsymbol{R}|^2}\, d^3R \ .$$

This suggests that the temporal correlation coefficients for the field and sources coincide: $K_u(\tau) = K_Q(\tau)$, i.e., the temporal coherence of the radiated field appears to be the same as that of the sources. This is only to be expected, since when the sources are fixed, the field variations are caused solely by the variation of $Q_j(t)$.

Turning to the spatial correlation of a radiated field, let us consider some special cases.

If a "cloud" of sources is small as compared with the mean radiation wavelength $\lambda_0 (L \sim V^{1/3} \ll \lambda_0)$, then we will regard the delay s/c in (2.126) as a constant quantity equal to $s_0/c = (|\boldsymbol{r}_1 - \boldsymbol{R}_0| - |\boldsymbol{r}_2 - \boldsymbol{R}_0|)/c = (r_1 - r_2)/c$, where $\boldsymbol{R}_0 = 0$ is the center of V. For observation points $\boldsymbol{r}_1$ and $\boldsymbol{r}_2$ lying outside the volume V, we then find

$$\Gamma_u(\tau; \boldsymbol{r}_1, \boldsymbol{r}_2) \approx \frac{N\psi_Q(\tau - s_0/c)}{(4\pi)^2 r_1 r_2} \ ,$$

where we have taken into consideration the fact that $\int \bar{n}(R)\, d^3R = N$. At $\tau = 0$ we thus obtain the spatial coherence function for a small radiation source

$$\Gamma_u(0; \boldsymbol{r}_1, \boldsymbol{r}_2) = \frac{N\psi_Q(-s_0/c)}{(4\pi)^2 r_1 r_2} \ , \qquad L \ll \lambda_0 \ . \tag{2.127}$$

According to (2.127), the field is spatially coherent inside a spherical frustrum of thickness $c\tau_c$, where τ_c is the correlation time for the sources. The spatial correlation radius l_c thus coincides here with the length of the coherent train, $c\tau_c$.

The situation is different for large (as compared with λ_0) volumes V. The appropriate variations are conveniently deduced using $\Gamma_u(\omega; \boldsymbol{r}_1, \boldsymbol{r}_2)$, i.e., the

Fourier transform of $\Gamma_u(\tau; \boldsymbol{r}_1, \boldsymbol{r}_2)$[11] in time. The coherence function (2.126) is [2.44]

$$\Gamma_u(\omega; \boldsymbol{r}_1, \boldsymbol{r}_2) = \frac{1}{2\pi}\int_{-\infty}^{\infty} \Gamma_u(\tau, \boldsymbol{r}_1, \boldsymbol{r}_2) e^{i\omega\tau} d\tau$$

$$= \frac{1}{4\pi^2} g_Q(\omega) \int \frac{\bar{n}(\boldsymbol{R}) e^{iks}}{|\boldsymbol{r}_1 - \boldsymbol{R}||\boldsymbol{r}_2 - \boldsymbol{R}|} d^3R \ , \qquad (2.128)$$

where $k = \omega/c$ and

$$g_Q(\omega) = \frac{1}{2\pi}\int_{-\infty}^{\infty} \psi_Q(\tau) e^{i\omega\tau} d\tau$$

is the spectral density of $Q(t)$ (see also [2.44]).

The very form of (2.128) suggests that if the size of the "cloud" of sources L is large as compared with λ, and if $\boldsymbol{r}_1$ and $\boldsymbol{r}_2$ lie within the cloud, then shifting any of $\boldsymbol{r}_1$ and $\boldsymbol{r}_2$ by λ will drastically change Γ_u, since the integrand contains the fast-oscillating function $\exp(iks)$. It follows that within a large $(L \gg \lambda)$ cloud of radiators, the spatial correlation radius is comparable to the wavelength[12]

$$l_u \sim \lambda \ .$$

As observation points move away from the cloud, the interference pattern should, clearly, become more ordered, if only because waves now come only from within a bounded solid angle. Away from the system of radiators this ordering manifests itself as an increase in the spatial correlation radius, just as, by the Van Cittert–Zernike theorem, l_u increases for the diffraction field of the hole (Sect. 2.4). The Van Cittert–Zernike theorem can also be formulated for the system of independent radiators, but its derivation turns out to be somewhat more complicated than in the problem of the diffraction of a random field.

We will only provide the final result here. Let $\boldsymbol{r}_1$ and $\boldsymbol{r}_2$ lie in the plane $z = \text{const}$, and $\boldsymbol{r}_{1,2} = (\boldsymbol{\varrho}_{1,2}, z)$. For the vector $\boldsymbol{R}$, too, we will separate the longitudinal (z') and transverse ($\boldsymbol{\varrho}'$) coordinates: $\boldsymbol{R} = (\boldsymbol{\varrho}', z')$, $d^3R = d^2\varrho'\, dz'$. If the distance to the cloud is large as compared with its size $(z \gg L)$ and the observation points lie not very far from the z-axis $(\varrho_{1,2} \ll z)$, then (2.128) takes the form

$$\Gamma_u(\omega; \boldsymbol{\varrho}, \boldsymbol{\varrho}_+, z) = \frac{g_Q(\omega)\exp\left(\dfrac{ik\boldsymbol{\varrho}\cdot\boldsymbol{\varrho}_+}{z}\right)}{(4\pi z)^2} \int \bar{n}(\boldsymbol{\varrho}', z') \exp\left(-\frac{ik\boldsymbol{\varrho}\cdot\boldsymbol{\varrho}'}{z}\right) d^2\varrho'\, dz' \ , \qquad (2.129)$$

where $\boldsymbol{\varrho} = \boldsymbol{\varrho}_1 - \boldsymbol{\varrho}_2$ and $\boldsymbol{\varrho}_+ = (\boldsymbol{\varrho}_1 + \boldsymbol{\varrho}_2)/2$. This expression bears a relation to (2.91),

[11] In optics this quantity is termed the *mutual spectral intensity* because it relates to the *mutual* coherence function $\Gamma_u(\tau; \boldsymbol{r}_1, \boldsymbol{r}_2)$ characterizing the statistical connection between the fields at two points.

[12] A simple illustration: a stone thrown into water produces waves of a regular circular form. If a handful of stones are thrown, the interference of the circular waves causes an irregular pattern of crests and troughs alternating in any direction, at intervals equal to about a wavelength.

if the reduced (to the plane $z' = 0$) intensity is understood to be

$$J(\boldsymbol{\varrho}') = \frac{1}{4\pi^2} g_Q(\omega)\bar{n}_s(\boldsymbol{\varrho}') \,, \quad \text{where} \tag{2.130}$$

$$\bar{n}_s(\boldsymbol{\varrho}') = \int \bar{n}(\boldsymbol{R})\,dz' = \int \bar{n}(\boldsymbol{\varrho}', z')\,dz'$$

is the number of radiators *per unit area* in the plane $z = 0$.

The Van Cittert–Zernike theorem also holds in the case of moving sources, but the spectral density $g_Q(\omega)$ appearing in the definition of the reduced intensity (2.130), as well as in (2.128, 129), should be averaged over all possible values of the Doppler frequency shift occurring due to source motion. In other words, in (2.128–130) it is necessary to change

$$g_Q(\omega) \to \langle g_Q(\omega - \Omega_d) \rangle \,, \tag{2.131}$$

where the angle brackets indicate averaging over the distribution of the Doppler shift $\Omega_d = \omega v_{||}/c$, the latter being proportional to the ray velocity $v_{||} = (\boldsymbol{m}, \boldsymbol{v})$, $\boldsymbol{m} = (\boldsymbol{r} - \boldsymbol{R})/|\boldsymbol{r} - \boldsymbol{R}|$.

The change (2.131) is legitimate on condition that the "free path" of the source, l_{free}, (i.e., the distance over which the radiator velocity changes noticeably, or the distance covered by the source during the correlation time for the radiated signal) is small as compared with the distance $|\boldsymbol{r} - \boldsymbol{R}|$ to the observation point: $l_{\text{free}} \ll |\boldsymbol{r} - \boldsymbol{R}|$. In the majority of cases of practical interest this condition is met.

A specific expression for the spectral density $g_Q(\omega)$ was obtained earlier (Exercise II.2.8.3 [2.16]) for the Poisson pulse process consisting of successive sections of exponentially decaying sinusoidal trains. This process may serve as a model for the radiation of atoms excited by collisions with their neighbors. The final equation for the spectral density, i.e., (II.2.154), has the form

$$g_Q(\omega) = \frac{n_1(\gamma + n_1)}{\pi(2\gamma + n_1)} \overline{A^2} \left\langle \frac{1}{(\bar{\omega}_0 + \Omega_d - \omega)^2 + (\gamma + n_1)^2} \right\rangle \tag{2.132}$$

where n_1 is the mean frequency of the collisions; γ is the radiation width of the emission lines and, simultaneously, the damping coefficient for pulses of the form $Q(t) = A\exp[-(\gamma + i\bar{\omega}_0)t + i\beta]$; A is the random amplitude with mean square $\overline{A^2}$; β is the random phase uniformly distributed in the interval $(0, 2\pi)$; and $\bar{\omega}_0$ is the radiation frequency in the reference frame in which the oscillator is at rest. For the Maxwell distribution of particle velocities, (2.132) gives either the Gaussian (for $\sigma_\omega \gg n_1 + \gamma$) or Lorentzian (at $\sigma_\omega \ll n_1 + \gamma$) emission line shape, depending on the relationship between the emission linewidth $n_1 + \gamma$ and the root-mean-square Doppler broadening $\sigma_\omega = \sqrt{\overline{\Omega_d^2}}$.

The Van Cittert–Zernike theorem permits of another generalization. The spectral density (2.132) corresponds to a model of atoms of only one sort radiating at only one frequency. But real hot gases may contain many sorts of atoms radiating at various frequencies, and in a dense cloud, many light pulses are absorbed by other atoms, thereby never reaching the observer. All these effects can be taken account of in the Van Cittert–Zernike theorem, if by $J(\varrho)$ we understand the *sum of reduced intensities* for various emission lines

$$J(\varrho) = \sum_m J_m(\varrho) = \frac{1}{4\pi^2} \sum_m g_Q^{(m)}(\omega) \bar{n}_s^{(m)}(\varrho) \; .$$

The spectrum of each of these emission lines is characterized by the natural frequency of the radiator $\bar{\omega}_0^{(m)}$, the oscillator strength $\overline{A_m^2}$, the radiation width γ_m, and the mean number of collisions $n_1^{(m)}$. The quantity $\bar{n}_s^{(m)}(\varrho)$ then determines the effective surface concentration of radiators whose pulses are capable of leaving the volume V. Clearly, if the radiators are distributed not over a volume, but over a surface, then the treatment should from the very beginning be based on the surface concentration $n_s(\varrho)$ and not on the volume concentration $n(\boldsymbol{R})$ (see, e.g., [2.45]).

2.6 Exercises

2.6.1 Express the mean density of the sound energy in a liquid $W = \varrho v^2/2 + \beta p^2/2$ and the mean energy-flux vector $\mathscr{S} = p\boldsymbol{v}$ in terms of the coherence function of the velocity potential and of the radiation intensity.

Solution. Expressing the pressure p and the velocity $\boldsymbol{v}$ through the velocity potential $\varphi(t, \boldsymbol{r})$

$$p = -\varrho \frac{\partial \varphi}{\partial t} \; , \qquad \boldsymbol{v} = \nabla \varphi \; ,$$

we have

$$W = \frac{\varrho}{2}\left[(\nabla \varphi)^2 + \frac{1}{c^2}\left(\frac{\partial \varphi}{\partial t}\right)^2 \right] , \qquad \mathscr{S} = -\varrho \frac{\partial \varphi}{\partial t} \nabla \varphi \; ,$$

where $c = (\beta \varrho)^{-1/2}$ is the sound velocity, ϱ is the mean density and β is the compressibility of the liquid. If $u(t, \boldsymbol{r})$ is the complex velocity potential (analytic signal) corresponding to the real potential $\varphi(t, \boldsymbol{r}) = \mathrm{Re}\{u\}$, then W and $\mathscr{S}$ are written in terms of u in the following way:

$$W = \frac{\varrho}{4}\left\{ |\nabla u|^2 + \frac{1}{c^2}\left|\frac{\partial u}{\partial t}\right|^2 + \mathrm{Re}\left[(\nabla u)^2 + \frac{1}{c^2}\left(\frac{\partial u}{\partial t}\right)^2 \right] \right\} ,$$

$$\mathscr{S} = -\frac{\varrho}{2} \mathrm{Re}\left\{ \frac{\partial u}{\partial t} \nabla u^* + \frac{\partial u}{\partial t} \nabla u \right\} .$$

Averaging of these expressions gives

$$\langle W \rangle = \frac{\varrho}{4}\left\{\left(\nabla_1 \cdot \nabla_2 + \frac{1}{c^2}\frac{\partial^2}{\partial t_1\,\partial t_2}\right)[\Gamma_u(1,2) + \mathrm{Re}\,\tilde{\Gamma}_u(1,2)]\right\}_{2\to 1},$$

$$\langle \mathscr{S} \rangle = -\frac{\varrho}{2}\mathrm{Re}\left\{\frac{\partial}{\partial t_1}\nabla_2[\Gamma_u(1,2) + \tilde{\Gamma}_u(1,2)]\right\}_{2\to 1}. \tag{2.133}$$

Symbol $2 \to 1$ here denotes the limiting process $t_2 \to t_1$, $\boldsymbol{r}_2 \to \boldsymbol{r}_1$ that follows after computing the temporal and spatial derivatives. Note that the conservation law $\mathrm{div}\,\mathscr{S} + \partial W/\partial t = 0$, which must also be fulfilled for the means $\langle W \rangle$ and $\langle \mathscr{S} \rangle$ when we substitute (2.133) into it, can be treated, as it were, as a conservation equation for Γ_u and $\tilde{\Gamma}_u$.

If the field u is *stationary*, then $\bar{u} = 0$, $\tilde{\Gamma} = 0$, and

$$\langle W \rangle = \frac{\varrho}{4}\left[\left(\nabla_1 \nabla_2 + \frac{1}{c^2}\frac{\partial^2}{\partial t_1 \partial t_2}\right)\Psi_u(1,2)\right]_{2\to 1},$$

$$\langle \mathscr{S} \rangle = -\frac{\varrho}{2}\mathrm{Re}\left\{\frac{\partial}{\partial t_1}\nabla_2 \Psi_u(1,2)\right\}_{2\to 1}.$$

If, in addition, the field u is *homogeneous*, then its covariance $\Psi_u(1,2)$ has the form (2.32), thus enabling $\langle W \rangle$ and $\langle \mathscr{S} \rangle$ to be expressed in terms of the radiation intensity

$$\langle W \rangle = \frac{\varrho}{2}\int_0^\infty k^2 d\omega \oint \mathscr{J}(\omega, \boldsymbol{n})\, do(\boldsymbol{n}),$$

$$\langle \mathscr{S} \rangle = \frac{c\varrho}{2}\int_0^\infty k^2\, d\omega \oint \boldsymbol{n}\, \mathscr{J}(\omega, \boldsymbol{n})\, do(\boldsymbol{n}).$$

For an isotropic random sound field, clearly, $\langle \mathscr{S} \rangle = 0$. For a quasi-monochromatic sound field (the spectrum is concentrated at ω_0), the factor k^2 in the above expressions can be brought outside the integration sign with the value $k_0^2 = \omega_0^2/c^2$.

2.6.2 Find the general form of the covariance of the isotropic, homogeneous, stationary scalar wave field.

Solution. The radiation intensity $\mathscr{J}(\omega, \boldsymbol{n})$ for the isotropic field is independent of the direction: $\mathscr{J}(\omega, \boldsymbol{n}) = \mathscr{P}(\omega)$, now, the covariance (2.32) is given by the single integral

$$\Psi_u(t, \boldsymbol{r}) = 4\pi \int_0^\infty \mathscr{P}(\omega)\frac{\sin kr}{kr}\mathrm{e}^{-\mathrm{i}\omega t} d\omega,$$

since

$$\oint \mathrm{e}^{\mathrm{i}k\boldsymbol{n}\cdot\boldsymbol{r}} do(\boldsymbol{n}) = 4\pi\frac{\sin kr}{kr}.$$

2.6.3 Estimate the longitudinal and transverse correlation radii of a wave field whose radiation intensity $\mathscr{J}(\omega, \boldsymbol{n})$ is concentrated within a narrow cone with $\theta \ll 1$.
Solution. Let the z-axis be the direction of maximum radiation intensity. We write the unit vector $\boldsymbol{n}$ in the form

$$\boldsymbol{n} = \boldsymbol{i}_z n_z + \boldsymbol{n}_\perp ,$$

and for $n_\perp \lesssim \theta \ll 1$ we have $n_z = \sqrt{1 - n_\perp^2} \approx 1 - n_\perp^2/2$, so that $(\boldsymbol{nr}) \approx z - n_\perp^2 z/2 + (\boldsymbol{n}_\perp, \boldsymbol{\varrho})$, $\mathscr{J}(\omega, \boldsymbol{n}) \approx \mathscr{J}(\omega, \boldsymbol{n}_\perp, 1)$ and the covariance (2.32) takes the form

$$\Psi(t, \boldsymbol{r}) = \int_0^\infty d\omega \exp(\mathrm{i}kz - \mathrm{i}\omega t) \oint \mathscr{J}(\omega, \boldsymbol{n}_\perp, 1) \exp\left(\mathrm{i}k\boldsymbol{n}_\perp \cdot \boldsymbol{\varrho} - \frac{\mathrm{i}}{2} k n_\perp^2 z \right) do(\boldsymbol{n}) \tag{2.134}$$

where $\boldsymbol{\varrho} = (x, y)$. The transverse $(l_\perp)$ and longitudinal $(l_\parallel)$ correlation radii are estimated from the conditions $|k\boldsymbol{n}_\perp \boldsymbol{\varrho}| \lesssim 1$, $|kn_\perp^2 z| \lesssim 1$, if we substitute $n_\perp \approx \theta$ into them. As a result

$$l_\perp \sim \frac{1}{k\theta} \sim \frac{\lambda}{\theta}, \qquad l_\parallel \sim \frac{1}{k\theta^2} \sim \frac{\lambda}{\theta^2} . \tag{2.135}$$

2.6.4 Find the relationship between the radiation intensity and the two-dimensional spectrum (2.34) in the case of a narrow angular spectrum [2.46].
Solution. In the narrow cone $|n_\perp| \lesssim \theta \ll 1$, the approximate relationship $do(\boldsymbol{n}) \approx dn_x dn_y$ holds. It can be represented in the equivalent form $do(\boldsymbol{n}) = k^{-2} d\kappa_x d\kappa_y = k^{-2} d^2\kappa_\perp$, if we introduce the two-dimensional wave vector $\kappa_\perp = k\boldsymbol{n}_\perp = (kn_x, kn_y)$. Equation (2.134) then takes the form of the two-dimensional spectral expansion

$$\Psi(t, \boldsymbol{r}) = \int_0^\infty d\omega \exp(\mathrm{i}kz - \mathrm{i}\omega t) \int_{-\infty}^\infty k^{-2} \mathscr{J}(\omega, \boldsymbol{n}_\perp, 1) \exp\left(\mathrm{i}\kappa_\perp \cdot \boldsymbol{\varrho} - \frac{\mathrm{i}\kappa_\perp^2 z}{2k} \right) d^2\kappa_\perp .$$

A comparison of this expansion with the expansion

$$\Psi(t, \boldsymbol{r}) = \int_0^\infty d\omega \int_{-\infty}^\infty F(\omega; \kappa_\perp, z) \exp(\mathrm{i}\kappa_\perp \cdot \boldsymbol{\varrho} - \mathrm{i}\omega t) d^2\kappa_\perp ,$$

inverse to (2.34), shows that

$$F(\omega; \kappa_\perp, z) = k^{-2} \mathscr{J}(\omega, \boldsymbol{n}_\perp, 1) \exp\left(\mathrm{i}kz - \frac{\mathrm{i}\kappa_\perp^2 z}{2k} \right) .$$

2.6.5 Express the mean density of the electromagnetic energy and the mean Poynting vector in a vacuum in terms of the radiation intensity matrix $\mathscr{J}_{\alpha\beta}$, which is introduced as in (2.31)

$$G_{\alpha\beta}^{(E)}(\omega, \boldsymbol{\kappa}) = \mathscr{J}_{\alpha\beta}(\omega, \boldsymbol{n}) k^{-2} \delta(\kappa - k), \qquad \boldsymbol{n} = \frac{\boldsymbol{\kappa}}{\kappa} , \qquad k \equiv \frac{\omega}{c} ;$$

and, as in (2.32),

$$\Psi^{(E)}_{\alpha\beta}(t, \boldsymbol{r}) = \int_0^\infty d\omega \oint \mathscr{I}_{\alpha\beta}(\omega, \boldsymbol{n}) \mathrm{e}^{\mathrm{i}(k\boldsymbol{n}\cdot\boldsymbol{r} - \omega t)} do(\boldsymbol{n}) \ . \tag{2.136}$$

Here $G^{(E)}_{\alpha\beta}$ is the spectral density matrix, which is the Fourier transform of the correlation matrix for the free, statistically homogeneous, electromagnetic field

$$\Psi^{(E)}_{\alpha\beta}(t, \boldsymbol{r}) = \langle \tilde{E}_\alpha(t_1, \boldsymbol{r}_1) \tilde{E}^*_\beta(t_2, \boldsymbol{r}_2) \rangle \ .$$

Solution. Taking the case of a stationary homogeneous electromagnetic field with zero mean and considering that in a vacuum $\langle |\boldsymbol{H}|^2 \rangle = \langle |\boldsymbol{E}|^2 \rangle$, we have

$$\langle W \rangle = \frac{1}{16\pi}(\langle |\boldsymbol{E}|^2 \rangle + \langle |\boldsymbol{H}|^2 \rangle) = \frac{1}{8\pi} \langle |\boldsymbol{E}|^2 \rangle \ .$$

But at $\bar{\boldsymbol{E}} = 0$, by (2.136),

$$\langle |\boldsymbol{E}|^2 \rangle = \sum_{\alpha=1}^{3} \Psi^{(E)}_{\alpha\beta}(0, 0) = \int_0^\infty d\omega \oint \mathscr{I}(\omega, \boldsymbol{n})\, do(\boldsymbol{n}) \ ,$$

where we have introduced the radiation intensity

$$\mathscr{I}(\omega, \boldsymbol{n}) = \sum_{\alpha=1}^{3} \mathscr{I}_{\alpha\alpha}(\omega, \boldsymbol{n}) \ . \tag{2.137}$$

As a result, we have

$$\langle W \rangle = \frac{1}{8\pi} \int_0^\infty d\omega \oint \mathscr{I}(\omega, \boldsymbol{n})\, do(\boldsymbol{n}) \ .$$

We can obtain the mean of the Poynting vector

$$\langle \mathscr{S} \rangle = \frac{c}{8\pi} \mathrm{Re} \langle [\boldsymbol{E} \cdot \boldsymbol{H}^*] \rangle$$

using the spectral representations for $\boldsymbol{E}$ and $\boldsymbol{H}$, and the relationship $\boldsymbol{H}_{\omega\kappa} = [\boldsymbol{n} \cdot \boldsymbol{E}_{\omega\kappa}]$

$$\langle \mathscr{S} \rangle = \frac{c}{8\pi} \mathrm{Re} \left\{ \iint_{-\infty}^{\infty} d^3\kappa'\, d^3\kappa'' \iint_0^\infty d\omega'\, d\omega'' \right.$$
$$\left. \times \langle [\boldsymbol{E}_{\omega'\boldsymbol{\kappa}'} [\boldsymbol{n}'' \boldsymbol{E}^*_{\omega''\boldsymbol{\kappa}''}]] \rangle \exp \langle \mathrm{i}(\boldsymbol{\kappa}' - \boldsymbol{\kappa}'') \cdot \boldsymbol{r} - \mathrm{i}(\omega' - \omega'')] \right\} \ , \tag{2.138}$$

where $\boldsymbol{n}'' = \boldsymbol{\kappa}''/\kappa''$. Expanding the double vector product gives

$$\langle \dots \rangle = \boldsymbol{n}'' \langle \boldsymbol{E}_{\omega'\boldsymbol{\kappa}'} \boldsymbol{E}^*_{\omega''\boldsymbol{\kappa}''} \rangle - \langle \boldsymbol{E}^*_{\omega''\boldsymbol{\kappa}''} (\boldsymbol{n}'' \cdot \boldsymbol{E}_{\omega'\boldsymbol{\kappa}'}) \rangle \ .$$

Since the spectral amplitudes are delta-correlated, the vector $\boldsymbol{n}''$ can be replaced by $\boldsymbol{n}' = \boldsymbol{\kappa}'/\kappa'$, but, because the electromagnetic field is transverse, $(\boldsymbol{n}' \cdot \boldsymbol{E}_{\omega'\boldsymbol{\kappa}'}) = 0$. As a result, by (2.137), we have

$$\langle \ldots \rangle = \sum_{\alpha=1}^{3} \boldsymbol{n}' G_{\alpha\alpha}(\omega', \boldsymbol{\kappa}')\delta(\omega' - \omega'')\delta(\boldsymbol{\kappa}' - \boldsymbol{\kappa}'')$$
$$= \boldsymbol{n}' \mathscr{J}(\omega, \boldsymbol{n})k^{-2}\delta(\omega' - \omega'')\delta(\boldsymbol{\kappa}' - \boldsymbol{\kappa}'')\delta(\kappa' - k) \ .$$

Substituting this into (2.138) gives

$$\langle \mathscr{S} \rangle = \frac{c}{8\pi} \int_0^{\infty} d\omega \oint \boldsymbol{n} \mathscr{J}(\omega, \boldsymbol{n})\, do(\boldsymbol{n}) \ . \tag{2.139}$$

We clearly see that the electromagnetic relations for $\langle W \rangle$ and $\langle \mathscr{S} \rangle$ differ from the acoustic expressions (see Exercise 2.6.1) only by the factor $1/8\pi$, instead of by $\varrho k^2/2$.

It follows from (2.139) that the quantity $c\boldsymbol{n}\mathscr{J}(\omega, \boldsymbol{n})/8\pi$ has the sense of an energy flux through a unit area into a unit solid angle per unit frequency interval. It is in exactly this way that the radiation intensity (brightness) is introduced into the phenomenological theory of radiation transport [2.47, 48]. The statistical and wave aspects of radiative transfer, including derivations of the relevant equations directly from the wave equation on the basis of a number of statistical hypotheses, are given in review [2.50] and monographs [2.51, 52].

2.6.6 Construct the correlation matrix $\Psi_{\alpha\beta}^{(E)}$ for the isotropic, homogeneous, stationary, electromagnetic field.

Solution. The matrix $\mathscr{J}_{\alpha\beta}$ appearing in (2.136) for the isotropic field should have the form

$$\mathscr{J}_{\alpha\beta}(\omega, \boldsymbol{n}) = \mathscr{P}(\omega)\delta_{\alpha\beta} + \mathscr{Q}(\omega)n_\alpha n_\beta \ .$$

Besides, since electromagnetic waves are transverse in free space, $\mathscr{J}_{\alpha\beta}$ should meet the condition

$$n_\alpha \mathscr{J}_{\alpha\beta} = n_\beta \mathscr{J}_{\alpha\beta} = 0 \ .$$

This is only possible if $\mathscr{Q} = -\mathscr{P}$, and then

$$\Psi_{\alpha\beta}^{(E)}(t, \boldsymbol{r}) = \int_0^{\infty} d\omega \mathscr{P}(\omega) \oint \mathrm{e}^{ik\boldsymbol{n}\cdot\boldsymbol{r} - i\omega t}(\delta_{\alpha\beta} - n_\alpha n_\beta)\, do(\boldsymbol{n})$$
$$= \int_0^{\infty} d\omega \mathscr{P}(\omega) \mathrm{e}^{-i\omega t}\left(\delta_{\alpha\beta} - \frac{1}{(ik)^2}\frac{\partial^2}{\partial r_\alpha \partial r_\beta}\right) \oint \mathrm{e}^{ik\boldsymbol{n}\cdot\boldsymbol{r}}\, do(\boldsymbol{n})$$
$$= \int_0^{\infty} d\omega \mathscr{P}(\omega) \mathrm{e}^{-i\omega t} k^{-2}\left\{\left[k^2 s + \frac{s'}{r}\right]\delta_{\alpha\beta} + \left[s'' - \frac{s'}{r}\right]\frac{r_\alpha r_\beta}{r^2}\right\} ,$$

where $s = 4\pi \sin kr/kr$, and the prime denotes differentiation with respect to r. The radiation intensity $\mathscr{J} = \Sigma \mathscr{J}_{\alpha\alpha}$ is here equal to $2\mathscr{P}(\omega)$, and hence it is independent of the direction. According to the results of Exercise 2.6.5, we thus arrive at

$$\langle \mathscr{S} \rangle = 0, \langle W \rangle = \int_0^\infty \mathscr{P}(\omega)\, d\omega \ .$$

2.6.7 Calculate the scintillation index in the far zone for the amplitude-phase screen with the log-normal probability distribution for the field.
Solution. Let $v = \exp(\chi + \mathrm{i}S)$, where the phase S and level χ are normally distributed, so that v is a log-normal field. Suppose that $\bar{S} = 0$ and $\langle S\chi \rangle = 0$. By virtue of the normalization (2.57), we have $\bar{\chi} = -\sigma_\chi^2$. As a result

$$v_0 = \langle \mathrm{e}^{\chi + \mathrm{i}S} \rangle = \mathrm{e}^{\bar{\chi} + 1/2 \langle (\tilde{\chi} + \mathrm{i}S)^2 \rangle} = \mathrm{e}^{-1/2(\sigma_S^2 + \sigma_\chi^2)} \ .$$

Accordingly, in the far zone

$$\beta^2 = \beta_\infty^2 = 1 - |v_0|^4 = 1 - \exp[-2(\sigma_S^2 + \sigma_\chi^2)] \ .$$

In the case of pure amplitude fluctuations ($\sigma_S^2 = 0$), the squared scintillation index in the far zone will be $\beta_\infty^2 = 1 - \exp(-2\sigma_\chi^2)$, whereas at the screen itself $\beta^2 = \beta_0^2 = \exp(4\sigma_\chi^2) - 1$. Clearly, $\beta_0^2 > 1 > \beta_\infty^2$, i.e., the amplitude fluctuations beyond the screen are smoothed out.

2.6.8 Let the field in the screen plane be real and normally distributed. Show that the squared scintillation index in the far zone is half that at the screen.
Solution. In the far zone, by (2.72), $\beta_\infty^2 = 1 - |v_0|^4$. Then, at the screen

$$\beta^2 = \beta_0^2 = \langle v^4 \rangle - 1 = v_0^4 + 6v_0^2 \langle \tilde{v}^2 \rangle + \langle \tilde{v}^4 \rangle \ .$$

But for the real normal random field $\langle \tilde{v}^4 \rangle = 3\langle \tilde{v}^2 \rangle = 3\sigma_v^2$. Therefore, taking account of the normalization $\sigma_v^2 + v_0^2 = 1$, we have

$$\beta_0^2 = v_0^4 + 6v_0^2(1 - v_0^2) + 3(1 - v_0^2)^2 = 2(1 - v_0^4) = 2\beta_\infty^2 \ ,$$

i.e., $\beta_\infty^2 = \beta_0^2/2$. In this and previous problems $\beta_\infty^2 < \beta_0^2$, since in addition to amplitude fluctuations, the field beyond the screen causes phase fluctuations which, clearly, have no effect on the degree of intensity fluctuations.

2.6.9 From the Fresnel approximation (2.22) derive the relationship (2.47), according to which the covariance beyond the infinite screen $\psi_\perp(\varrho)$ coincides with that for the boundary field, $\psi_v(\varrho)$ for the incident plane wave.
Solution. In the Fresnel approximation

$$\psi_\perp(\varrho_1, \varrho_2, z) = \left(\frac{k}{2\pi z}\right)^2 \iint_{-\infty}^{\infty} \psi_v(\varrho' - \varrho'') \exp\left\{\frac{ik}{2z}[(\varrho_1 - \varrho')^2 - (\varrho_2 - \varrho'')^2]\right\} \times d^2\varrho'\, d^2\varrho'' \ .$$

Introducing the notation $\varrho = \varrho_1 - \varrho_2$, $\varrho_+ = (\varrho_1 + \varrho_2)/2$, $\xi = (\varrho' - \varrho'')$,

$\boldsymbol{\eta} = (\boldsymbol{\varrho}' + \boldsymbol{\varrho}'')/2$, we arrive at

$$\psi_\perp(\boldsymbol{\varrho}, \boldsymbol{\varrho}_+, z) = \left(\frac{k}{2\pi z}\right)^2 \iint\limits_{-\infty}^{\infty} \psi_v(\boldsymbol{\xi}) \exp\left\{\frac{ik}{z}[\boldsymbol{\varrho}\cdot\boldsymbol{\varrho}_+ + \boldsymbol{\xi}\cdot\boldsymbol{\eta} - \boldsymbol{\eta}\cdot\boldsymbol{\varrho} - \boldsymbol{\xi}\cdot\boldsymbol{\varrho}_+]\right\} \times d^2\xi \, d^2\eta \ .$$

The integral with respect to $\boldsymbol{\eta}$ is taken here to be precisely equal to $(2\pi z/k)^2 \delta(\boldsymbol{\xi} - \boldsymbol{\varrho})$, thus leading to the relation $\psi_\perp(\boldsymbol{\varrho}, z) = \psi_v(\boldsymbol{\varrho})$.

2.6.10 Find the law for the variation of the mean and covariance of a spherical wave that has passed through an infinite, statistically homogeneous, chaotic screen. *Solution.* Suppose that on the left of the plane $z = 0$, where the screen is located, there is a source producing a spherical wave

$$\exp\{ik|\boldsymbol{r} - \boldsymbol{r}_0|\}/|\boldsymbol{r} - \boldsymbol{r}_0| \ ,$$

where $\boldsymbol{r}_0 = (0, -z_0)$ (Fig. 2.8). If $f(\boldsymbol{\varrho})$ is the complex transmittance function for the screen, then the boundary field at $z = +0$ will be

$$v(\boldsymbol{\varrho}') = f(\boldsymbol{\varrho}') \frac{\exp(ik\sqrt{z_0^2 + \varrho'^2})}{\sqrt{z_0^2 + \varrho'^2}} \ .$$

Next, we substitute this expression for the field into the Green's function (2.17), fix the observation point at $\boldsymbol{r} = (\boldsymbol{\varrho}, z)$, and expand the exponent in the deviation $\boldsymbol{\delta} = \boldsymbol{\varrho}' - \boldsymbol{\varrho}'_{\rm st}$ of the variable $\boldsymbol{\varrho}'$ from the stationary point $\boldsymbol{\varrho}'_{\rm st} = \boldsymbol{\varrho} z/(z + z_0)$ lying on the ray connecting the observation point to the source. We then replace all the factors, except $f(\boldsymbol{\varrho})$, at the exponential function by their stationary values, i.e., the values at $\boldsymbol{\varrho}' = \boldsymbol{\varrho}'_{\rm st}$.

The resulting approximate formula

$$u(\boldsymbol{\varrho}, z) \approx \frac{k e^{ikR_0}(z + z_0)^3}{2\pi i R_0^3 z z_0} \int\limits_{-\infty}^{\infty} f(\boldsymbol{\varrho}_{\rm st} + \boldsymbol{\delta}) \exp\left(\frac{ik}{2L}\left[\delta^2 - \frac{(\boldsymbol{\delta}\cdot\boldsymbol{\varrho})^2}{R_0^2}\right]\right) d^2\delta \qquad (2.140)$$

may be called the Fresnel approximation for the spherical wave. Here

$$R = |\boldsymbol{r} - \boldsymbol{r}_0| = \sqrt{(z + z_0)^2 + \varrho^2}, \quad L = R_0 z z_0/(z + z_0)^2 \ .$$

From (2.140) we find the mean field

$$\bar{u}(\boldsymbol{r}) = \bar{f} e^{ikR_0}/R_0 \ .$$

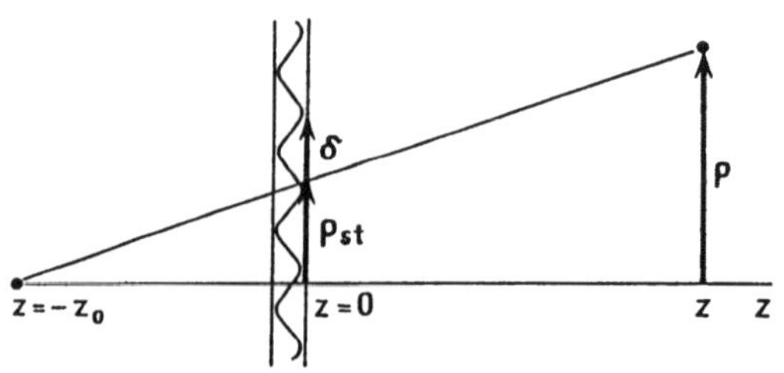

Fig. 2.8. Passage of a spherical wave through a chaotic screen

This is a spherical wave with amplitude $\bar{f}$, and the covariance of the field in the plane $z = \text{const}$

$$\psi_u(\boldsymbol{\varrho}_1, \boldsymbol{\varrho}_2, z) = \psi_u(\boldsymbol{\varrho}, \boldsymbol{\varrho}_+, z) = \frac{\exp\left(ik\frac{\boldsymbol{\varrho}\cdot\boldsymbol{\varrho}_+}{R_0}\right)}{R_0^2}\psi_f\left(\boldsymbol{\varrho}\frac{z}{z+z_0}\right),$$

where $\boldsymbol{\varrho} = \boldsymbol{\varrho}_1 - \boldsymbol{\rho}_2, \boldsymbol{\varrho}_+ = (\boldsymbol{\varrho}_1 + \boldsymbol{\varrho}_2)/2$. A similar expression is also obtained for the coherence function.

Unlike the plane wave considered in Sect. 2.3, the covariance of the spherical wave beyond the screen is not conserved. Specifically, the correlation radius in the plane $z = \text{const}$ increases linearly beyond the screen with the distance from the source

$$l_u = l_f\frac{z+z_0}{z_0},$$

where l_f is the correlation radius of the transmittance function f. In this problem, the role of wave parameter is played by the quantity $D_{\text{sph}} = 2L/kl_f^2$, which for $z \to \infty$ tends to a finite limit $D_{\max} = 2z_0/kl_f^2$. It is easily seen that for $D_{\max} \lesssim 1$, when the radius of the first Fresnel zone $\sqrt{\lambda z z_0/(z+z_0)}$ at any z is less than the size of the inhomogeneities at the screen l_{f}, the field of the spherical wave beyond the screen, even at infinity, is not normalizable.

2.6.11 Find the intensity distribution for the field in the far zone of a bounded chaotic screen.

Solution. The field in the Fraunhofer zone is given by (2.86), where v should be understood to signify the boundary field $v(\boldsymbol{\varrho}) = f(\boldsymbol{\varrho})u_{\text{p}}$ produced as the plane wave $u_{\text{p}} = \exp(ikz)$ strikes the chaotic screen with a transmittance function $f(\boldsymbol{\varrho})$. The intensity distribution in the far zone can be computed from (2.86), as for the field (2.111) excited by an antenna with fluctuating currents in Sect. 2.5. If we consider that (2.111) goes over into (2.86) when we substitute $-2ikv$ for q, by analogy with the intensity distribution (2.119), we can write

$$\tilde{I}_u(\boldsymbol{\varrho}_+, z) = \left(\frac{k}{2\pi z}\right)^2 \left[|\bar{v}|^2 S^2 \left|\mathscr{F}_M\left(\frac{k\boldsymbol{\varrho}_+}{z}\right)\right|^2 + \sigma_v^2 S_v S \mathscr{F}_v\left(\frac{k\boldsymbol{\varrho}_+}{z}\right)\right], \tag{2.141}$$

where S_v and $\mathscr{F}_v$ are introduced as in (2.116).

Note that a similar intensity distribution (but with $z \to F$) is obtained in the focal plane of a lens placed directly beyond the bounded chaotic screen. In both cases, the intensity distribution varies as shown in Fig. 2.6 in passing over from weak ($\sigma_v \ll \bar{v}$) to strong ($\sigma_v \gg \bar{v}$) fluctuations.

2.6.12 Find the mean directive gain for a parabolic reflector antenna with smooth irregularities.

Solution. Let $z = \varrho^2/2F$ be the equation for the unperturbed paraboloid, and $z = \varrho^2/2F + \zeta(\boldsymbol{\varrho})$ that for the reflector with irregularities $\zeta(\boldsymbol{\varrho})$. A radiator located at the paraboloid focus (Fig. 2.9) sends onto the reflector a spherical wave which, on being reflected, becomes a distorted plane wave. With smooth (in the scale of λ) irregularities of the reflector, the field distortion in the aperture can be taken account of by introducing the phase factor $\exp[\mathrm{i}S(\varrho)] = \exp[2\mathrm{i}k\zeta(\varrho)]$ ($2k\zeta$ is an additional phase progression due to the irregularities). This reduces the problem to one of calculating the diffraction of the field with the boundary value at the aperture $v(\varrho) = \exp[2\mathrm{i}k\zeta(\varrho)]$.

If the height of the irregularities $\zeta(\varrho)$ is distributed according to the normal law, then

$$\bar{v} = \exp(-2k^2\sigma_\zeta^2) \ , \quad \sigma_v^2 = 1 - |\bar{v}|^2 = 1 - \mathrm{e}^{-4k^2\sigma_\zeta^2} \ .$$

For small irregularities

$$\sigma_v^2 \approx 4k^2\sigma_\zeta^2 \ll 1, \quad l_v \sim l_\zeta \quad (k\sigma_\zeta \ll 1) \ ,$$

and for large ones

$$\sigma_v^2 \approx 1, \quad l_v \sim l_\zeta/k\sigma_\zeta \ll l_\zeta \quad (k\sigma_\zeta \gg 1) \ ,$$

where l_ζ is the correlation radius for the irregularities, and the quantity $4k^2\sigma_\zeta^2$ plays the same role as the phase variance σ_S^2 for the phase screen.

To calculate the directive gain we use (2.122), where, considering the previous problem, we can substitute S_v, σ_v^2, $|\bar{v}|^2$, and $\sigma_v^2 + |\bar{v}|^2 = 1$ for S_q, σ_q^2, $|\bar{q}|^2$, and $Q^2 = \sigma_q^2 + |\bar{q}|^2$, respectively. We obtain

$$\bar{G} = G_0\left\{1 - \left(1 - \frac{S_v}{S}\right)\sigma_v^2\right\} = G_0\left\{\frac{S_v}{S} + \left(1 - \frac{S_v}{S}\right)\mathrm{e}^{-4k^2\sigma_\zeta^2}\right\} \ .$$

(In addition, this takes into account that $S_A = S$ at $\bar{v} = \text{const.}$) In the limiting cases we have

$$\bar{G} = G_0(1 - \sigma_v^2) \approx G_0(1 - 4k^2\sigma_\zeta^2) \quad (k\sigma_\zeta \ll 1) \ ,$$

$$\bar{G} \approx G_0\left(\frac{S_v}{S} + \mathrm{e}^{-4k^2\sigma_\zeta^2}\right) \qquad (k\sigma_\zeta \gg 1) \ , \tag{2.142}$$

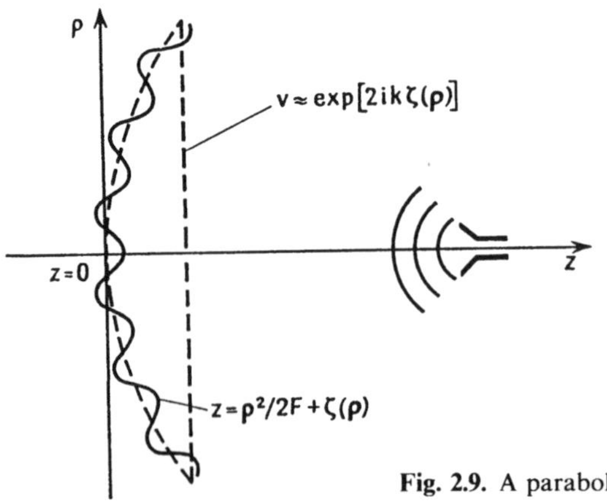

Fig. 2.9. A parabolic antenna

or, since $S_v \sim l_v^2 \sim l_\zeta^2/(k\sigma_\zeta)^2$,

$$\bar{G} \approx G_0\left(e^{-4k^2\sigma_\zeta^2} + \frac{l_\zeta^2}{Sk^2\sigma_\zeta^2}\right) \quad (k\sigma_\zeta \gg 1) . \tag{2.143}$$

Figure 2.10 shows the qualitative behavior of the normalized mean directive gain $\bar{G}/G_0$ as a function of the root-mean-square phase progression $k\sigma_\zeta$. The initial portion of the curve corresponds to the first of (2.142), and the final portion to (2.143). It is seen that the directive gain is halved already at $k\sigma_\zeta \sim 1/2$, i.e., at $\sigma_\zeta \sim 1/2k = \lambda/2\pi$. This coincides with the engineering criterion of reflector smoothness $\sigma_\zeta < \lambda/8$. For large reflector antennas of present-day radio telescopes, the quantity σ_ζ may be made as small as a centimeter. In accordance with the criterion $\sigma_\zeta < \lambda/8$, the effectiveness of such antennas is sharply reduced in going over from a centimeter to a millimeter wave range.

2.6.13 Estimate the longitudinal correlation radius of the field produced by a cloud of statistically independent sources.
Solution. Place the observation points $\boldsymbol{r}_1$ and $\boldsymbol{r}_2$ on the z-axis: $\boldsymbol{r}_1 = (0, 0, z + \zeta)$, $\boldsymbol{r}_2 = (0, 0, z)$. The path difference s in (2.128) will be

$$s = |\boldsymbol{r}_1 - \boldsymbol{R}| - |\boldsymbol{r}_2 - \boldsymbol{R}| = \sqrt{(z + \zeta - z')^2 + \varrho'^2} - \sqrt{(z - z')^2 + \varrho'^2} , \tag{2.144}$$

where the integration variables $\boldsymbol{\varrho}'$ and z' vary within the confines of the volume V occupied by the sources: $|\varrho'| \lesssim L, |z'| \lesssim L$. We will consider the longitudinal correlation in three characteristic zones.

In the *near* zone ($z \lesssim L$) we may expect that $l_\parallel$ will be small as compared with L, which enables us to expand (2.144) into a power series in ζ and to keep only the linear term

$$s \approx \zeta|z - z'|/\sqrt{|z - z'|^2 + \varrho'^2} , \tag{2.145}$$

For all values of ϱ' and z', the factor at ζ is not larger than unity. Therefore, the presence of $\exp(iks)$ in the integrand in (2.138) will only be felt for $k\zeta \gg 1$. Accordingly, the longitudinal correlation radius can be estimated from the condition $k\zeta \sim 1$, i.e.,

$$l_\parallel \sim \frac{1}{k} \sim \lambda \quad (z \lesssim L) . \tag{2.146}$$

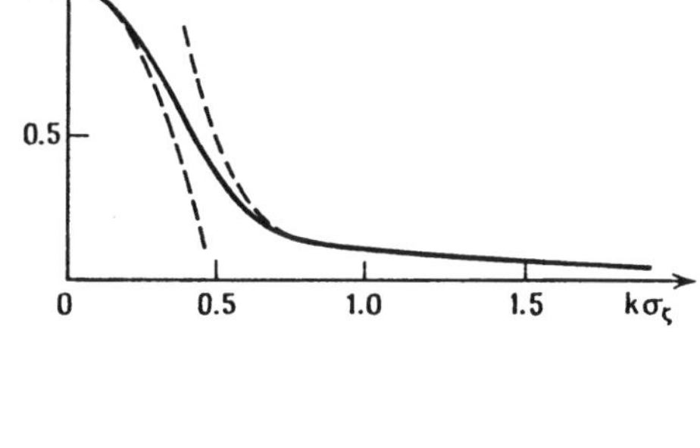

Fig. 2.10. Variation of $\bar{G}/G_0$ with $k\sigma_\zeta$ for an antenna with irregularities

In the *intermediate* zone ($L \ll z \ll kL^2$), the ratios z'/z and ϱ'/z are small as compared with unity, with the result that we can simplify (2.146) to obtain

$$s \approx \zeta(1-(\varrho'^2/2z^2)) \ . \tag{2.147}$$

The first term in (2.147) is independent of the integration variables and, since $|\exp(ik\zeta)| = 1$, can be ignored when examining the correlation. The second term does not exceed $\zeta L^2/2z^2$ in modulus. From the condition $k\zeta L^2/z^2 \sim 1$

$$l_{\parallel} \sim \lambda(z^2/L^2) \quad (L \ll z \ll kL^2) \ . \tag{2.148}$$

This estimate is similar to the expression (2.135) for the longitudinal correlation radius of a homogeneous random field with angular spectrum width $\theta \sim a/z$. In that case, the angular spectrum width θ may be said to be governed by the apparent angular size of the source cloud, $\gamma \sim L/z$.

In the *far* zone ($z \gg kL^2$), for all ϱ' and z' the quantity ks differs from $k\zeta$ by no more than π. Therefore, in the far zone $\exp(iks) \approx \exp(ik\zeta)$, and the longitudinal correlation radius tends to infinity

$$l_{\parallel} = \infty \quad (z \gg kL^2) \ , \tag{2.149}$$

which corresponds to a fully developed radiation field pattern in the Fraunhofer zone.

2.6.14 The degree of determinism $D(Y|Z)$ of an observable field $Y(\mathbf{r}, t)$ relative to a model (hypothetic or predictable) field $Z(\mathbf{r}, t)$ was defined in [2.52] as the correlation coefficient of Y and Z:

$$D(Y|Z) = \langle YZ\rangle/\langle Y^2\rangle^{1/2}\langle Z^2\rangle^{1/2}.$$

If $D = 1$, one may speak of the perfectly deterministic (predictable) behavior of field Y, while a small value for D, $|D| \ll 1$, corresponds to a random (perfectly unpredictable) behavior of Y, with all intermediate values, $0 < |D| < 1$, standing for a partially deterministic field Y.

Let $Y(x)$ be an inhomogeneous speckle field measured at an aperture $|x| \leqslant l/2$. Estimate the degree of determinism of field $Y(x)$ outside the aperture, i.e., when $|x| > l/2$, for the model field being calculated using the field source recovery technique.

Solution. Let us represent the model field $Z(x)$—which is hopefully useful for predicting the behavior of $Y(x)$ outside the aperture—be a sum of plane waves $Z(x) = \Sigma A_n \exp(i\kappa_n x)$ and let us estimate A_n and κ_n from the Fourier transforms of $Y(x)$ in the interval $|x| \leqslant l/2$. Neglecting amplitude fluctuations A_n and supposing the random deviations of wave numbers $\tilde{\kappa}_n = \kappa_n - \bar{\kappa}_n$ from their mean values $\bar{\kappa}_n$ to be normally distributed with variation σ_n^2, which depends on the signal-to-noise ratio, we obtain [2.53]:

$$D(Y|Z) = \Sigma A_n \exp\{-\tfrac{1}{2}\sigma_n^2(|x| - l/2)^2\}/[\Sigma A_n^2]^{1/2} \ .$$

It is essential that the length $x_{\text{det}} \sim 1/\sigma_n$ of the deterministic (predictable) behavior of field Y can remarkably exceed the correlation radius of speckle field $x_c \sim 1/|\kappa_n - \kappa_m|$.

3. Thermal Electromagnetic Fields

This chapter is concerned with fluctuations of an electromagnetic field that, at the microscopic level, are caused by the thermal motion of electrically charged microparticles. The theory, however, is constructed at the level of macroscopic electrodynamics, which assumes the field fluctuations to be thermal fluctuations of the volume densities of the charge and current. Although we deal here with so-called equilibrium fluctuations, i.e., random deviations from a state of thermal equilibrium, the results are to some degree applicable to nonequilibrium conditions, namely to those cases where transport phenomena are not yet noticeable. We will refer to the general correlation theory of such fluctuations as fluctuational electrodynamics. It derives, on the one hand, from (stochastic) Maxwell's equations, and on the other hand, from the fluctuation-dissipation theorem (FDT).

Fluctuational electrodynamics is based on the general equations of electrodynamics, and so we need impose no constraints on the relation between the wavelength λ of interest and the characteristic size l of bodies. Also, we may thus use a unified approach to both limiting cases: the classical theory of thermal radiation, based on geometrical optics ($\lambda \ll l$), and thermal noise in quasi-stationary electric networks ($\lambda \gg l$) discussed in Sect. II.3.5.

These and other aspects of fluctuational electrodynamics are covered in Sect. 3.1. Section 3.2 deals with stochastic Maxwell's equations that contain random (Langevin) sources. We formulated FDT earlier for quasi-stationary networks and now, when we consider the general case of electromagnetic fields, we will have to extend FDT to cover random fields (Sect. 3.3) and, specifically, electromagnetic fields (Sect. 3.4). We then use the reciprocity theorem and Lorentz lemma to arrive at the main result of fluctuational electrodynamics, the generalized Kirchhoff law (Sect. 3.5).

Subsequent sections of the chapter give a description of the various forms of the generalized Kirchhoff law and some of its applications, the nature of the forms and applications being clear from the sections headings.

3.1 Preliminary Remarks

A thermal electromagnetic field is produced by the chaotic thermal motion of charged microparticles (electrons, ions, etc.) of which bodies are made up. This alone suggests that the intensity of the thermal field should grow with temperature, as do Brownian movement or thermal noise in electric circuits. But in macroscopic electrodynamics, the electromagnetic field is treated not as a sum of fields of individual microcharges, but rather as a field produced by sources that are macroscopic as well and are described, for instance, by the volume densities of the charge and current. As applied to thermal electromagnetic fields, this means that its sources are *space-time fluctuations of the charge and current* in the physically infinitesimal volume elements of bodies. These elements should be small as compared both with the wavelength in question (in spectral treatment) and with the size of the body and its inhomogeneities, the separations between the bodies, and so on. If the physical conditions and geometry of a problem allow us to break down the body into such infinitesimal volume elements, we can use the equations of macroscopic electrodynamics, i.e., Maxwell's equations.

We have already considered the thermal fluctuations of electric quantities, but in the *quasi-stationary* region, i.e., for electric systems whose sizes (l) are very much smaller than the wavelength λ: $l \ll \lambda$.[1] The state of such systems may be described by a finite number of *integral* functions of time: currents through cross-sections of conductors, charges at lumped capacitances, and so forth. Using the Nyquist theorem and its generalization, the fluctuation-dissipation theorem (FDT), in Sect. II.3.5 we obtained expressions for the second moments of *thermodynamically equilibrium* fluctuations of such integral quantities. We are thus already acquainted with the correlation theory of electric thermal fluctuations in quasi-stationary networks, i.e., in the case of $l \ll \lambda$.

But there is also quite another domain of phenomena for which the theory of thermal electromagnetic fields has long been developed, but for the opposite condition where all the characteristic sizes of the bodies l are *large* as compared with the wavelength: $l \gg \lambda$. This is the classical theory of thermal radiation due to Kirchhoff, Boltzmann and a number of other prominent physicists of the late 19th century. The theory was completed in 1900 by Planck who discovered a law describing the spectral distribution of equilibrium thermal radiation confined in a fairly large cavity ($l \gg \lambda$). Other laws of this theory (e.g., Kirchhoff's laws) also rely on the asymptotic description of the electromagnetic field in the *geometrical optics approximation*, which is only suitable for sufficiently short waves. Historically, this is due to the fact that the classical theory of thermal radiation was developed primarily for the *optical* range of λ.

It is obvious that the general fluctuational electrodynamics of fields of thermal origin based on the Maxwell equations (clearly understood as *stochastic* equations) should be free of any constraints concerning the relationship between the body size l and the wavelength λ. In addition to the Nyquist theory of thermal noise in quasi-stationary networks ($l \ll \lambda$) and the classical theory of thermal radiation ($l \gg \lambda$), the theory should cover the intermediate region $l \sim \lambda$, in which both limits are no longer valid. Such a general theory of thermal electromagnetic fields has been worked out fairly recently [3.1], and the present chapter sketches its fundamentals. For simplicity, we will from now on refer to the random electromagnetic field of thermal origin simply as the *fluctuational* field.

[1] Quasi-stationarity in this *electrodynamic* sense should not be confused with the *statistical* quasi-stationarity of random processes.

Before we go on to discuss the theory systematically, we will dwell upon two questions that might be provoked by what has just been said. One of these might be posed in the following way: why does the classical theory of thermal radiation deal not with the thermal fluctuational field in general, but only with the *radiation*, i.e., with the *wave* field alone?

This question is quite justified because the charge and current fluctuations in each volume element of a body engender not only a diverging electromagnetic wave, but also the so-called *near* quasi-stationary field. The latter, however, is known to decay with distance R from its "point" source much faster than the wave field: not as $1/R$ but as $1/R^2$ or $1/R^3$, stretching only over the distance of the order of λ. Besides, the net energy flux of this near field over the period is zero. It is for this reason that the classical theory of thermal radiation, which is concerned only with the transport of "radiant" energy by sufficiently short waves ($\lambda \ll l$), simply does not take into account the near fluctuational field. It is, therefore, unable to describe adequately the fluctuational field, e.g., in a cavity that is not large, or even small, as compared with the wavelength. Equally, it ignores the peculiarities of the thermal field at small distances ($l \lesssim \lambda$) from the surface of a heated body. Nowadays, these peculiarities and their origin are readily understood.

Incessant thermal fluctuations of the charge and current densities or, to borrow the pictorial phrase of Leontovich, incessant electric "swarming", occur in each element of the volume inside a heated body. The total *wave* field caused by these sources outside the body constitutes the thermal electromagnetic radiation described by Kirchhoff's law. But the total of *near* fields (from the volume elements at the surface) constitutes the *quasi-stationary* thermal field that is, as it were, a *lining on the surface* of any body. This field is concentrated near the surface within a thickness of about λ, so that it can be ignored for $l \gg \lambda$ but, as $l \to 0$, its strength grows *indefinitely*. There is no paradox here, of course, since at distances of the order of atomic spacing, a macroscopic treatment of the field is no longer valid.

Although the quasi-stationary fluctuational field does not contribute to the energy flux, it manifests itself in quite an observable way, e.g., in the molecular forces of cohesion between bodies (see Sect. 3.9) and in some other phenomena.

The use of the geometrical optics approximation imposes constraints on the classical theory of thermal radiation not only in that this theory does not take into account the quasi-stationary fluctuational field, but also in that it ignores *diffractional* phenomena in the radiation itself.

Another legitimate question is whether the region $l \sim \lambda$, (i.e., where the size of the bodies is comparable to the wavelength) is of interest.

In its earliest days radio engineering made use of meter, or even centimeter waves emerging from oscillatory spark discharges. Quite soon, however, it was forced to go over to very long (kilometer) waves. But after the invention of vacuum tubes for oscillators and amplifiers, a long process of real engineering, using increasingly shorter waves, began. This path has gradually led to decimeter, centimeter and millimeter waves and, accordingly, to radiotechnical apparatus in which all, or some, of the dimensions are comparable with the wavelength, i.e., to non-quasi-stationary *distributed wave* systems. Such devices (wave guides, cavity resonators, horn and reflector antennas, etc.) form the subject-matter of so-called microwave electrodynamics. As the sensitivity and accuracy of radiotechnical techniques improved, it was only natural that for distributed systems, too, the issue of thermal noise should come to the fore. Thus, as far as the needs of radio engineering are concerned, the problem of thermal noise in the microwave range lead to the development of general fluctuational electrodynamics. Radiotechnical applications, however, do not exhaust the physical results of fluctuational electrodynamics. For example, one of its major consequences is the generalization of such a fundamental law as Kirchhoff's radiation law to the general case of an arbitrary relationship between body size and wavelength (Sect. 3.5).

It should be noted in passing that, strictly speaking, Kirchhoff's law already refers to *nonequilibrium* conditions. The point is that a heated body radiates into a colder environment, i.e., the temperature is not the same over the entire space. If the energy lost by the body is not compensated for, then the process will be nonstationary: the body loses energy and cools down. Clearly, stationary conditions are also not very difficult to achieve if the body temperature is maintained at a constant level, the energy being supplied by some external source as is the case, say, in the incandescent lamp. It is, however, clear that such a stationary state is thermodynamically not in equilibrium anyway, since in the body energy is being converted in a one-way manner, and transport phenomena occur (e.g., through thermal conduction). Under these conditions, the use of equilibrium laws (including FDT) is no longer quite rigorous, but still justified if, as is often the case, the role of the transport phenomena is as yet insignificant.

Before leaving the subject, a remark about the probability distribution functions for fluctuational fields is in order. It might seem that finding these functions is a more complex problem than calculating the moments, but in many cases this is not so. A fluctuational field is composed of a multitude of *microfields* created by the charge and current fluctuations in the various volume elements. If these contributions can be thought of as statistically independent and similar (e.g., their share in the total variance is similar), then by the central limit theorem of probability theory, the total field will be *Gaussian*. Hence, its complete statistical description only requires a knowledge of the covariance of its strengths (as the means of the strengths are zero).

3.2 Stochastic Maxwell Equations

It is well known that Maxwell's original equations

$$\operatorname{curl} \boldsymbol{H} = \frac{1}{c}\frac{\partial \boldsymbol{D}}{\partial t} + \frac{4\pi}{c}\boldsymbol{j}, \quad \operatorname{curl} \boldsymbol{E} = -\frac{1}{c}\frac{\partial \boldsymbol{B}}{\partial t} \tag{3.1}$$

must be supplemented by the so-called *material equations* relating the vectors of the electric displacement $\boldsymbol{D}$ and magnetic induction $\boldsymbol{B}$ to the strengths $\boldsymbol{E}$ and $\boldsymbol{H}$. These equations describe the electrodynamic behavior of a medium, and in the general case may be nonlinear. However, if the problem of thermal fluctuations is formulated using *macroscopic* equations (as is the case in fluctuational electrodynamics), and if we are interested only in thermodynamically equilibrium fluctuations, then the macroscopic nonlinearity of the system is of no importance, i.e., linearized equations will suffice. This point was clarified earlier for discrete systems (Sect. II.3.5), but it also holds for continuous systems. In other words, we can begin with *linearized* material equations without loss of generality. These latter may be the most general, i.e., allowing for inhomogeneities and the anisotropy of a medium, the temporal and spatial dispersion, as well as for the motion of the medium. However, to simplify the treatment and bring out the principal points, we will confine our discussion to the case of the fixed and initially isotropic medium featuring only a temporal nonlocality (frequency dispersion). In this particular case, the medium can be described by material equations of the form (see footnote 20, Chap. 3)

$$\boldsymbol{D}(t,\boldsymbol{r}) = \int_{-\infty}^{+\infty} \varepsilon(t-t',\boldsymbol{r})\boldsymbol{E}(t',\boldsymbol{r})dt' \ ,$$

$$\boldsymbol{B}(t,\boldsymbol{r}) = \int_{-\infty}^{+\infty} \mu(t-t',\boldsymbol{r})\boldsymbol{H}(t',\boldsymbol{r})dt' \ , \tag{3.2}$$

where ε is the electric permittivity, and μ the magnetic permeability.

Turning to the stochastic Maxwell equations, we now introduce on the right of (3.1) random external sources that "cause" thermal fluctuations of the field and of all the related electrodynamic quantities. These continuous sources play here the same auxiliary role of equivalent Langevin "forces", as the Nyquist integral e.m.f.'s in electric networks with lumped parameters (Sect. II.3.5).

Fluctuation sources can be chosen in a number of ways, specifically, as the volume densities of the electric ($\boldsymbol{j}_e$) and magnetic ($\boldsymbol{j}_m$) currents[2]. The stochastic Maxwell equations will then take the form

$$\operatorname{curl}\boldsymbol{H} = \frac{1}{c}\frac{\partial \boldsymbol{D}}{\partial t} + \frac{4\pi}{c}\boldsymbol{j} + \frac{4\pi}{c}\boldsymbol{j}_e, \quad \operatorname{curl}\boldsymbol{E} = -\frac{1}{c}\frac{\partial \boldsymbol{B}}{\partial t} - \frac{4\pi}{c}\boldsymbol{j}_m \ . \tag{3.3}$$

In place of the currents, we may use the external electric displacement $\mathfrak{D}$ and magnetic induction $\mathfrak{B}$, letting

$$\boldsymbol{j}_e = \frac{1}{4\pi}\frac{\partial \mathfrak{D}}{\partial t}, \quad \boldsymbol{j}_m = \frac{1}{4\pi}\frac{\partial \mathfrak{B}}{\partial t} \tag{3.4}$$

which, as we will see later, is advisable in certain cases. We will regard the fields of the fluctuation sources $\boldsymbol{j}_e$ and $\boldsymbol{j}_m$ (like $\mathfrak{D}$ and $\mathfrak{B}$) as *stationary* in time.

We now turn to the energy balance equation (the Poynting theorem) that follows from (3.3) and write it for the *complete* field[3]. To this end, we multiply the

---

[2] The use of magnetic currents is, of course, only a formal trick, but it is convenient in many problems (see. e.g., [3.2]).

[3] It will be recalled that in electrodynamics, the *complete* field is one that is zero beyond a certain *closed* surface Σ. Such a surface can be constructed either when there are ideally reflecting envelopes, or when the field at infinity decays rather fast (the surface, or part of it, may be infinitely remote).

first of (3.1) by $\boldsymbol{E}$, the second by $\boldsymbol{H}$, add up the results and take the integral over the volume V bounded by the surface Σ. We thus obtain

$$\int_V (\boldsymbol{E}\,\mathrm{curl}\,\boldsymbol{H} - \boldsymbol{H}\,\mathrm{curl}\,\boldsymbol{E})d^3r = \frac{4\pi}{c}\left\{\frac{dU}{dt} + Q + \int_V (\boldsymbol{E}\cdot\boldsymbol{j}_\mathrm{e} + \boldsymbol{H}\cdot\boldsymbol{j}_\mathrm{m})d^3r\right\} ,$$

where

$$\frac{dU}{dt} = \frac{1}{4\pi}\int_V\left(\boldsymbol{E}\cdot\frac{\partial\boldsymbol{D}}{\partial t} + \boldsymbol{H}\cdot\frac{\partial\boldsymbol{B}}{\partial t}\right)d^3r, \quad Q = \int_V \boldsymbol{j}\cdot\boldsymbol{E}\,d^3r ,$$

dU/dt being the rate of the variation of the electromagnetic energy in V, and Q the power dissipated in this volume. The integral on the left-hand side is proportional to the energy flux (the Poynting vector flux) through Σ, so that for a *complete* field it is zero. Averaging over the ensemble of the random "forces" $\boldsymbol{j}_\mathrm{e}$ and $\boldsymbol{j}_\mathrm{m}$ and considering that for the *complete stationary* field $d\langle U\rangle/dt = 0$, we obtain for the mean dissipated power

$$\langle Q\rangle = -\int_V \langle \boldsymbol{E}\cdot\boldsymbol{j}_\mathrm{e} + \boldsymbol{H}\cdot\boldsymbol{j}_\mathrm{m}\rangle\, d^3r . \tag{3.5}$$

We are interested in the spectral amplitudes of electrodynamic quantities stationary in time, i.e., in the spatial fields of the Fourier transforms

$$\boldsymbol{E}(\omega, \boldsymbol{r}) = \frac{1}{2\pi}\int_{-\infty}^{\infty} \boldsymbol{E}(t, \boldsymbol{r})\mathrm{e}^{\mathrm{i}\omega t}\,dt ,$$

and, likewise, for $\boldsymbol{H}$, $\boldsymbol{D}$, $\boldsymbol{B}$, $\boldsymbol{j}$, $\boldsymbol{j}_\mathrm{e}$, and $\boldsymbol{j}_\mathrm{m}$. For the spectral amplitudes, (3.3) takes the form

$$\mathrm{curl}\,\boldsymbol{H} = -\mathrm{i}k\boldsymbol{D} + \frac{4\pi}{c}\boldsymbol{j}_\mathrm{e}, \quad \mathrm{curl}\,\boldsymbol{E} = \mathrm{i}k\boldsymbol{B} - \frac{4\pi}{c}\boldsymbol{j}_\mathrm{m} , \tag{3.6}$$

where $k = \omega/c$ and the material equations (3.3) take the form

$$\boldsymbol{D}(\omega, \boldsymbol{r}) = \varepsilon(\omega, \boldsymbol{r})\boldsymbol{E}(\omega, \boldsymbol{r}) ,$$

$$\boldsymbol{B}(\omega, \boldsymbol{r}) = \mu(\omega, \boldsymbol{r})\boldsymbol{H}(\omega, \boldsymbol{r}) , \quad \text{where} \tag{3.7}$$

$$\left.\begin{matrix}\varepsilon(\omega, \boldsymbol{r})\\ \mu(\omega, \boldsymbol{r})\end{matrix}\right\} = \int_0^\infty \left.\begin{matrix}\varepsilon(\tau, \boldsymbol{r})\\ \mu(\tau, \boldsymbol{r})\end{matrix}\right\}\mathrm{e}^{\mathrm{i}\omega\tau}\,d\tau . \tag{3.8}$$

Note that in (3.6) we combined the terms $4\pi\boldsymbol{j}/c$ and $\mathrm{i}k\boldsymbol{D}$, thereby incorporating the electric conductivity into ε, or rather into its imaginary part $\varepsilon''(\varepsilon = \varepsilon' + \mathrm{i}\varepsilon'')$. By introducing μ $(\mu = \mu' + \mathrm{i}\mu'')$, we can also make allowances for magnetic losses.

The set of inhomogeneous linear equations (3.6, 7) uniquely defines the spectral amplitudes of the strengths and inductions of a fluctuational field with given boundary conditions and, if necessary, with conditions at infinity and specified random sources $\boldsymbol{j}_\mathrm{e}(\omega, \boldsymbol{r})$ and $\boldsymbol{j}_\mathrm{m}(\omega, \boldsymbol{r})$. This is, therefore, a problem with *random sources* [scheme (1) according to the classification introduced in Sect. 2.1], whose stochastic behavior should be known. Since the equation is linear, the second (nth) moments of the fluctuational field can be found given a knowledge of the second (nth) moments of their sources. These second moments, if related to the fluctuations about a thermodynamic equilibrium, are completely defined (as in the case of discrete systems) by the fluctuation-dissipation theorem, which can easily be generalized to *continuous* systems (fields). Therefore, we will leave Maxwell's equations for the moment and turn to this generalization of the FDT.

3.3 Equilibrium Thermal Fluctuations in Continuous Dissipative Systems

In Sect. II.3.5, we formulated the Fluctuation–Dissipation Theorem for *discrete* systems described by a finite number n of generalized coordinates $q_j(t)$. Using this discrete form of the FDT, we can obtain results for continuous systems, either indirectly, as in problem 1, or by artificially discretizing a continuous system, e.g., by breaking down the space occupied by a field into small cells and substituting difference for differential operators. It is with the help of the latter technique that the discrete FDT can be applied to stochastic Maxwell equations in [Ref. 3.3, Chap. 13]. It is interesting, however, to obtain the *general* form of the FDT for *continuous* systems described by random fields — functions of time and space. Such a generalization, which can be implemented when applying the theorem to the fields of equilibrium thermal fluctuations of any physical nature — electromagnetic, mechanic (in aero- and hydrodynamics, elasticity theory), temperature, enthropy, etc. — is realizable in the following way [3.4].

Consider the fluctuations, described by a stationary homogeneous field $\xi(t, \boldsymbol{r})$, in a continuous system occupying a volume V. For simplicity, we take the case of the one-component field[4]. The generalization to the multicomponent field $\xi^{(i)}(t, \boldsymbol{r})$ ($i = 1, 2, \ldots$) is easily obtained, and the result is given below. Let $f(t, \boldsymbol{r})$ be the volume density of the field of a generalized force that is conjugate, in the Lagrangian sense, to the generalized coordinate $\xi(t, \boldsymbol{r})$, the spectral amplitudes $\xi(\omega, \boldsymbol{r})$ and $f(\omega, \boldsymbol{r})$,being related to each other by the linear spatial operators $\hat{A}$ and $\hat{A}^{-1}$

$$\xi(\omega, \boldsymbol{r}) = \hat{A} f(\omega, \boldsymbol{r}), \quad f(\omega, \boldsymbol{r}) = \hat{A}^{-1} \xi(\omega, \boldsymbol{r}) \ , \tag{3.9}$$

where $\hat{A}$ and $\hat{A}^{-1}$ are mutually inverse operators, i.e., their product is equal to the unit operator $\hat{E}(\hat{A}\hat{A}^{-1} = \hat{A}^{-1}\hat{A} = \hat{E})$, which leaves the function unchanged $[\hat{E}f(\boldsymbol{r}) = f(\boldsymbol{r})]$. In practice, $\hat{A}^{-1}$ is mostly a differential operator.

[4] That is described by one function ξ.

If, for instance, we are dealing with the wave equation

$$\Delta\xi - \frac{1}{c^2}\frac{\partial^2\xi}{\partial t^2} = -4\pi f(t, \boldsymbol{r}) ,$$

whose partial solution is known to be

$$\xi(t, \boldsymbol{r}) = \int f\left(t - \frac{R}{c}, \boldsymbol{r}'\right)\frac{d^3r'}{R} \quad (R = |\boldsymbol{r} - \boldsymbol{r}'|) ,$$

then for the spectral amplitudes, we have ($k = \omega/c$)

$$(\Delta + k^2)\xi(\omega, \boldsymbol{r}) = -4\pi f(\omega, \boldsymbol{r}) ,$$

$$\xi(\omega, \boldsymbol{r}) = \int f(\omega, \boldsymbol{r}')\frac{e^{ikR}}{R}\,d^3r' .$$

Thus,

$$\hat{A}^{-1} = -\frac{\Delta + k^2}{4\pi}, \quad \hat{A}\{\dots\} = \int\{\dots\}\frac{e^{ikR}}{R}\,d^3\boldsymbol{r}' .$$

Assume now that we have a full orthonormalized system of real functions $\varphi_j(\boldsymbol{r})$, defined for V as

$$\int_V \varphi_j(\boldsymbol{r})\varphi_k(\boldsymbol{r})\,d^3r = \delta_{jk} . \tag{3.10}$$

Expanding $\xi(t, \boldsymbol{r})$ and $f(t, \boldsymbol{r})$ in $\varphi_j(\boldsymbol{r})$ gives

$$\xi(t, \boldsymbol{r}) = \sum_j \xi_j(t)\varphi_j(\boldsymbol{r}) , \tag{3.11a}$$

$$f(t, \boldsymbol{r}) = \sum_j f_j(t)\varphi_j(\boldsymbol{r}) , \tag{3.11b}$$

or, in the spectral representation,

$$\xi(\omega, \boldsymbol{r}) = \sum_j \xi_j(\omega)\varphi_j(\boldsymbol{r}) , \tag{3.12a}$$

$$f(\omega, \boldsymbol{r}) = \sum_j f_j(\omega)\varphi_j(\boldsymbol{r}) , \quad \text{where}$$

$$\xi_j(\omega) = \frac{1}{2\pi}\int_{-\infty}^{+\infty} \xi_j(t)e^{i\omega t}\,dt , \tag{3.12b}$$

and likewise for $f_j(\omega)$. Note that, by (3.10, 11), the mean power dissipated in V

under the action of the force $f(t, \boldsymbol{r})$ is written as

$$\langle Q\rangle = \int \left\langle f \frac{\partial \xi}{\partial t} \right\rangle d^3r = -\int \left\langle \xi \frac{\partial f}{\partial t} \right\rangle d^3r = -\sum_j \left\langle \xi_j(t) \frac{\partial f_j(t)}{\partial t} \right\rangle . \tag{3.13}$$

If we now take $\xi_j(t)$ to be those discrete variables that describe fluctuations in the system in question, we can naturally extend the FDT to include the spectral amplitudes $\xi_j(\omega)$ and $f_j(\omega)$. We need only assume here that the FDT remains valid for an *infinite* set of the discrete variables $\xi_j(t)$. Proceeding from this assumption, we will now find the spatial covariances for $\xi(\omega, \boldsymbol{r})$ and $f(\omega, \boldsymbol{r})$.

How are $\xi_j(\omega)$ and $f_j(\omega)$ related? Substituting (3.11a, b) into the first of (3.9) gives

$$\sum_k \xi_k(\omega)\varphi_k(\boldsymbol{r}) = \sum_k f_k(\omega)\hat{A}\varphi_k(\boldsymbol{r}) .$$

Next, we multiply this equality on the left-hand side by $\varphi_j(\boldsymbol{r})$ and integrate over V. Taking into account (3.10) we obtain

$$\xi_j(\omega) = \sum_k f_k(\omega) \int_V \varphi_j(\boldsymbol{r})\hat{A}\varphi_k(\boldsymbol{r})d^3r \equiv \sum_k \alpha_{jk}(\omega)f_k(\omega) , \tag{3.14a}$$

where

$$\alpha_{jk}(\omega) = \int_V \varphi_j(\boldsymbol{r})\hat{A}\varphi_k(\boldsymbol{r})d^3r \tag{3.15a}$$

are, clearly, the coefficients of expansion of $\hat{A}\varphi_k(\boldsymbol{r})$ in the basis functions $\varphi_j(\boldsymbol{r})$, i.e.,

$$\hat{A}\varphi_k(\boldsymbol{r}) = \sum_j \alpha_{jk}\varphi_j(\boldsymbol{r}) . \tag{3.16a}$$

Using the second of the equations (3.5), the same operations lead to the inverse of (3.14a)

$$f_j(\omega) = \sum_k \alpha_{jk}^{-1}(\omega)\xi_k(\omega) , \quad \text{where} \tag{3.14b}$$

$$\alpha_{jk}^{-1}(\omega) = \int_V \varphi_j(\boldsymbol{r})\hat{A}^{-1}\varphi_k(\boldsymbol{r})d^3r , \tag{3.15b}$$

i.e., α_{jk}^{-1} are the coefficients of expansion of $\hat{A}^{-1}\varphi_k(\boldsymbol{r})$ in $\varphi_j(\boldsymbol{r})$

$$\hat{A}^{-1}\varphi_k(\boldsymbol{r}) = \sum_j \alpha_{jk}^{-1}\varphi_j(\boldsymbol{r}) . \tag{3.16b}$$

Quantities (3.15a,b), which are the coefficients in the equations (3.14a,b) relating $\xi_j(\omega)$ and $f_j(\omega)$, thus form the elements of the infinite-dimensional *generalized susceptibility* matrix $\{\alpha_{jk}^{-1}\}$ and the inverse matrix $\{\alpha_{jk}\}$.

The discrete FDT gives the following expressions for the spectral densities of $\xi_j(t)$ and $f_j(t)$ [(II.3.118) and (II.3.122)]. Note that the right-hand sides now have a negative sign since we used the spectral expansion with exp(iωt) earlier, but in this volume we use the expansion with exp($-$iωt):

$$\langle \xi_j(\omega)\xi_k^*(\omega)\rangle = -\frac{\mathrm{i}\Theta(\omega, T)}{2\pi\omega}(\alpha_{jk} - \alpha_{kj}^*) , \tag{3.17a}$$

$$\langle f_j(\omega)f_k^*(\omega)\rangle = -\frac{\mathrm{i}\Theta(\omega, T)}{2\pi\omega}(\alpha_{kj}^{-1*} - \alpha_{jk}^{-1}) , \tag{3.17b}$$

where $\Theta(\omega, T)$ is the mean energy of the quantum oscillator

$$\Theta(\omega, T) = \hbar\omega\left\{\frac{1}{\mathrm{e}^{\hbar\omega/\kappa T} - 1} + \frac{1}{2}\right\} = \frac{\hbar\omega}{2}\coth\frac{\hbar\omega}{2\kappa T} . \tag{3.18}$$

This is where the expressions (3.17a,b) come into play.

According to (3.12a) and (3.17a),

$$\begin{aligned}\langle \xi(\omega, \boldsymbol{r})\xi^*(\omega, \boldsymbol{r}')\rangle &= \sum_{j,k}\langle \xi_j(\omega)\xi_k^*(\omega)\rangle\varphi_j(\boldsymbol{r})\varphi_k(\boldsymbol{r}')\\ &= -\frac{\mathrm{i}\Theta(\omega, T)}{2\pi\omega}\sum_{j,k}(\alpha_{jk} - \alpha_{kj}^*)\varphi_j(\boldsymbol{r})\varphi_k(\boldsymbol{r}') .\end{aligned}$$

Using (3.16a), we can write this expression as

$$\langle \xi(\omega, \boldsymbol{r})\xi^*(\omega, \boldsymbol{r}')\rangle = -\frac{\mathrm{i}\Theta(\omega, T)}{2\pi\omega}\sum_j\{\varphi_j(\boldsymbol{r}')\hat{A}\varphi_j(\boldsymbol{r}) - \varphi_j(\boldsymbol{r})\hat{A}^*\varphi_j(\boldsymbol{r})\} .$$

Here, $\hat{A}$ operates on the functions of $\boldsymbol{r}$, and $\hat{A}^*$ on the functions of $\boldsymbol{r}'$, so that these can be factored out. The sum is now the expansion of the delta-function of $\boldsymbol{r} - \boldsymbol{r}'$ in the basis functions

$$\sum_j \varphi_j(\boldsymbol{r})\varphi_j(\boldsymbol{r}') = \delta(\boldsymbol{r} - \boldsymbol{r}') .$$

As a result,

$$\langle \xi(\omega, \boldsymbol{r})\xi^*(\omega, \boldsymbol{r}')\rangle = -\frac{\mathrm{i}\Theta(\omega, T)}{2\pi\omega}(\hat{A} - \hat{A}^*)\delta(\boldsymbol{r} - \boldsymbol{r}') . \tag{3.19a}$$

With the help of (3.12b), (3.17b), and (3.16b), we can obtain the spatial covariance for the spectral amplitude of force $f(\omega, \boldsymbol{r})$ in exactly the same way:

$$\langle f(\omega, \boldsymbol{r})f^*(\omega, \boldsymbol{r}')\rangle = -\frac{\mathrm{i}\Theta(\omega, T)}{2\pi\omega}(\hat{A}^{-1*} - \hat{A}^{-1})\delta(\boldsymbol{r} - \boldsymbol{r}') . \tag{3.19b}$$

Since the delta-function is dependent only on $\boldsymbol{\varrho} = \boldsymbol{r} - \boldsymbol{r}'$, we can express $\hat{A}$ and $\hat{A}^*$ in terms of the components of $\boldsymbol{\varrho}$. Furthermore we can represent the differential operators $\hat{A}^{-1}$ and $\hat{A}^{-1*}$ (the first contains ∇, the second ∇') *via* the operator $\nabla_\varrho = \nabla = -\nabla'$.

Equations (3.19a,b), in which the configuration and size of the volume V are in no way reflected, represent the FDT for the one-component homogeneous field $\xi(t, \boldsymbol{r})$ and the related field of the Langevin force $f(t, \boldsymbol{r})$.

The generalization of this result to the multicomponent field, i.e., to the system of homogeneous, uniformly interrelated fields $\xi^{(j)}(t, \boldsymbol{r})$, is almost self-evident. Instead of equations (3.9), we will now have the *system* of linear equations

$$\begin{aligned} \xi^{(j)}(\omega, \boldsymbol{r}) &= \sum_k \hat{A}_{jk} f^{(k)}(\omega, \boldsymbol{r}) \ , \\ f^{(j)}(\omega, \boldsymbol{r}) &= \sum_k \hat{A}_{jk}^{-1} \xi^{(k)}(\omega, \boldsymbol{r}) \ , \end{aligned} \tag{3.20}$$

where $\hat{A}_{jk}$ and $\hat{A}_{jk}^{-1}$ are elements of the direct and inverse *operator matrices*, and, instead of the expression (3.13) for the power developed by the forces $f^{(j)}(t, \boldsymbol{r})$, we will have

$$\begin{aligned} \langle Q \rangle &= \sum_j \int_V \left\langle f^{(j)}(t, \boldsymbol{r}) \frac{\partial \xi^{(j)}(t, \boldsymbol{r})}{\partial t} \right\rangle d^3 r \\ &= -\sum_j \int_V \left\langle \xi^{(j)}(t, \boldsymbol{r}) \frac{\partial f^{(j)}(t, \boldsymbol{r})}{\partial t} \right\rangle d^3 r \ . \end{aligned} \tag{3.21}$$

Lastly, the final equations (3.19a,b) are replaced by correlation *matrices* of the spectral amplitudes $\xi^{(j)}(\omega, \boldsymbol{r})$ and $f^{(j)}(\omega, \boldsymbol{r})$

$$\langle \xi^{(j)}(\omega, \boldsymbol{r}) \xi^{(k)*}(\omega, \boldsymbol{r}') \rangle = -\frac{\mathrm{i}\Theta(\omega, T)}{2\pi\omega} (\hat{A}_{jk} - \hat{A}_{kj}^*) \delta(\boldsymbol{\varrho}) \ , \tag{3.22a}$$

$$\langle f^{(j)}(\omega, \boldsymbol{r}) f^{(k)*}(\omega, \boldsymbol{r}') \rangle = -\frac{\mathrm{i}\Theta(\omega, T)}{2\pi\omega} (\hat{A}_{kj}^{-1*} - \hat{A}_{jk}^{-1}) \delta(\boldsymbol{\varrho}) \ , \tag{3.22b}$$

where, as earlier, $\boldsymbol{\varrho} = \boldsymbol{r} - \boldsymbol{r}'$.

Consequently, in continuous as in discrete systems, the FDT expresses the covariances (matrices) of the ω-amplitudes[5] in a perfectly regular manner. These functions are uniquely defined by the form of $\hat{A}_{jk}$ and $\hat{A}_{jk}^{-1}$, i.e., by the *linearized macroscopic equations* for the system in question.

If these are Maxwell's equations, then the formulas (3.22a,b) allow us to find the spatial correlation of a fluctuational field, and those external fields (inductions or currents) that "create" the fluctuational electromagnetic field. If we are dealing, say,

[5] And hence of the $\omega\kappa$-amplitudes, if the ω-amplitudes are represented as spatial Fourier integrals (see Exercise 6).

with the equations of elasticity theory, then, using once again (3.22a,b), we can determine the correlation of the thermal fluctuations of strain and temperature, and also of the corresponding external stresses and heat sources [3.5], and so on. More specifically, the correlation matrices of fluctuational and external thermal fields are determined by the *skew-Hermitian parts* of the appropriate linear operators, or, to put it another way, they are conditioned by the *dissipative* behavior of the system. In the absence of energy absorption, i.e., without dissipation, the operators or operator matrices will be Hermitian and the differences on the right-hand sides of any of the equations expressing the FDT vanish. This means that in a lossless medium (or its part), there are no external sources of fluctuations, and hence no contribution from such a medium to a fluctuational field. In (3.19a) and (3.22a) *only this contribution* is taken into account and in this case, too, it vanishes. But a fluctuational field may be produced by sources localized *outside* this nonabsorbing part of the medium, i.e., it may enter this volume from the outside. Therefore, the covariances of a *fluctuational field* are in general, nonzero also in those regions of space where there is no absorption.

3.4 Correlation of External Thermal Sources in Electrodynamics

We now return to the spectral Maxwell equations (3.6) and, using the material equations (3.7), eliminate the displacement $\boldsymbol{D}$ and induction $\boldsymbol{B}$

$$\operatorname{curl} \boldsymbol{H} = -\mathrm{i}k\varepsilon \boldsymbol{E} + \frac{4\pi}{c}\boldsymbol{j}_{\mathrm{e}}, \quad \operatorname{curl} \boldsymbol{E} = \mathrm{i}k\mu \boldsymbol{H} - \frac{4\pi}{c}\boldsymbol{j}_{\mathrm{m}} . \tag{3.23}$$

The system under consideration—the fluctuational electromagnetic field—is described by the six-vector $\xi^{(j)} \equiv \{\boldsymbol{E}, \boldsymbol{H}\}$. What is here, according to Lagrange, the six-vector of the external forces $f^{(j)}$ conjugate to $\xi^{(j)}$?

According to the general expression (3.21), the mean dissipating power $\langle Q \rangle$ is equal (with the minus sign) to the sum of the products of the coordinates by the variation *rates* of the forces. Comparing (3.5) with (3.21), we can easily see that the external *currents* $\boldsymbol{j}_{\mathrm{e}}$ and $\boldsymbol{j}_{\mathrm{m}}$ are not conjugate "forces" for the "coordinates" $\xi^{(j)} = \{\boldsymbol{E}, \boldsymbol{H}\}$. If, on the other hand, we make use of the external *fields* $\mathfrak{D}$ and $\mathfrak{B}$ and substitute (3.4) into (3.5), we then obtain for $\langle Q \rangle$ a formula that has precisely the form (3.21):

$$\langle Q \rangle = -\frac{1}{4\pi}\int_V \left\langle \boldsymbol{E}\cdot\frac{\partial \mathfrak{D}}{\partial t} + \boldsymbol{H}\cdot\frac{\partial \mathfrak{B}}{\partial t}\right\rangle d^3r .$$

Consequently, for $\xi^{(j)} = \{\boldsymbol{E}, \boldsymbol{H}\}$ the "forces" are $f^{(j)} = \{\mathfrak{D}/4\pi,\ \mathfrak{B}/4\pi\}$, and the general equations (3.22a,b), expressing the FDT, should be applied just to such $\xi^{(j)}$ and $f^{(j)}$.

It is obvious that, having obtained the correlation matrix for the spectral amplitudes of the external displacement $\mathfrak{D}$ and magnetic induction $\mathfrak{B}$, we will also be able to write this matrix for the spectral amplitudes of the external currents, because, by (3.4),

$$\boldsymbol{j}_{\mathrm{e}} = -\frac{\mathrm{i}\omega}{4\pi}\mathfrak{D}, \quad \boldsymbol{j}_{\mathrm{m}} = -\frac{\mathrm{i}\omega}{4\pi}\mathfrak{B} . \tag{3.24}$$

Substituting (3.24) into (3.23), we write (3.23) in a component form

$$\frac{\mathfrak{D}_1}{4\pi} = -\frac{\varepsilon E_1}{4\pi} + \frac{1}{4\pi i k}(\nabla_2 H_3 - \nabla_3 H_2)\circlearrowleft ,$$

$$\frac{\mathfrak{B}_1}{4\pi} = -\frac{\mu H_1}{4\pi} - \frac{1}{4\pi i k}(\nabla_2 E_3 - \nabla_3 E_2)\circlearrowleft ,$$

where $k = \omega/c$ and the sign $\circlearrowleft$ denotes the cyclic permutation of the subscripts 1, 2, 3. We now compare these six equations with the second group of equations (3.20) where, as will be recalled, we put

$$\xi^{(j)} = \{\boldsymbol{E}, \boldsymbol{H}\}, \quad f^{(j)} = \{\mathfrak{D}/4\pi, \mathfrak{B}/4\pi\} \quad (j = 1, \ldots, 6) .$$

We thus arrive at the matrix of the operators $\hat{A}_{jk}^{-1}(\nabla)$

$$\hat{A}_{jk}^{-1}(\nabla) = \left\{\begin{array}{c|c} -\dfrac{\varepsilon}{4\pi}\delta_{\alpha\beta} & \dfrac{1}{4\pi i k}\begin{pmatrix} 0 & \nabla_3 & -\nabla_2 \\ -\nabla_3 & 0 & \nabla_1 \\ \nabla_2 & -\nabla_1 & 0 \end{pmatrix} \\ \hline -\dfrac{1}{4\pi i k}\begin{pmatrix} 0 & \nabla_3 & -\nabla_2 \\ -\nabla_3 & 0 & \nabla_1 \\ \nabla_2 & -\nabla_1 & 0 \end{pmatrix} & -\dfrac{\mu}{4\pi}\delta_{\alpha\beta} \end{array}\right\} .$$

$$(\alpha, \beta = 1, 2, 3).$$

The transposed complex-conjugate matrix dependent on ∇' will now be

$$\hat{A}_{kj}^{-1*}(\nabla') = \left\{\begin{array}{c|c} -\dfrac{\varepsilon^*}{4\pi}\delta_{\alpha\beta} & -\dfrac{1}{4\pi i k}\begin{pmatrix} 0 & \nabla'_3 & -\nabla'_2 \\ -\nabla'_3 & 0 & \nabla'_1 \\ \nabla'_2 & -\nabla'_1 & 0 \end{pmatrix} \\ \hline \dfrac{1}{4\pi i k}\begin{pmatrix} 0 & \nabla'_3 & -\nabla'_2 \\ -\nabla'_3 & 0 & \nabla'_1 \\ \nabla'_2 & -\nabla'_1 & 0 \end{pmatrix} & -\dfrac{\mu^*}{4\pi}\delta_{\alpha\beta} \end{array}\right\} .$$

$$(\alpha, \beta = 1, 2, 3).$$

If we take into account that $\nabla' = -\nabla$ (since the operators act on a function of $\boldsymbol{r}-\boldsymbol{r}'$), we obtain

$$\hat{A}_{kj}^{-1*}(\nabla') - \hat{A}_{jk}^{-1}(\nabla) = \left\{\begin{array}{c|c} \frac{\varepsilon - \varepsilon^*}{4\pi}\delta_{\alpha\beta} & 0 \\ \hline 0 & \frac{\mu-\mu^*}{4\pi}\delta_{\alpha\beta} \end{array}\right\} \quad (\alpha, \beta = 1, 2, 3) \ .$$

Finally, substituting this result into (3.22b), i.e., the FDT for the Langevin forces, gives the covariances of the external fields $\mathfrak{D}$ and $\mathfrak{B}$

$$\begin{aligned} \langle \mathfrak{D}_\alpha(\omega, \boldsymbol{r}) \mathfrak{D}_\beta^*(\omega, \boldsymbol{r}') \rangle &= -\frac{2\mathrm{i}\Theta(\omega, T)}{\omega}(\varepsilon - \varepsilon^*)\delta_{\alpha\beta}\delta(\boldsymbol{r}-\boldsymbol{r}') \ , \\ \langle \mathfrak{B}_\alpha(\omega, \boldsymbol{r}) \mathfrak{B}_\beta^*(\omega, \boldsymbol{r}') \rangle &= -\frac{2\mathrm{i}\Theta(\omega, T)}{\omega}(\mu - \mu^*)\delta_{\alpha\beta}\delta(\boldsymbol{r}-\boldsymbol{r}') \ . \\ \langle \mathfrak{D}_\alpha(\omega, \boldsymbol{r}) \mathfrak{B}_\beta^*(\omega, \boldsymbol{r}') \rangle &= 0 \ . \end{aligned} \tag{3.25}$$

In the general case of an *anisotropic* or gyrotropic medium described by the permeability *tensors* $\varepsilon_{\alpha\beta}$ and $\mu_{\alpha\beta}$, we would obtain the same relations (3.25), but with the diagonal tensors $\varepsilon\delta_{\alpha\beta}$ and $\mu\delta_{\alpha\beta}$ replaced by $\varepsilon_{\alpha\beta}$ and $\mu_{\alpha\beta}$, $\varepsilon^*\delta_{\alpha\beta}$ and $\mu^*\delta_{\alpha\beta}$ by $\varepsilon_{\beta\alpha}^*$ and $\mu_{\beta\alpha}^*$.

Using (3.24,25), it is a straightforward exercise to write the covariances of the external currents. We will now give them for the anisotropic medium[6]

$$\begin{aligned} \langle j_{\mathrm{e}\alpha}(\omega, \boldsymbol{r}) j_{\mathrm{e}\beta}^*(\omega, \boldsymbol{r}') \rangle &= -\frac{\mathrm{i}\omega\Theta(\omega, T)}{8\pi^2}(\varepsilon_{\alpha\beta} - \varepsilon_{\beta\alpha}^*)\delta(\boldsymbol{r}-\boldsymbol{r}') \ , \\ \langle j_{\mathrm{m}\alpha}(\omega, \boldsymbol{r}) j_{\mathrm{m}\beta}^*(\omega, \boldsymbol{r}') \rangle &= -\frac{\mathrm{i}\omega\Theta(\omega, T)}{8\pi^2}(\mu_{\alpha\beta} - \mu_{\beta\alpha}^*)\delta(\boldsymbol{r}-\boldsymbol{r}') \ . \\ \langle j_{\mathrm{e}\alpha}(\omega, \boldsymbol{r}) j_{\mathrm{m}\beta}^*(\omega, \boldsymbol{r}') \rangle &= 0 \ . \end{aligned} \tag{3.26}$$

A less formal derivation of these formulas is given in Exercise 2.

Thus, the electric and magnetic sources of a fluctuational field are not spatially correlated, and the radius of the spatial correlation of both is zero [delta-function $\delta(\boldsymbol{r}-\boldsymbol{r}')$]. For media without spatial dispersion, to which we confine our discussion here, the last result appears to be self-evident. In fact, the actual correlation radius of thermal field *sources* in such media may only be a microquantity (of the order of interatomic distances, say); therefore, in the macroscopic theory, which views media as continuous, it must be zero. On the contrary, in media with *spatial*

[6] Equations (3.26) are due to Landau and Lifshits [3.3] who derived them with the help of the discrete form of the FDT. Relations of the form (3.26), based on the continuous FDT, were obtained in [3.4]. Later, independent of the general continuous form of FDT derived in [3.4], equations of the form (3.26) were obtained by many authors (see, e.g. [References 23, 55–62], Ref. 3.6).

dispersion, material equations are nonlocalized not only in t, but also in $\boldsymbol{r}$, i.e., instead of (3.2), these equations contain operators both in t and $\boldsymbol{r}$[7]. For the sources of the fluctuational field, this gives a nonzero correlation radius which has the same order of magnitude as the size of the nonlocality region in the material equations (see [Ref. 3.6, Sect. 4]).

Piecewise homogeneous media, i.e., those with distinct boundaries between the media, are described by (3.25, 26) (due to delta-correlation) up to the boundaries. But in the presence of a spatial nonlocality, the situation changes: near boundaries, namely within a layer whose thickness is of the order of the correlation radius, the fields of the external sources are inhomogeneous (their covariances are individually dependent on $\boldsymbol{r}$ and $\boldsymbol{r}'$ and not on the difference $\boldsymbol{r}-\boldsymbol{r}'$) and anisotropic, even if the medium itself is isotropic.

We were able to write immediately the covariances of the *sources* of the fluctuational field using the theorem (3.22b) only because the operators $\hat{A}_{\alpha\beta}^{-1}$ in (3.22b) are known to us from Maxwell's equations. We face a different situation with the correlation of "coordinates", i.e., of the *strengths* of the fluctuational electromagnetic fields $\boldsymbol{E}$ and $\boldsymbol{H}$. To write the covariances for $\boldsymbol{E}$ and $\boldsymbol{H}$, we have, by (3.22a), to know the *inverse* operators $\hat{A}_{\alpha\beta}$, i.e., to "invert" Maxwell's equations by expressing $\boldsymbol{E}$ and $\boldsymbol{H}$ in terms of $\boldsymbol{j}_e$ and $\boldsymbol{j}_m$ (or $\mathfrak{D}$ and $\mathfrak{B}$). In other words, we have to *solve* the inhomogeneous Maxwell equations, regarding the distribution of the sources $\boldsymbol{j}_e$ and $\boldsymbol{j}_m$ as arbitrary but specified, and imposing the boundary conditions appropriate for a given problem.

If the boundary problem is solved, we will thereby obtain $\hat{A}_{\alpha\beta}$ (normally in the form of integrals covering regions where $\boldsymbol{j}_e$ and $\boldsymbol{j}_m$ are nonzero), and will be able to apply the theorem (3.22a). We may also proceed differently and construct average bilinear combinations of interest to us of the solutions obtained for $\boldsymbol{E}$ and $\boldsymbol{H}$. The integrands of multiple integrals, which express the second moments of the strengths, will include covariances of the external currents $\boldsymbol{j}_e$ and $\boldsymbol{j}_m$, i.e., expressions (3.26), which we have already obtained. Both techniques are quite legitimate, and a number of fluctuational electrodynamic problems have been solved in exactly this way [3.1, 7, 8].

Both techniques are tedious and uneconomic, however, because for each specific problem, one has to repeat the process anew. We will, therefore, look at what results we arrive when the FDT is applied to the solution of the Maxwell equations written in the *general integral* form, say, in terms of *Green's functions*.

3.5 Generalized Kirchhoff's Law

Green's functions are solutions of a boundary problem with elementary *point* field sources. Since we are concerned with electrodynamics, we will introduce them in the following way.

---

[7] Accordingly, equations (3.7) will also not be algebraic, but will contain spatial operators. Only equations for $\omega\kappa$-amplitudes will be algebraic.

Let $\boldsymbol{j}_{0\mathrm{e}}$ be a deterministic electric current concentrated at a point $A\,(\boldsymbol{r}=\boldsymbol{r}_A)$ and directed along a constant unit vector $\boldsymbol{l}$

$$\boldsymbol{j}_{0\mathrm{e}} = \boldsymbol{l}\delta(\boldsymbol{r}-\boldsymbol{r}_A)\ . \tag{3.27}$$

$\boldsymbol{j}_{0\mathrm{e}}$ may be regarded as a polarization current, i.e., the source may be thought of as a point *electric dipole* with a moment $\boldsymbol{p} = -\boldsymbol{l}/\mathrm{i}\omega$[8]. The solution to Maxwell's equations, i.e., the electromagnetic field produced by the source (3.27) in a given system of bodies and media, will be denoted by $\boldsymbol{E}_{0\mathrm{e}}(\boldsymbol{r},\boldsymbol{r}_A,\boldsymbol{l})$, $\boldsymbol{H}_{0\mathrm{e}}(\boldsymbol{r},\boldsymbol{r}_A,\boldsymbol{l})$, and this *deterministic* field will, for the sake of brevity, be called *diffractional*. It is the fields $\boldsymbol{E}_{0\mathrm{e}}$ and $\boldsymbol{H}_{0\mathrm{e}}$ that are Green's functions.

We wish to express $\boldsymbol{E}$ and $\boldsymbol{H}$ for the fluctuational field caused by the randomly distributed currents $\boldsymbol{j}_\mathrm{e}$ and $\boldsymbol{j}_\mathrm{m}$ in terms of these currents and Green's functions $\boldsymbol{E}_{0\mathrm{e}}$ and $\boldsymbol{H}_{0\mathrm{e}}$. This can conveniently be done using the electrodynamic *reciprocity theorem* that interrelates the strengths and sources of two different fields, 1 and 2, for the same system of bodies and media.

Consider two distributions of external electric and magnetic currents $\boldsymbol{j}_{1\mathrm{e}}, \boldsymbol{j}_{1\mathrm{m}}$ and $\boldsymbol{j}_{2\mathrm{e}}, \boldsymbol{j}_{2\mathrm{m}}$ that, other things being equal, produce the fields $\boldsymbol{E}_1$, $\boldsymbol{H}_1$ and $\boldsymbol{E}_2$, $\boldsymbol{H}_2$, respectively. The reciprocity theorem is derived in a fairly simple way (without strong restrictions) from two sets of Maxwell's equations: one for the field and sources 1, and other for the field and sources 2 (see, e.g., [3.6]). It states that

$$\int\limits_V (\boldsymbol{E}_1\cdot\boldsymbol{j}_{\mathrm{e}2}-\boldsymbol{H}_1\cdot\boldsymbol{j}_{\mathrm{m}2})\,d^3r = \int\limits_V (\boldsymbol{E}_2\cdot\boldsymbol{j}_{\mathrm{e}1}-\boldsymbol{H}_2\cdot\boldsymbol{j}_{\mathrm{m}1})\,d^3r\ , \tag{3.28}$$

where V is the volume of the complete field. In actual fact, the integrals (3.28) cover only those regions of space where the external currents are nonzero.

In (3.28), we now identify the desired fluctuational field with field 1

$$\boldsymbol{j}_{\mathrm{e}1}=\boldsymbol{j}_\mathrm{e},\quad \boldsymbol{j}_{\mathrm{m}1}=\boldsymbol{j}_\mathrm{m},\quad \boldsymbol{E}_1=\boldsymbol{E},\quad \boldsymbol{H}_1=\boldsymbol{H}\ , \tag{3.29}$$

and the auxiliary diffractional field with field 2

$$\boldsymbol{j}_{\mathrm{e}2}=\boldsymbol{l}\delta(\boldsymbol{r}-\boldsymbol{r}_A),\quad \boldsymbol{j}_{\mathrm{m}2}=0,\quad \boldsymbol{E}_2=\boldsymbol{E}_{0\mathrm{e}},\quad \boldsymbol{H}_2=\boldsymbol{H}_{0\mathrm{e}}\ .$$

Then, (3.28) gives

$$\begin{aligned}\int\limits_V \boldsymbol{E}(\boldsymbol{r})\cdot\boldsymbol{l}\delta(\boldsymbol{r}-\boldsymbol{r}_A)\,d^3r &= E_l(\boldsymbol{r}_A)\\ &= \int\limits_V \{\boldsymbol{E}_{0\mathrm{e}}(\boldsymbol{r},\boldsymbol{r}_A,\boldsymbol{l})\cdot\boldsymbol{j}_\mathrm{e}(\boldsymbol{r})-\boldsymbol{H}_{0\mathrm{e}}(\boldsymbol{r},\boldsymbol{r}_A,\boldsymbol{l})\cdot\boldsymbol{j}_\mathrm{m}(\boldsymbol{r})\}\,d^3r\ ,\end{aligned} \tag{3.30}$$

i.e., what is required. The component along the direction $\boldsymbol{l}$ of the electric strength of the fluctuational field at point A, is expressed through the distribution of random

[8] The polarization current density is $\partial\boldsymbol{P}/\partial t$, where $\boldsymbol{P}$ is the volume polarization density. For the point dipole with moment $\boldsymbol{p}$ we have $\boldsymbol{P}=\boldsymbol{p}\delta(\boldsymbol{r}-\boldsymbol{r}_A)$, so that $\boldsymbol{j}_{0\mathrm{e}}=(\partial\boldsymbol{p}/\partial t)\ \delta(\boldsymbol{r}-\boldsymbol{r}_A)$, and for the spectral amplitudes $\boldsymbol{j}_{0\mathrm{e}}=-\mathrm{i}\omega\boldsymbol{p}\delta(\boldsymbol{r}-\boldsymbol{r}_A)$. At $\boldsymbol{p}=-\boldsymbol{l}/\mathrm{i}\omega$ we arrive at (3.27).

currents and Green's functions—the deterministic field strengths produced by an electric dipole oriented along $\boldsymbol{l}$ and located at A.

In order to obtain a similar expression for the component of the magnetic fluctuational field it is necessary, while retaining (3.29), to utilize another field 2 in the reciprocity theorem (3.28), namely the diffractional field $\boldsymbol{E}_{0\mathrm{m}}$, $\boldsymbol{H}_{0\mathrm{m}}$ produced by the *magnetic* point current

$$\boldsymbol{j}_{0\mathrm{m}} = \boldsymbol{l}\delta(\boldsymbol{r}-\boldsymbol{r}_A) \ ,$$

or, similarly, by the *magnetic dipole* with moment $\boldsymbol{m} = -\boldsymbol{l}/\mathrm{i}\omega$, placed at A

$$\boldsymbol{j}_{\mathrm{e}2} = 0, \quad \boldsymbol{j}_{\mathrm{m}2} = \boldsymbol{l}\delta(\boldsymbol{r}-\boldsymbol{r}_A), \quad \boldsymbol{E}_2 = \boldsymbol{E}_{0\mathrm{m}}, \quad \boldsymbol{H}_2 = \boldsymbol{H}_{0\mathrm{m}} \ .$$

From (3.28) we obtain

$$H_l(\boldsymbol{r}_A) = -\int_V \{\boldsymbol{E}_{0\mathrm{m}}(\boldsymbol{r},\boldsymbol{r}_A,\boldsymbol{l})\cdot\boldsymbol{j}_{\mathrm{e}}(\boldsymbol{r}) - \boldsymbol{H}_{0\mathrm{m}}(\boldsymbol{r},\boldsymbol{r}_A,\boldsymbol{l})\cdot\boldsymbol{j}_{\mathrm{m}}(\boldsymbol{r})\}\, d^3r \ . \tag{3.31}$$

From the expressions (3.30, 31) which are linear in $\boldsymbol{j}_e$ and $\boldsymbol{j}_{\mathrm{m}}$, we can now obtain any moments of the fluctuational field, if we know the moments of the same order for $\boldsymbol{j}_{\mathrm{e}}$ and $\boldsymbol{j}_{\mathrm{m}}$. Specifically, we can work out *a fortiori* the second moments of the fluctuational field components, since the second moments of the external currents are given by (3.26).

Let us, for example, find the mean of the product $E_{l_1}(\boldsymbol{r}_1)\,H^*_{l_2}(\boldsymbol{r}_2)$. We put $\boldsymbol{l} = \boldsymbol{l}_1$, $\boldsymbol{r}_A = \boldsymbol{r}_1$ in (3.30) and write the scalar products of the vectors in components ($\boldsymbol{ab} = a_\alpha b_\alpha$, where the summation is over the double subscripts from 1 to 3). We thus arrive at

$$E_{l_1}(\boldsymbol{r}_1) = \int \{E_{0\mathrm{e}\alpha}(\boldsymbol{r},\boldsymbol{r}_1,\boldsymbol{l}_1)\cdot j_{\mathrm{e}\alpha}(\boldsymbol{r}) - H_{0\mathrm{e}\alpha}(\boldsymbol{r},\boldsymbol{r}_1,\boldsymbol{l}_1)\cdot j_{\mathrm{m}\alpha}(\boldsymbol{r})\}\, d^3r \ . \tag{3.32}$$

Setting $\boldsymbol{l} = \boldsymbol{l}_2$ and $\boldsymbol{r}_A = \boldsymbol{r}_2$ in (3.31) gives

$$H_{l_2}(\boldsymbol{r}_2) = -\int \{E_{0\mathrm{m}\beta}(\boldsymbol{r}',\boldsymbol{r}_2,\boldsymbol{l}_2) j_{\mathrm{e}\beta}(\boldsymbol{r}') - H_{0\mathrm{m}\beta}(\boldsymbol{r}',\boldsymbol{r}_2,\boldsymbol{l}_2) j_{\mathrm{m}\beta}(\boldsymbol{r}')\}\, d^3r' \ . \tag{3.33}$$

In the following, we will omit $\boldsymbol{r}_1$, $\boldsymbol{l}_1$ in $\boldsymbol{E}_{0\mathrm{e}}$, $\boldsymbol{H}_{0\mathrm{e}}$ and $\boldsymbol{r}_2$, $\boldsymbol{l}_2$ in $\boldsymbol{E}_{0\mathrm{m}}$, $\boldsymbol{H}_{0\mathrm{m}}$. Next, we multiply (3.32) by the expression that is complex-conjugate to (3.33) and average over the equilibrium ensemble of the random sources $\boldsymbol{j}_{\mathrm{e}}$ and $\boldsymbol{j}_{\mathrm{m}}$:

$$\begin{aligned}\langle E_{l_1}(\boldsymbol{r}_1) H^*_{l_2}(\boldsymbol{r}_2)\rangle = -\int\!\!\int \{ & E_{0\mathrm{e}\alpha}(\boldsymbol{r}) E^*_{0\mathrm{m}\beta}(\boldsymbol{r}') \langle j_{\mathrm{e}\alpha}(\boldsymbol{r}) j^*_{\mathrm{e}\beta}(\boldsymbol{r}')\rangle \\ & + H_{0\mathrm{e}\alpha}(\boldsymbol{r}) H^*_{0\mathrm{m}\beta}(\boldsymbol{r}') \langle j_{\mathrm{m}\alpha}(\boldsymbol{r}) j^*_{\mathrm{m}\beta}(\boldsymbol{r}')\rangle \\ & - E_{0\mathrm{e}\alpha}(\boldsymbol{r}) H^*_{0\mathrm{m}\beta}(\boldsymbol{r}') \langle j_{\mathrm{e}\alpha}(\boldsymbol{r}) j^*_{\mathrm{m}\beta}(\boldsymbol{r}')\rangle \\ & - H_{0\mathrm{e}\alpha}(\boldsymbol{r}) E_{0\mathrm{m}\beta}(\boldsymbol{r}') \langle j_{\mathrm{m}\alpha}(\boldsymbol{r}) j_{\mathrm{e}\beta}(\boldsymbol{r}')\rangle\}\, d^3r\, d^3r' \ .\end{aligned}$$

By (3.26), the last two terms of the integrand are zero because the electric and magnetic external currents are uncorrelated, whereas the covariances of the currents in the first two terms contain $\delta(\boldsymbol{r}-\boldsymbol{r}')$, thus allowing us to carry out the integration with respect to $\boldsymbol{r}'$. As a result

$$\langle E_{l_1}(\boldsymbol{r}_1) H^*_{l_2}(\boldsymbol{r}_2)\rangle = \frac{\mathrm{i}\omega\Theta(\omega,T)}{8\pi^2}\int\{E_{0\mathrm{e}\alpha}E^*_{0\mathrm{m}\beta}(\varepsilon_{\alpha\beta}-\varepsilon^*_{\beta\alpha}) + H_{0\mathrm{e}\alpha}H^*_{0\mathrm{m}\beta}(\mu_{\alpha\beta}-\mu^*_{\beta\alpha})\}\,d^3r \ .$$

But

$$q_{\mathrm{em}*}(\boldsymbol{r}) = -\frac{\mathrm{i}\omega}{16\pi}\{E_{0\mathrm{e}\alpha}E^*_{0\mathrm{m}\beta}(\varepsilon_{\alpha\beta}-\varepsilon^*_{\beta\alpha}) + H_{0\mathrm{e}\alpha}H^*_{0\mathrm{m}\beta}(\mu_{\alpha\beta}-\mu^*_{\beta\alpha})\} \tag{3.35}$$

is nothing other than the spectral volume density of the *mixed thermal* losses (electric plus magnetic) at a point $\boldsymbol{r}$. Before proceeding any further we would like to clarify the notion of "mixed" losses and take as an example an isotropic medium having only electric losses, when

$$\varepsilon_{\alpha\beta} = \left(\varepsilon' + \mathrm{i}\frac{4\pi\sigma}{\omega}\right)\delta_{\alpha\beta} \ , \qquad \mu_{\alpha\beta} = \mu'\delta_{\alpha\beta} \ ,$$

where ε' and μ' are the real permittivity and permeability, respectively, and σ' is the electric conductivity. For this medium, (3.35) becomes

$$q_{\mathrm{em}*}(\boldsymbol{r}) = \tfrac{1}{2}\sigma\,\boldsymbol{E}_{0\mathrm{e}}\cdot\boldsymbol{E}^*_{0\mathrm{m}}, \quad q_{\mathrm{e}*\mathrm{m}} = (q_{\mathrm{e}*\mathrm{m}})^*$$

"Mixed" losses owe their name to fields caused by two different sources. Using the above example, we will see how these losses come about. It is common knowledge that the instantaneous volume density of the Joule heat is $q(t,\boldsymbol{r}) = \sigma(\boldsymbol{r})\boldsymbol{E}^2(t,\boldsymbol{r})$. For the harmonic field $\boldsymbol{E}(t,\boldsymbol{r}) = [\boldsymbol{E}(\omega,\boldsymbol{r})\exp(-\mathrm{i}\omega t) + \boldsymbol{E}^*(\omega,\boldsymbol{r})\times\exp(\mathrm{i}\omega t)]/2$, the average over the period $T = 2\pi/\omega$ is

$$q(\omega,\boldsymbol{r}) = \tfrac{1}{2}\sigma(\boldsymbol{r})|E(\omega,\boldsymbol{r})|^2 \ .$$

If $\boldsymbol{E} = \boldsymbol{E}_1 + \boldsymbol{E}_2$, then, clearly,

$$q(\omega,\boldsymbol{r}) = q_{11*} + q_{12*} + q_{1*2} + q_{22*} \ ,$$

where $q_{11*}(\omega,\boldsymbol{r}) = \sigma|\boldsymbol{E}_1|^2/2$, $q_{22*}(\omega,\boldsymbol{r}) = \sigma|\boldsymbol{E}_2|^2/2$ are the losses of each of the fields $\boldsymbol{E}_1$ and $\boldsymbol{E}_2$, respectively, and $q_{12*}(\omega,\boldsymbol{r}) = \sigma\boldsymbol{E}_1\boldsymbol{E}^*_2/2$, $q_{1*2}(\omega,\boldsymbol{r}) = \sigma\boldsymbol{E}^*_1\boldsymbol{E}_2/2$ are the mixed losses of $\boldsymbol{E}_1$ and $\boldsymbol{E}_2$.

The integral of $q_{\mathrm{em}*}(\boldsymbol{r})$ with respect to $\boldsymbol{r}$ is the spectral density of the losses over the entire volume of a medium (actually, in all the regions with energy dissipation, i.e., $\varepsilon_{\alpha\beta} - \varepsilon^*_{\beta\alpha} \neq 0$ and/or $\mu_{\alpha\beta} - \mu^*_{\beta\alpha} \neq 0$)

$$Q_{\mathrm{em}*}(\boldsymbol{r}_1,\boldsymbol{l}_1;\boldsymbol{r}_2,\boldsymbol{l}_2) = \int q_{\mathrm{em}*}(\boldsymbol{r})\,d^3r \ . \tag{3.36}$$

Clearly, $q_{\mathrm{em}*}$ and $Q_{\mathrm{em}*}$ vary with the coordinates and directions of the point sources of the auxiliary diffractional fields: in this case the electric dipole with moment $\boldsymbol{p} = -\boldsymbol{l}_1/\mathrm{i}\omega$ at point $\boldsymbol{r}_1$, and the magnetic dipole with moment

$\boldsymbol{m} = -\,\boldsymbol{l}_2/\mathrm{i}\omega$ at $\boldsymbol{r}_2$. It follows from (3.34–36) that

$$\langle E_{l_1}(\boldsymbol{r}_1) H^*_{l_2}(\boldsymbol{r}_2)\rangle = -\frac{2}{\pi}\,\Theta(\omega, T)\, Q_{\mathrm{em}*}(\boldsymbol{r}_1, \boldsymbol{l}_1; \boldsymbol{r}_2, \boldsymbol{l}_2)\ . \tag{3.37}$$

A similar treatment for the covariances of $\boldsymbol{E}$ and $\boldsymbol{H}$ given by (3.31, 32) and (3.26), respectively, yields

$$\langle E_{l_1}(\boldsymbol{r}_1) E^*_{l_2}(\boldsymbol{r}_2)\rangle = \frac{2}{\pi}\,\Theta(\omega, T)\, Q_{\mathrm{ee}*}(\boldsymbol{r}_1, \boldsymbol{l}_1; \boldsymbol{r}_2, \boldsymbol{l}_2)\ , \tag{3.38}$$

$$\langle H_{l_1}(\boldsymbol{r}_1) H^*_{l_2}(\boldsymbol{r}_2)\rangle = \frac{2}{\pi}\,\Theta(\omega, T)\, Q_{\mathrm{mm}*}(\boldsymbol{r}_1, \boldsymbol{l}_1; \boldsymbol{r}_2, \boldsymbol{l}_2)\ . \tag{3.39}$$

In equation (3.38), mixed losses of the diffractional fields occur due to two electric dipoles ($\boldsymbol{p}_1 = -\boldsymbol{l}_1/\mathrm{i}\omega$ at $\boldsymbol{r}_1$ and $\boldsymbol{p}_2 = -\boldsymbol{l}_2/\mathrm{i}\omega$ at $\boldsymbol{r}_2$), and in (3.39) due to two magnetic dipoles ($\boldsymbol{m}_1 = -\boldsymbol{l}_1/\mathrm{i}\omega$ at $\boldsymbol{r}_1$ and $\boldsymbol{m}_2 = -\boldsymbol{l}_2/\mathrm{i}\omega$ at $\boldsymbol{r}_2$). Not the mixed, but proper losses of the diffractional field of one source determine the *root-mean-square value* of any *one* component of the fluctuational field at the *same* point. For instance, letting $\boldsymbol{r}_1 = \boldsymbol{r}_2 = \boldsymbol{r}$ and $\boldsymbol{l}_1 = \boldsymbol{l}_2 = \boldsymbol{l}$ in (3.38) gives

$$\langle |E_l(\boldsymbol{r})|^2\rangle = \frac{2}{\pi}\,\Theta(\omega, T)\, Q_{\mathrm{ee}*}(\boldsymbol{r}, \boldsymbol{l}; \boldsymbol{r}, \boldsymbol{l})\ . \tag{3.40}$$

This expression thus includes the proper losses of the diffractional field produced by the electric dipole with moment $\boldsymbol{p} = -\,\boldsymbol{l}/\mathrm{i}\omega$ located at $\boldsymbol{r}$.

Equations (3.37–39) can be termed the *Kirchhoff form* of the FDT, as they are a *direct generalization of Kirchhoff's law* in the classical theory of thermal radiation. This law is known to relate the intensity of the thermal radiation of a body in any direction to the absorption in that body when exposed to a plane wave propagating in the opposite direction. (It may be said that such a plane wave comes from a point source at infinity). The generalization thus concerns at once three aspects of the situation.

First, we can now find the means of the products of *any* components of $\boldsymbol{E}$ and $\boldsymbol{H}$, and not only of those that define the energy density and flux (the Poynting vector flux)—the two quantities that are the sole concern of the classical theory of radiation.

Second, we can compute not only the mean products of components at the *same* point ($\boldsymbol{r}_1 = \boldsymbol{r}_2$), which are required to work out the density and energy flux, but also the spatial *covariances* of the fluctuational field ($\boldsymbol{r}_1 \neq \boldsymbol{r}_2$).

Third, and most important, (3.37–39) contain no constraints on the relationships between the wavelength λ and characteristic scale l of the problem (the size of the bodies, the curvature radii of their surfaces, the distances from the body to an observation point, etc.). In other words, unlike the classical theory of thermal radiation, which is bound by the constraints of geometrical optics, we can now calculate the second moments of a fluctuational field, that is to say both the wave part (taking into account all the diffraction phenomena), and the nonwave (quasi-stationary) part for any λ vs l ratio.

It is worth discussing here the computational advantages provided by the generalized Kirchhoff's law. Admittedly, in order to find auxiliary diffractional fields (Green's functions), we will still need to solve the appropriate boundary electrodynamic problems using conventional methods. But these problems are much simpler than those mentioned at the end of the previous section. For (3.37–39), it is necessary to solve the homogeneous Maxwell's equations that have dipole singularities at specified points $\boldsymbol{r}_1$ and $\boldsymbol{r}_2$, and not the inhomogeneous equations with a specified, but arbitrary distribution of external currents $\boldsymbol{j}_e$ and $\boldsymbol{j}_m$. In addition, it is possible in many cases to employ known Green's functions, i.e., the solutions to problems with points sources obtained earlier (e.g., the classical Sommerfeld problem concerning the field of a dipole located above the plane boundary of an absorbing medium).

Lastly, it is often not necessary to compute the *strengths* of the diffractional fields. In the final analysis, (3.37–39) include *thermal losses* of the diffractional fields, and not these strengths. In many cases of practical interest, these losses can be derived to a sufficient accuracy by *approximate* methods. These cases include, for instance, well-conducting bodies (strong skin-effect); bodies large (or, on the contrary, small) as compared with λ (if only in some of their dimensions, as is the case with thin conductors). We will now go on to furnish several examples of the use of the generalized Kirchhoff's law, and at the same time illustrate the passing to the classical theory of thermal radiation in the limiting case. This section concludes with two general remarks.

The first remark refers to the generalization of the law (3.37–39) to the case of *nonuniformly heated* bodies. If temperature gradients are small enough for a function $\Theta = \Theta(\boldsymbol{r})$ to be so well-behaved that the contribution of nonequilibrium processes is still negligible, then the inclusion of nonuniform heating will reduce to having to retain $\Theta(\boldsymbol{r})$ under the volume integration sign. In (3.37–39), under such quasi-homogeneous conditions, it is necessary to write $\int \Theta\, dQ_0$ instead of $\Theta\, Q_0$, where $dQ_0 = q_0 d^3r$ is the thermal losses of the diffractional field in the volume element d^3r of the body under consideration.

The second remark refers to *zero oscillations*. All the equations of the correlation theory of equilibrium and quasi-equilibrium fluctuational fields include as a factor the mean energy of the oscillator $\Theta(\omega, T)$ containing the summand $\hbar\omega/2$—the energy of the so-called *zero oscillations* [see (3.18)]. The term in (3.18) that varies with T and vanishes at $T = 0$ corresponds to the so-called *black-body radiation*. It is only this that is generally considered when dealing with radiation, i.e., the energy *flux*, since it alone is an observable quantity in these cases. But relations of the form (3.37) also contain the energy flux of the zero oscillations. When and why should this flux be ignored?

The fact is that in deriving relations of the form (3.37–39), it is tacitly assumed that the body in question is the *only* source of the fluctuational field. However, zero oscillations exist in the absence of this body as well, as these are produced by *all* bodies, including ones at absolute zero ($T = 0$). It may be said that as far as zero oscillations are concerned, an equilibrium state always occurs i.e., zero oscillations are always *standing* waves. Accordingly, any energy flux of these oscillations is nullified by a return flux of the same intensity[9]. Therefore, in any formulas referring to energy *flux* (but not to energy *density*!) only that part of $\Theta(\omega, T)$ should be kept that concerns black-body radiation, and the energy flux of zero oscillations should be subtracted because it is always compensated for in any surroundings of the body.

3.6 Examples of the Use of the Generalized Kirchhoff's Law

As a first example, consider a half-space $z < 0$ filled with an absorbing, homogeneous, isotropic medium with permittivity ε_1 and permeability μ_1, above which the medium is also homogeneous and isotropic but transparent (ε and μ are real) (Fig. 3.1). We already know that to find the second moments of a fluctuational field at a point in a transparent medium requires a knowledge of the diffractional field $\boldsymbol{E}_{0\mathrm{e}}$, $\boldsymbol{H}_{0\mathrm{e}}$ and $\boldsymbol{E}_{0\mathrm{m}}$, $\boldsymbol{H}_{0\mathrm{m}}$ of the elementary electric and magnetic dipoles with moments $\boldsymbol{p} = -\boldsymbol{l}_1/\mathrm{i}\omega$ and $\boldsymbol{m} = -\boldsymbol{l}_2/\mathrm{i}\omega$ placed at this point ($\boldsymbol{l}_1$ and $\boldsymbol{l}_2$ are unit vectors). The solution to this classical problem may be found in any textbook on radio wave propagation (for a more detailed treatment see, e.g., [3.9, 10]). It is more convenient to think of the losses of the diffractional field $\boldsymbol{E}_0 = \boldsymbol{E}_{0\mathrm{e}} + \boldsymbol{E}_{0\mathrm{m}}$, $\boldsymbol{H}_0 = \boldsymbol{H}_{0\mathrm{e}} + \boldsymbol{H}_{0\mathrm{m}}$ in the absorbing half-space $z < 0$ as an energy flux through the boundary $z = 0$ into this half-space

$$Q_0 = -\frac{c}{16\pi}\int_0^{\infty} r\,dr \int_0^{2\pi} d\varphi\,\{[\boldsymbol{E}_0\boldsymbol{H}_0^*]_z + [\boldsymbol{E}_0^*\boldsymbol{H}_0]_z\}_{z=0}\ . \tag{3.41}$$

(As usual $[\boldsymbol{E}\boldsymbol{B}] = \boldsymbol{E} \times \boldsymbol{B}$.) In the plane $z = 0$, we introduced in (3.41) the polar coordinates r and φ, so that the dipole lies at a height h above the origin $r = 0$. Clearly, Q_0 will only be dependent on h and, of course, on the dipole orientations $\boldsymbol{l}_1$ and $\boldsymbol{l}_2$. By (3.37–40), various expressions for Q_0 corresponding to various orientations $\boldsymbol{l}_1$ and $\boldsymbol{l}_2$ will give the means of the products of the appropriate components of the fluctuational field. If, say, we direct $\boldsymbol{l}_1$ along the x-axis and put $\boldsymbol{m} = 0$, we will

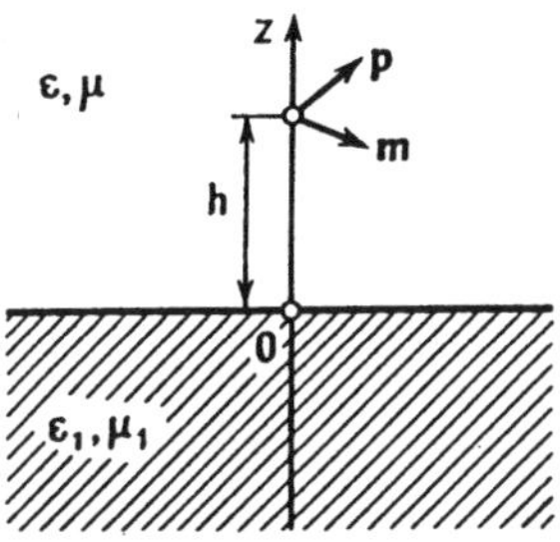

Fig. 3.1. Auxiliary moments $\boldsymbol{m}$ and $\boldsymbol{p}$ above the absorbing half-space

[9] Of course, it is nullified *on average*, i.e., the mean values of opposite fluxes cancel out. As for *fluctuations* of the energy flux, these are also influenced by zero oscillations.

obtain $\langle |E_x|^2 \rangle$; if we direct $\boldsymbol{l}_1$ along the x-axis and $\boldsymbol{l}_2$ along the y-axis, then the appropriate Q_0 will give $\langle E_x H_y^* \rangle$, and so forth. Only the following moments will appear nonzero[10]:

$$\begin{aligned} &\langle |E_x|^2 \rangle = \langle |E_y|^2 \rangle, \ \langle |H_x|^2 \rangle = \langle |H_y|^2 \rangle \ , \\ &\langle |E_z|^2 \rangle = \langle |H_z|^2 \rangle, \ \langle E_x H_y^* \rangle = -\langle E_y H_x^* \rangle \ . \end{aligned} \tag{3.42}$$

They now depend only on the height h of the observation point above the boundary.

Knowing (3.42), we can easily derive expressions for the spectral densities (in $\omega > 0$) of the *electric and magnetic energies* of a fluctuational field in a transparent medium (with ε and μ independent of ω, i.e., without dispersion), and for the z-component of the *Poynting vector*

$$\begin{aligned} &u_{e\omega}(h) = \frac{\varepsilon}{4\pi} \langle |E|^2 \rangle, \quad u_{m\omega}(h) = \frac{\mu}{4\pi} \langle |H|^2 \rangle \ , \\ &\mathscr{S}_{\omega z}(h) = \frac{c}{4\pi} \langle E_x H_y^* - E_y H_x^* + E_x^* H_y - E_y^* H_x \rangle \ . \end{aligned} \tag{3.43}$$

Like (3.42), (3.43) are given by single integrals from 0 to ∞ with respect to a certain variable p, the distance h of the observation point from the boundary $z = 0$ entering into the integrals only through the exponential function $\exp[-(q+q^*)h]$, where $q = \sqrt{p^2 - k^2}$ and k is the wave number in a transparent medium ($k = k_0 \sqrt{\varepsilon\mu}$, $k_0 = \omega/c$). Thereby, the integrals with respect to p, expressing (3.43), decompose into two essentially different summands:

(1) Integrals over the interval of p from 0 to k. Here q is purely imaginary, $q + q^* = 0$, and this part of $\mathscr{S}_{\omega z}$ is *independent of* h in (3.43). This is the *wave* part of the fluctuational field.

(2) Integrals over the interval of p from k to ∞, where q is real and positive. The appropriate contributions to $u_{e\omega}(h)$ and $u_{m\omega}(h)$ fall off with distance from $z = 0$, and do not enter at all into the energy flux $\mathscr{S}_{\omega z}$, since the integrand also contains the factor $q - q^*$ which vanishes at real q. This is the *quasi-stationary* fluctuational field which makes no contribution to the energy flux.

As to the *wave* field, if for the integration variable we substitute the angle θ between the wave vector $\boldsymbol{k}$ and the plane $z = 0$ for p, so that $p = k \sin\theta$, (3.43) will become

$$u_{e\omega}^{w} = u_{m\omega}^{w} = \frac{u_\omega^w}{2} = \frac{u_{0\omega} n^3}{4} \int_0^{\pi/2} [1 - \mathscr{R}(\theta)] \sin\theta \, d\theta \ , \tag{3.44}$$

$$\mathscr{S}_{\omega z}^{w} = 2\pi \mathscr{T}_{0\omega} n^2 \int_0^{\pi/2} [1 - \mathscr{R}(\theta)] \cos\theta \, \sin\theta \, d\theta \ . \tag{3.45}$$

Here $n = \sqrt{\varepsilon\mu}$ is the refractive index for a transparent medium, $u_{0\omega}$ and

---

[10] A detailed treatment is given in [Ref. 3.6, Sect. 6].

$\mathcal{T}_{0\omega} = cu_{0\omega}/4\pi$ are the energy density and the intensity of equilibrium radiation in a *vacuum*, respectively; $\mathcal{R}(\theta)$ is the half-sum of the Fresnel reflection coefficients (in energy) for a plane wave incident at an angle θ with two polarizations (the electric vector parallel and perpendicular to the incidence plane)

$$\mathcal{R}(\theta) = \tfrac{1}{2}[\mathcal{R}_{\|}(\theta) + \mathcal{R}_{\perp}(\theta)] \ ,$$

$$\mathcal{R}_{\|}(\theta) = \left| \frac{\dfrac{\varepsilon_1}{\varepsilon}\cos\theta - \sqrt{N^2 - \sin^2\theta}}{\dfrac{\varepsilon_1}{\varepsilon}\cos\theta + \sqrt{N^2 - \sin^2\theta}} \right|^2 ,$$

$$\mathcal{R}_{\perp}(\theta) = \left| \frac{\dfrac{\mu_1}{\mu}\cos\theta - \sqrt{N^2 - \sin^2\theta}}{\dfrac{\mu_1}{\mu}\cos\theta + \sqrt{N^2 - \sin^2\theta}} \right|^2 , \tag{3.46}$$

where $N = \sqrt{\varepsilon_1\mu_1/\varepsilon\mu}$ is the relative refractive index. Equations (3.46) include the half-sum of $\mathcal{R}_{\|}$ and $\mathcal{R}_{\perp}$, because in this problem both polarizations are completely equivalent.

A comparison of (3.45) with the general expression for the normal component of the Poynting vector $\mathcal{S}_{\omega z}$ in terms of the intensity $\mathcal{T}_{\omega}(\theta, \varphi)$

$$\mathcal{S}_{\omega z} = \int_{\theta \leqslant \pi/2} \mathcal{T}_{\omega}(\theta, \varphi)\cos\theta \, do = \int_0^{2\pi} d\varphi \int_0^{\pi/2} \mathcal{T}_{\omega}(\theta, \varphi)\cos\theta \sin\theta \, d\theta$$

indicates that here, i.e., in the case of the radiation of a half-space filled with a homogeneous, isotropic, absorbing medium, $\mathcal{T}_{\omega}$ is independent of φ and is (*Kirchhoff's law*)

$$\mathcal{T}_{\omega}(\theta) = \mathcal{T}_{0\omega} n^2 [1 - \mathcal{R}(\theta)] \ . \tag{3.47}$$

By simply putting $\mathcal{R} = 0$, we can go over to the case of an infinite space filled with a transparent medium with refractive index n. We will arrive at the Clausius law relating the equilibrium intensities in a transparent medium to those in a vacuum

$$\mathcal{T}_{\omega} = \mathcal{T}_{0\omega} n^2 \ .$$

By (3.44), the same transition to $\mathcal{R} = 0$ gives

$$u_{\omega}^{\mathrm{w}} = \tfrac{1}{2} u_{0\omega} n^3 \ ,$$

i.e., *half* the energy density $u_{0\omega} n^3$ of the equilibrium radiation in a transparent medium, as is to be expected since we separated out of this radiation the *unidirectional* energy flux (in the direction $z > 0$).

For the *quasi-stationary* field, quadratures expressing the energy density $u_{e\omega}^{\mathrm{q.st}}$ and $u_{m\omega}^{\mathrm{q.st}}$ are more complicated and permit only of approximate estimates.

If kh is much larger than

$$\left|\frac{\varepsilon}{\varepsilon_1}\sqrt{N^2-1}\right| \quad \text{and} \quad \left|\frac{\mu}{\mu_1}\sqrt{N^2-1}\right|, \quad \text{then}$$

$$u_{e\omega}^{q.st} \approx u_{m\omega}^{q.st} \approx \frac{u_{\omega}^{q.st}}{2} \approx \frac{u_{0\omega}n}{8(kh)^2}\operatorname{Re}\left\{\frac{\varepsilon_1\mu+\mu_1\varepsilon}{\sqrt{N^2-1}}\right\},$$

i.e., the energy densities fall off with distance from the boundary (with increasing h), obeying the law $1/h^2$. At distances h, for which

$$(kh)^2 \gtrsim \frac{1}{4n^2}\operatorname{Re}\left\{\frac{\varepsilon_1\mu+\mu_1\varepsilon}{\sqrt{N^2-1}}\right\}, \tag{3.48}$$

the energy density of the quasi-stationary part of the thermal field is negligible as compared with the energy density of the wave part of the field (3.44), which is constant for $z > 0$. On the contrary, at small h the quasi-stationary field predominates, so that small gaps and cavities, for which (3.48) is not met, are filled precisely with quasi-stationary fluctuational fields.

For small kh, keeping terms up to order $1/(kh)^3$, we get

$$\begin{aligned} u_{e\omega}^{q.st.} &\approx \frac{u_{0\omega}n^3\varepsilon}{4(kh)^2}\left(\frac{\varepsilon_1''}{|\varepsilon+\varepsilon_1|^2 kh}+\operatorname{Re}\left\{\frac{\varepsilon_1\sqrt{N^2-1}}{(\varepsilon+\varepsilon_1)^2}\right\}\right), \\ u_{m\omega}^{q.st.} &\approx \frac{u_{0\omega}n^3\mu}{4(kh)^2}\left(\frac{\mu_1''}{|\mu+\mu_1|^2 kh}+\operatorname{Re}\left\{\frac{\mu_1\sqrt{N^2-1}}{(\mu+\mu_1)^2}\right\}\right). \end{aligned} \tag{3.49}$$

This suggests that for electric losses alone ($\mu_1'' = \operatorname{Im}\{\mu_1\} = 0$), the electric energy density, $u_{e\omega}^{q.st.}$; grows with decreasing h as $1/h^3$, and the magnetic energy density, $u_{m\omega}^{q.st.}$; grows only as $1/h^2$ (however, if $\varepsilon_1'' = \operatorname{Im}\{\varepsilon_1\} = 0$, the converse is true). The integral of $u_{\omega}^{q.st.} = u_{e\omega}^{q.st.} + u_{m\omega}^{q.st.}$ over any finite volume adjacent to the boundary also diverges. As was said in Sect. 3.4, this is the result of the assumed delta-correlation of external (Langevin) sources in an absorbing medium. If we took a nonlocal medium in which the external fluctuational currents had a nonzero correlation radius a, then the density $u_{\omega}^{q.st.}$ would remain finite as $h \to 0$, reaching values of order $u_{0\omega}/(ka)^3$.

Present-day microwave amplifiers enable a direct experiment to be carried out to demonstrate the strong growth of quasi-stationary fluctuational fields near the surface of a heated conducting body, but unfortunately such an experiment has not yet been performed. Incidentally, in a somewhat different problem concerned with heat transfer through a thin gap between two differently heated bodies, the influence of the quasi-stationary thermal field has already been studied theoretically and, in a way, supported experimentally [3.7].

In the above example we obtained the *classical* Kirchhoff law (3.47) for the wave part of the field, since for the infinite half-space it holds for any wavelength. True, at sufficiently small distances h from the surface of a medium, some effects have been observed that lie beyond the scope of the classical theory of thermal radiation (the presence of the quasi-stationary fluctuational field). But of especial interest are applications of the general theory to problems where the classical theory breaks down, since body sizes are of comparable size to the wavelength. The wave part of the thermal field should then also exhibit a new feature — the dependence of wave field characteristics on the ratio l/λ, where l is the characteristic dimension. (Earlier this was referred to as the distance h from the boundary). To illustrate this statement, we will take as an example the thermal radiation of a uniformly heated ball.

The general technique remains as before: we seek the auxiliary (diffraction) field of dipoles with moments $\boldsymbol{p} = -\boldsymbol{l}_1/\mathrm{i}\omega$ and $\boldsymbol{m} = -\boldsymbol{l}_2/\mathrm{i}\omega$ located at a point, in a transparent medium (real ε and μ) surrounding a sphere containing a medium with complex ε_1 and μ_1. The solution to this boundary problem (with the usual continuity conditions of the tangential components of $\boldsymbol{E}_0$ and $\boldsymbol{H}_0$ at the sphere surface, and the radiation condition at infinity) is readily obtainable as a series in fundamental vector spherical harmonics. The losses Q_0 of the diffraction field within the sphere can then be computed as the energy flux input into the sphere through its surface[11]. Recall that in all of those, essentially nonequilibrium, problems (the radiation of a heated body) it is supposed that a quasi-stationary state, specifically a constant temperature, is somehow maintained inside a body.

We will provide only one result here — the variation of the mean Poynting vector for the fluctuational field of a heated ball with radius a. For a sphere with good electric conductivity and no magnetic losses, Fig. 3.2 illustrates the variation of the power p_ω/p_∞ (with the specific power p_ω that is radiated by a unit surface of the sphere) with the parameter $\alpha = k_0 a = \omega a/c = 2\pi a/\lambda$. As $\lambda/a \to 0$ ($\alpha \to \infty$), the power tends to a constant value p_∞ corresponding to the classical Kirchhoff law. As α increases the specific power of radiation is seen to go up at first and then fall off with weakly pronounced maxima that are somewhat shifted to the right of the values of α corresponding to the natural frequencies ω of electric oscillations of the ball[12]. For a small ($\alpha \sim 1$) well-conducting ball the radiation per unit surface is

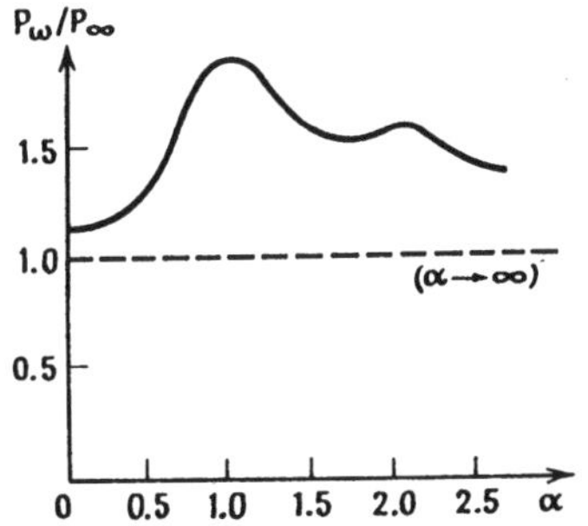

Fig. 3.2. The mean Poynting vector for the thermal field of a hot ball as a function of the dimensionless radius

[11] A detailed treatment is given in [Ref. 3.6, Sect. 7], where this technique is used to solve the problem for the infinite round cylinder as well.

[12] If the ball only had magnetic losses, the maxima would lie at the natural frequencies of magnetic type oscillations. With both types of losses, the picture would be more complicated.

1.5–2 times more intense than for a large one ($\alpha \gg 1$). The variation of p_ω with λ/a is just that new element which is provided by fluctuational electrodynamics.

The total energy flux radiated by the ball at ω is $P_\omega = 4\pi a^2 p_\omega$ and, of course, it is the same for any closed surface surrounding the ball. But the mean energy density u_ω of the fluctuational field behaves differently. At a distance from the ball, u_ω falls off as $1/R^2$. This is the radiation (wave field) density. Near the ball surface u_ω grows fast and at $R = a$ becomes infinite. Clearly, this growth of u_ω is also due to the quasi-stationary part of the fluctuational field.

It will be noticed that the solution of the auxiliary diffraction problem concerned with fields of dipoles $\boldsymbol{p}$ and $\boldsymbol{m}$ is substantially simplified for well-conducting bodies (strong skin-effect). For such bodies it is possible to use the solution to the diffraction problem with so-called *impedance* boundary conditions, i.e., it is possible to ignore the field inside a body. In a first approximation, the diffraction field is calculated for ideally conducting bodies of the same form and configuration as those at hand (owing to the simplicity of the boundary conditions, the case of ideally conducting bodies enjoys greatest favor among diffraction theorists). Thermal losses are then determined by integrating the *surface density* of the Joule heat over the body surface. An element $d\Sigma$ of the body surface absorbs the heat

$$dQ_0 = \frac{c\zeta'}{8\pi} |H_{0t}|^2 \, d\Sigma, \quad \zeta' = \sqrt{\frac{\mu_1 \omega}{8\pi\sigma_1}}, \tag{3.50}$$

where ζ' is the real part of the impedance, and H_{0t} (the tangential component of $\boldsymbol{H}_0$) is borrowed from the solution to the diffraction problem for an ideally conducting body.

Another approximate method of calculating the losses Q_0 of a diffraction field is applied in cases where (1) all the characteristic dimensions of a body (including the curvature radii of its surface) are large as compared to the wavelength λ and the penetration depth of the field inside the body, and (2) the moments of the fluctuational field are only sought away from the body—at distances significantly larger than λ. Accordingly, the point sources of the diffraction field are also placed at such distances. Moreover, the body is assumed to have a shape such that in the geometrical optics approximation, the rays coming out of a point source are subjected only to single reflections at the boundary.

These conditions warrant the assumption that in the immediate vicinity of each body surface element, the field of a remote point source is a plane wave; consequently, the total diffraction field at the surface can be found using the *Fresnel reflection coefficients* (for each of the polarizations, and with the local value of the incidence angle of a primary wave). As it has been assumed that there are no multiple reflections and that the refracted wave is absorbed without reaching other areas of the body surface, the losses Q_0 of the diffraction field are simply the total power of the waves refracted into the body, the integral being taken over the "exposed" part of the surface.

Both approximate techniques of calculating Q_0 (i.e., the use of the strong skin-effect and Fresnel reflection equations) are illustrated by a number of examples in [3.6].

We will now turn to the case where the radiation of a body in free space (for simplicity, in a vacuum) is only of interest to us in the wave zone of the body, i.e., for $R \gg l^2/\lambda$, where l is the body size. Since the dipoles $\boldsymbol{p}$ and $\boldsymbol{m}$ are placed at such a large distance, we may assume that the incoming wave is plane in the region occupied by the body. If the moments of $\boldsymbol{p}$ and $\boldsymbol{m}$ (i.e., the unit vectors $\boldsymbol{l}_1$ and $\boldsymbol{l}_2$) are taken to be mutually orthogonal, forming an orthogonal right-hand screw connection with $\boldsymbol{R}$ (Fig. 3.3), then the fields in the incident wave will only differ in sign

$$\boldsymbol{E}_{0\mathrm{e}} = -\boldsymbol{E}_{0\mathrm{m}} \equiv \boldsymbol{E}_0, \quad \boldsymbol{H}_{0\mathrm{e}} = -\boldsymbol{H}_{0\mathrm{m}} \equiv \boldsymbol{H}_0 \ , \tag{3.51}$$

i.e., the incident wave is linearly polarized (in $\boldsymbol{l}_1$), and its amplitude (at $p = 1/\mathrm{i}\omega$) is $|E_0| = 2\pi/c\lambda R$.

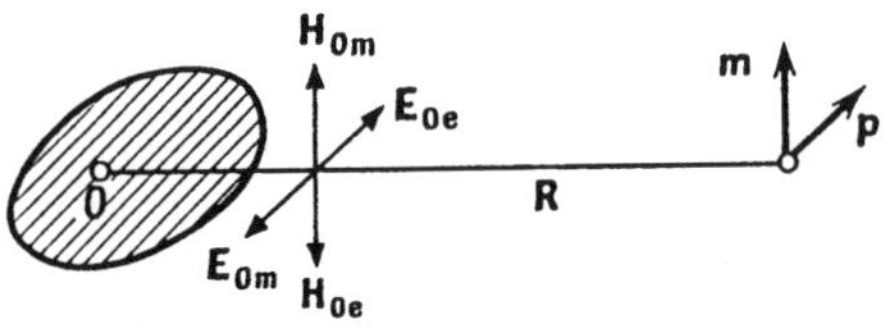

Fig. 3.3. Auxiliary dipole moments $\boldsymbol{m}$ and $\boldsymbol{p}$ that are required to calculate the thermal field of a body

By virtue of (3.51), formulas (3.30, 31) give for the strengths of a thermal field at a point $\boldsymbol{R}$

$$E_1 = H_2 = \int_V (\boldsymbol{E}_0 \cdot \boldsymbol{j}_\mathrm{e} - \boldsymbol{H}_0 \cdot \boldsymbol{j}_\mathrm{m})\, d^3r \ , \tag{3.52}$$

where V is the body volume, the subscripts 1 and 2 denoting the components along $\boldsymbol{l}_1$ and $\boldsymbol{l}_2$. By (3.37–39), we thus obtain

$$\langle |E_1|^2 \rangle = \langle |H_2|^2 \rangle = \langle E_1 H_2^* \rangle = \frac{2}{\pi}\, \Theta(\omega, T) Q_0 \ .$$

Accordingly, the energy flux (the Poynting vector) of the body radiation (with the polarization $\boldsymbol{l}_1$ and in the direction of $\boldsymbol{R}$) will be

$$\mathscr{S}_R = \frac{c}{4\pi} \langle E_1 H_2^* + E_1^* H_2 \rangle = \frac{c}{2\pi} \langle |E_1|^2 \rangle = \frac{c}{\pi^2}\, \Theta(\omega, T) Q_0 \ , \tag{3.53}$$

and the flux within the solid angle do, i.e., through the area $R^2\, do$, at the distance R from the body is

$$dP_\omega = \mathscr{S}_R R^2\, do = \frac{c}{\pi^2}\, \Theta(\omega, T)\, Q_0 R^2\, do \ . \tag{3.54}$$

The subscript ω indicates, as usual, that this refers to the spectrum for $\omega \geqslant 0$.

We will now express Q_0 in the body in terms of the so-called *effective absorption cross-section*

$$\sigma_{\text{eff}} = \frac{Q_0}{\dfrac{c}{8\pi}|E_0|^2} .$$

Thus, σ_{eff} is the area of the front of an incident plane wave, such that the energy flux through it equals the power Q_0 absorbed by the body (for the polarization $\boldsymbol{l}_1$). Substituting $|E_0|$ into σ_{eff} gives

$$Q_0 = \sigma_{\text{eff}} \frac{\pi}{2c\lambda^2 R^2} .$$

Lastly, inserting this into (3.54) and doubling the result (to account for both independent polarizations) we get

$$dP_\omega = \frac{\Theta(\omega, T)}{\pi\lambda^2} \sigma_{\text{eff}}\, do . \tag{3.55}$$

We have thus obtained the energy flux of the thermal radiation of a body in the far zone into the solid angle do in the direction of $\boldsymbol{R}$ at a frequency ω.

If the body is a plate normal to R and its dimensions are much larger than λ, then $\sigma_{\text{eff}} = A\Sigma$, where Σ is the plate area and A its absorption coefficient. Then

$$dP_\omega = \frac{\Theta(\omega, T)}{\pi\lambda^2} A\Sigma\, do = \frac{\omega^2\Theta(\omega, T)}{4\pi^3 c^2} A\Sigma\, do , \tag{3.56}$$

in agreement with the classical theory of thermal radiation.

3.7 Waveguide Form of Kirchhoff's Law

When electromagnetic signals are transmitted over waveguides, the spectral intensity of thermal "noise", i.e., the power of the thermal radiation transmitted over a waveguide within the spectral interval $(\omega, \omega + d\omega)$ becomes of interest. A thermal field may be produced by the walls of the waveguide itself, antenna arrangements to which the waveguide is connected (slot, horn), reflectors, diaphragms, attenuators, and so on. We will refer to any system of such elements as a *radiator*. Clearly, the fluctuational *wave* field in the waveguide can also be represented as a superposition of *travelling* (subcritical) natural waves of the waveguide, or the so-called natural modes or eigenmodes.

It is not difficult to predict from (3.37–39) that the thermal radiation power sent out into a waveguide by any radiator in the nth mode and at frequency ω will be determined by the *absorption coefficient* of this radiator, when exposed to the nth natural wave of frequency ω. An appropriate relation called the *waveguide form of*

Kirchhoff's law [Ref. 3.1, Sect. 17] is readily obtainable using both the reciprocity theorem, and the principle of detailed balance [Ref. 3.6, Sect. 9]. We will only give the second technique here, which *interalia* does not assume that the reciprocity theorem is satisfied.

It was shown in Sect. II.3.5 that an arbitrary radiator matched to the line (waveguide, coaxial line) for $\omega \geqslant 0$, beams into it, within the frequency range $(\omega, \omega + d\omega)$, the power[13]

$$P_\omega d\omega = \frac{\Theta(\omega, T)}{2\pi} d\omega \ .$$

If the matched radiator only excites one (nth) proper wave, this relationship must be obeyed for this wave as well, i.e., for each mode and frequency for which the radiator is matched we have

$$P_{\omega n} = \frac{\Theta}{2\pi} \ . \tag{3.57}$$

Consider an equilibrium system in which there is an unmatched radiator M (the mismatch, generally speaking, being different on either side of M) between two cross-sections of a regular waveguide, as shown in Fig. 3.4.

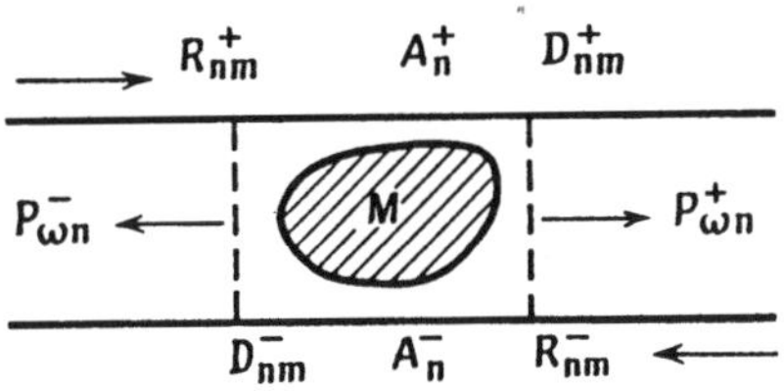

Fig. 3.4. Parameters of the thermal radiation of a body in a waveguide

We will introduce for M absorption coefficients of the nth wave, $A_n^\pm$, and coefficients of its transformation $(n \to m)$ in reflecting from $M(R_{nm}^\pm)$ and in passing through the region occupied by $M(D_{nm}^\pm)$, where the superscripts *plus* and *minus* denote the incidence of the primary nth wave on M from the *left* and *right*, respectively. From the energy conservation law we have

$$\begin{aligned} 1 &= \sum_m{}' (R_{nm}^+ + D_{nm}^+) + A_n^+ \ , \\ 1 &= \sum_m{}' (R_{nm}^- + D_{nm}^-) + A_n^- \ . \end{aligned} \tag{3.58}$$

The prime at the summation sign implies that the summation is only over those m-waves that at frequency ω are travelling (subcritical).

---

[13] Of course, $\Theta(\omega, T)$ is understood here to be the energy of black-body radiation only (Sect. 3.5). In the nonquantum region $\hbar\omega \ll kT$, $\Theta(\omega, T) = kT$ without reservations.

Suppose that on either side of M there are black-body radiators, i.e., bodies that have the same temperature as M and are matched to the waveguide (at a given ω) for each of the travelling waves under consideration. Denote by $P^{\pm}_{\omega n}$ the powers radiated by M at the nth wave to the right and left (Fig. 3.4). The principle of detailed balance requires that the opposite energy fluxes at this wave be equal (on either side of M)

$$\begin{aligned} \frac{\Theta}{2\pi} &= P^{+}_{\omega n} + \frac{\Theta}{2\pi} \sum_m{}' (R^{-}_{mn} + D^{+}_{mn}) \,, \\ \frac{\Theta}{2\pi} &= P^{-}_{\omega n} + \frac{\Theta}{2\pi} \sum_m{}' (R^{+}_{mn} + D^{-}_{mn}) \,. \end{aligned} \tag{3.59}$$

Multiplying (3.58) by $\Theta/2\pi$ and subtracting the results from the appropriate equalities (3.59) gives

$$\begin{aligned} P^{+}_{\omega n} &= \frac{\Theta}{2\pi} \left\{ A^{-}_{n} + \sum_m{}' (R^{-}_{nm} - R^{-}_{mn} + D^{-}_{nm} - D^{+}_{mn}) \right\} , \\ P^{-}_{\omega n} &= \frac{\Theta}{2\pi} \left\{ A^{+}_{n} + \sum_m{}' (R^{+}_{nm} - R^{+}_{mn} + D^{+}_{nm} - D^{-}_{mn}) \right\} , \end{aligned} \tag{3.60}$$

which is the desired result.

If the *reciprocity principle is met*, the matrices R and D feature the symmetry

$$R^{\pm}_{nm} = R^{\mp}_{nm}, \quad D^{\pm}_{nm} = D^{\mp}_{nm} \,,$$

and we then obtain from (3.60)

$$P^{+}_{\omega n} = \frac{\Theta}{2\pi} A^{-}_{n}, \quad P^{-}_{\omega n} = \frac{\Theta}{2\pi} A^{+}_{n} \,, \tag{3.61}$$

i.e., the power radiated by M to the right (left) is determined by its absorption coefficient for the nth wave striking from the right (left).

In the absence of transformation of wave types, when the matrices R and D are diagonal, we obtain from (3.60)

$$\begin{aligned} P^{+}_{\omega n} &= \frac{\Theta}{2\pi} (A^{-}_{n} + D^{-}_{n} - D^{+}_{n}) \,, \\ P^{-}_{\omega n} &= \frac{\Theta}{2\pi} (A^{+}_{n} + D^{+}_{n} - D^{-}_{n}) \,. \end{aligned} \tag{3.62}$$

If the reciprocity principle is satisfied ($D^{+}_{n} = D^{-}_{n}$) or M is totally opaque ($D^{\pm}_{n} = 0$), we again obtain (3.61).

Note that since $P^{\pm}_{\omega n}$ are nonnegative, it follows from (3.62) that for any body or system M

$$1 - R^{\mp}_{n} = A^{\mp}_{n} + D^{\mp}_{n} \geqslant D^{\pm}_{n} \,,$$

i.e., the proportion of energy absorbed and transmitted by M is never smaller than the share of energy transmitted through M when its opposite side is exposed.

If material equations for bodies constituting M are local, then we can go over to the quasi-equilibrium case, taking the temperature T to be a function of co-ordinates, and can make use of the mean $\Theta(\omega, T)$ weighted by the local value of the Joule heat losses inside the body

$$\bar{\Theta} = \frac{1}{Q_{0n}(\omega)} \int \Theta(\omega, T)\, dQ_{0n} \,,$$

then (3.61) will be written in a more general form

$$P_{\omega n}^{\pm} = \frac{\bar{\Theta}}{2\pi} A_n^{\mp}(\omega) \,. \tag{3.63}$$

Clearly, $A_n^{\pm}(\omega)$ are determined by the entire structure of the field produced in M when it is struck by the nth wave, and not just by its absorbing elements (see problem 4).

According to (3.61), a body satisfying the reciprocity principle radiates in a waveguide the following *total* powers in the frequency interval $d\omega$:

$$P_{\omega}^{\pm}\, d\omega = \frac{\Theta}{2\pi} \sum_n{}' A_n^{\mp}(\omega)\, d\omega \,. \tag{3.64}$$

The summation is over all the subcritical waves, whose number varies at different values of ω. Figure 3.5 illustrates (3.64) for the unidirectional radiation of a well-conducting transverse partition in a rectangular waveguide. Power P_ω (on an arbitrary scale) is shown as a function of $\xi = 2a/\lambda$, where a is the shorter side of the rectangular cross-section. With increasing ξ (and hence ω), an ever increasing number of proper E- and H-waves become travelling waves. The partition having only electric, and no magnetic losses, the absorption coefficients of the E- and H-waves behave differently beyond the critical frequencies: for H-waves these increase smoothly (from zero at a critical frequency), and for E-waves they form sharp peaks just after the critical frequency. It is these peaks that are responsible for

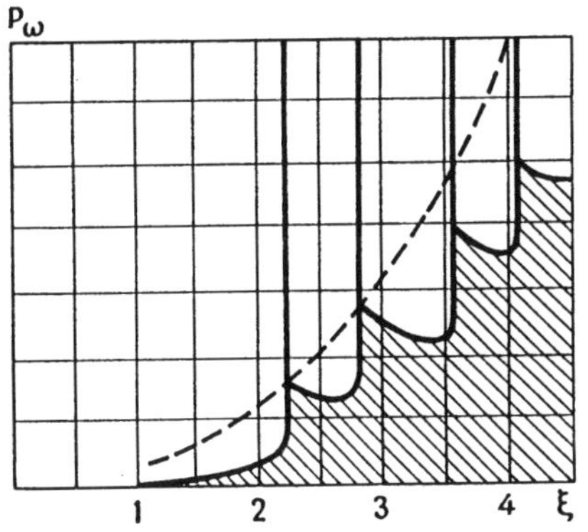

Fig. 3.5. Variation of P_ω with ξ in a rectangular waveguide

the strong spikes of P_ω that, on the scale adopted, reach far beyond the confines of the figure. The dashed lines in the figure show the variation of $P_\omega(\xi)$ that would be obtained in extrapolating the classical Kirchhoff law to the region of small ξ. Clearly, this asymptotic law is inapplicable.

It is to be noted that for any number of waves propagating in any *one direction* (e.g., radially in the case of a ball) the power radiated can be represented in the form (3.64), i.e., as a sum over mutually orthogonal modes. Consequently, in calling the equations of the form (3.64) the *waveguide* form of Kirchhoff's law, we restrict their applicability somewhat. This is, however, justified since waveguides open up real possibilities of separating waves of individual types (E and H) and numbers, whereas with bodies radiating into free space, only the sum that enters in (3.64) is of interest. This sum is infinite (as there are no critical frequencies).

The waveguide form of Kirchhoff's law holds not only for the infinite regular waveguide, but also for a radiator (system of bodies) M located at the end of a *semi-infinite* waveguide. Consider, for instance, a waveguide to whose end a horn is connected, before and/or inside which any bodies M_α ($\alpha = 1, 2, \ldots$) with temperatures T_α are located (Fig. 3.6). By radiator M we understand here all the system of bodies M_α confined within the closed surface bounding the left half-space and sufficiently remote from the waveguide end, i.e., crossing it where any number of one-dimensional mutually orthogonal waves of the regular waveguide already exist (section Σ in Fig. 3.6). The power of the thermal radiation sent out by M into the waveguide at the nth natural wave will be written in the same form (3.63)

$$P^+_{\omega n} = \frac{1}{2\pi}\sum_\alpha \Theta_\alpha A^-_{\alpha n}(\omega) \,, \tag{3.65}$$

where the summation covers all the bodies M_α (including, of course, the horn if it has some heat losses).

It will be recalled that $A^-_{\alpha n}(\omega)$ is the proportion of the power of the nth wave incident from the right through Σ that is absorbed by M_α, i.e., when the apparatus is in the *transmit* condition. In that case, some of the power is radiated into the surrounding space. A straightforward exercise, however, shows that this proportion can also be included into (3.65). In fact, suppose that the medium filling the space bounded by a closed surface and enveloping the bodies M_α features an *arbitrarily small absorption*. If, under these conditions, all the power radiated in the transmit operation is eventually absorbed by the medium, the latter may be

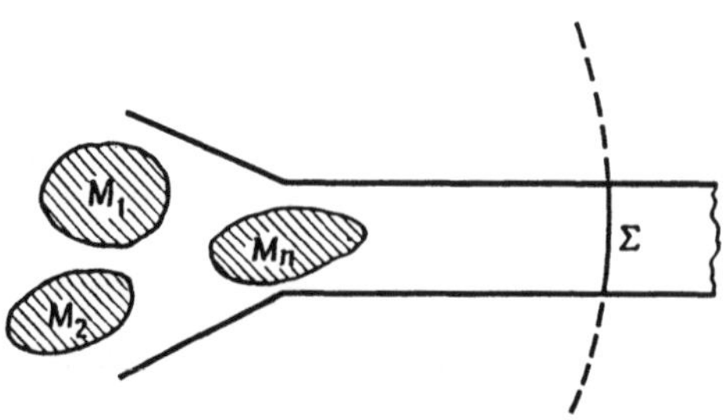

Fig. 3.6. Hot bodies radiating into a waveguide

regarded as just *one of* M_α. One of the summands in (3.65) will thus refer to the *medium*, and its A_n^- will play the role of the so-called *energy radiation coefficient*, i.e., the proportion of the power of the nth wave coming from the right *radiated* by the system in question into a *nonabsorbing* medium.

3.8 Thermal Radiation and Antennas

There are many types of antennas. For example, there are *thin* (wire) antennas widely used in broadcasting and communications at short and longer waves, reflector antennas used in radio astronomy and communications at ultra-short and even shorter waves, and so forth. The material of any antenna is subject to thermal fluctuations responsible for the antenna's thermal self-radiation. Of interest in many cases are not only self-noise and self-radiation, but also the action of the fluctuational fields of external heated bodies and media on the antenna (and hence on subsequent channels). We will concentrate our attention on this *induced* thermal noise in *thin* antennas.[14]

An external fluctuational field induces in antennas currents that are of special interest because it is generally these that are the directly measured quantities. In principle, if we know the covariance of incident thermal radiation, the moments of the induced current are obtainable using the appropriate relationships of the theory of antenna excitation. These calculations, however, involve difficulties that are primarily due to the fact that a fluctuational field has no delta-correlation along the antenna wires. These difficulties can to a large extent be circumvented, if we again make use of Green's function and the reciprocity theorem.

In the theory of thin antennas it is more convenient to deal not with external currents, but with external fields and e.m.f.'s. It is therefore advisable to modify the reciprocity theorem appropriately. We write the "electric part" of the theorem (3.28)

$$\int \boldsymbol{E}\cdot\boldsymbol{j}_{0\mathrm{e}}\,d^3r = \int \boldsymbol{E}_0\cdot\boldsymbol{j}_\mathrm{e}\,d^3r \tag{3.66}$$

and replace the external currents $\boldsymbol{j}_\mathrm{e}$ and $\boldsymbol{j}_{0\mathrm{e}}$—the *sources* of the fluctuational ($\boldsymbol{E}$) and auxiliary diffraction ($\boldsymbol{E}_0$) fields—by the *external fields* $\boldsymbol{K}$ and $\boldsymbol{K}_0$, respectively. Since in the spectral representation $\boldsymbol{j}_\mathrm{e} = -\mathrm{i}\omega\mathfrak{D}/4\pi$, and $\mathfrak{D} = \varepsilon\boldsymbol{K}$, where $\varepsilon = \varepsilon' + \mathrm{i}4\pi\sigma/\omega$ (ε' is the real permittivity and σ the conductivity), we have

$$\boldsymbol{j}_\mathrm{e} = -\frac{\mathrm{i}\omega\varepsilon}{4\pi}\boldsymbol{K}, \quad \boldsymbol{j}_{0\mathrm{e}} = -\frac{\mathrm{i}\omega\varepsilon}{4\pi}\boldsymbol{K}_0\ , \tag{3.67}$$

where the second relation—for the diffraction current $\boldsymbol{j}_{0\mathrm{e}}$—is clearly analogous to

[14] A theory that enables the thermal self-radiation to be calculated for reflector antennas is presented in [Ref. 3.6, Sects. 10 and 11], and for thin antennas in [Ref. 3.6, Sect. 12]. Section 13 of this reference is devoted to induced thermal noise in antennas and gives a more detailed discussion than in the rest of this section.

the first. If we designate the current densities *induced* in the antenna by $\boldsymbol{E}$ and $\boldsymbol{E}_0$ as $\boldsymbol{j}$ and $\boldsymbol{j}_0$, respectively, then

$$\boldsymbol{j} = -\frac{\mathrm{i}\omega\varepsilon}{4\pi}\boldsymbol{E}, \quad \boldsymbol{j}_0 = -\frac{\mathrm{i}\omega\varepsilon}{4\pi}\boldsymbol{E}_0 \ . \tag{3.68}$$

It follows from (3.67, 68) that $\boldsymbol{E}\boldsymbol{j}_{0\mathrm{e}} = \boldsymbol{K}_0\boldsymbol{j}$ and $\boldsymbol{E}_0\boldsymbol{j}_\mathrm{e} = \boldsymbol{K}\boldsymbol{j}_0$, so that the theorem (3.66) takes the form[15]

$$\int \boldsymbol{K}_0(\boldsymbol{r}')\cdot\boldsymbol{j}(\boldsymbol{r}')\,d^3r' = \int \boldsymbol{K}(\boldsymbol{r})\cdot\boldsymbol{j}_0(\boldsymbol{r})\,d^3r \ . \tag{3.69}$$

This form of the reciprocity theorem—with external fields and induced currents—is most often used in the case of thin conductors (including the theory of wire antennas), and also in the quasi-stationary regions (including the theory of networks with lumped parameters).

In order to make use of (3.69), we require a knowledge of the covariance of the components of the external fluctuational field $\boldsymbol{K}(\boldsymbol{r})$. This is easily derived from the first relation (3.67) and the first formula of the FDT (3.26), which confines the problem to the case of the isotropic material of wires, when $\varepsilon_{\alpha\beta} = \varepsilon\delta_{\alpha\beta}$,

$$\begin{aligned}\langle K_\alpha(\boldsymbol{r})K_\beta^*(\boldsymbol{r}')\rangle &= \frac{16\pi^2}{\omega^2|\varepsilon|^2}\langle j_{\mathrm{e}\alpha}(\boldsymbol{r})j_{\mathrm{e}\beta}^*(\boldsymbol{r}')\rangle \\ &= -\frac{2\mathrm{i}}{\omega}\Theta(\omega,T)\left(\frac{1}{\varepsilon^*}-\frac{1}{\varepsilon}\right)\delta_{\alpha\beta}\delta(\boldsymbol{r}-\boldsymbol{r}') \ .\end{aligned} \tag{3.70}$$

Furthermore, if, as in this case, we have metal wires, then we can normally neglect the displacement currents inside the wires and consider that $\varepsilon \approx \mathrm{i}4\pi\sigma/\omega$. Then

$$\langle K_\alpha(\boldsymbol{r})K_\beta^*(\boldsymbol{r}')\rangle = \frac{\Theta(\omega,T)}{\pi\sigma}\delta_{\alpha\beta}\delta(\boldsymbol{r}-\boldsymbol{r}') \ . \tag{3.71}$$

Now we would like to find the induced fluctuational current I through a section of a wire in the antenna array. Generally speaking, the cross-section Σ of this thin (quasi-linear, in the language of electrodynamics) conductor may vary over the length s of the wire, but is assumed that all its linear dimensions are small as compared to the wavelength λ in the surrounding nonconducting medium. Besides the coordinate s along the wire axis, we introduce the radius vector $\boldsymbol{\varrho}$ in the wire cross-section. Thus, $d^3r' = d^2\varrho'\,ds'$.

In order to be able to express $I(s)$ in terms of $\boldsymbol{K}$, we assume that the auxiliary external field $\boldsymbol{K}_0(\boldsymbol{r}')$ only acts in the wire cross-section with the coordinate $s' = s$, is constant over the entire cross-section and directed along the wire axis (unit vector $\boldsymbol{s}$)

$$\boldsymbol{K}_0(\boldsymbol{\varrho}', s') = \mathscr{E}_0\,\boldsymbol{s}\,\delta(s'-s) \ . \tag{3.72}$$

[15] In the following it is more convenient to denote the integration variable on the left-hand side differently, which is why we have introduced $\boldsymbol{r}'$.

Clearly, $\mathscr{E}_0$ is the integral external e.m.f. applied across the section at s

$$\int \boldsymbol{K}_0 \cdot \boldsymbol{s}\, ds' = \mathscr{E}_0 \ .$$

Since $\boldsymbol{K}_0$ has the form (3.72), the integral on the left-hand side of (3.69) will be

$$\int \boldsymbol{K}_0(\boldsymbol{\varrho}', s')\cdot \boldsymbol{j}(\boldsymbol{\varrho}', s')\, d^2\varrho'\, ds'$$

$$= \mathscr{E}_0 \int_{\Sigma} d^2\varrho' \int j_s(\boldsymbol{\varrho}', s')\,\delta(s'-s)\, ds'$$

$$= \mathscr{E}_0 \int_{\Sigma} d^2\varrho' j_s(\boldsymbol{\varrho}', s) = \mathscr{E}_0 I(s) \ ,$$

and the reciprocity theorem (3.69) gives

$$\mathscr{E}_0 I(s) = \int \boldsymbol{K}(\boldsymbol{r})\cdot \boldsymbol{j}_0(\boldsymbol{r})\, d^3r \ . \tag{3.73}$$

Using this formula, we easily obtain the mean square modulus of the fluctuational current $I(s)$. Multiplying (3.73) by the complex conjugate equality and averaging, we get

$$|\mathscr{E}_0|^2 \langle |I(s)|^2 \rangle = \int\int \langle K_\alpha(\boldsymbol{r}) K_\beta^*(\boldsymbol{r}') \rangle j_{0\alpha}(\boldsymbol{r}) j_{0\beta}^*(\boldsymbol{r}')\, d^3r\, d^3r' \ .$$

Now we substitute the covariance (3.71) of the external thermal field $\boldsymbol{K}$ to find

$$|\mathscr{E}_0|^2 \langle |I(s)|^2 \rangle = \frac{2}{\pi} \int \Theta(\omega, T)\, dQ_0 \ , \quad \text{where} \tag{3.74}$$

$$dQ_0 = \frac{\sigma}{2} |\boldsymbol{E}_0|^2 d^3r = \frac{|j_0|^2}{2\sigma} d^3r \tag{3.75}$$

is the Joule losses due to the auxiliary e.m.f. $\mathscr{E}_0$ concentrated in the cross-section s. It is quite obvious that (3.74) is the modified form of generalized Kirchhoff's law (3.40). It enables us to do without the inhomogeneous boundary problem of fluctuational field $\boldsymbol{E}$ in the antenna material but reduces finding $\langle |I(s)|^2 \rangle$ to the calculation of quadrature, if the solution to the problem of the antenna excited by the auxiliary field $\boldsymbol{E}_0$ in the form of a delta-function is known, i.e., the problem of the field excited by the e.m.f. $\mathscr{E}_0$ applied across the appropriate cross-section. If $Z_{\text{in}}(s)$ is the input impedance of the antenna relative to the e.m.f. connected across s, then $\mathscr{E}_0 = Z_{\text{in}}(s) I_0(s)$, where

$$I_0(s) = \int j_{0s}(\boldsymbol{\varrho}, s)\, d^2\varrho$$

is the total "diffraction" current through this section. In an analogous way we can introduce the equivalent thermal e.m.f. $\hat{\mathscr{E}}(s)$ across s, defining it in terms of the impedance $Z_{\text{in}}(S)$ and the total fluctuational current

$$\hat{\mathscr{E}}(s) = Z_{\text{in}}(s) I(s) \ . \tag{3.76}$$

Thus, $\hat{\mathscr{E}}(s) = \mathscr{E}_0 I(s)/I_0(s)$, i.e., by (3.73),

$$\hat{\mathscr{E}}(s) = \frac{1}{I_0(s)} \int \boldsymbol{K}(\boldsymbol{r}) \cdot \boldsymbol{j}_0(\boldsymbol{r}) d^3 r \ . \tag{3.77}$$

This is the equivalent thermal e.m.f. at s in the general case where the external field $\boldsymbol{K}(\boldsymbol{r})$ produced by the former is arbitrarily distributed over the volume of the antenna wires. By virtue of (3.74), the mean square of the modulus of $\hat{\mathscr{E}}(s)$ is

$$\langle |\hat{\mathscr{E}}(s)|^2 \rangle = \frac{2}{\pi |I_0(s)|^2} \int \Theta(\omega, T) dQ_0 \ . \tag{3.78}$$

It should be stressed that $\hat{\mathscr{E}}(s)$ is not a fluctuational e.m.f. distributed along the wire. Equation (3.76) introduces (in essence) for each s, an e.m.f. $\hat{\mathscr{E}}(s)$ that basically provides an adequate value of the fluctuational field through this section.

In the case of the *quasi-stationary* network, however, where in *each branch* the total current I and input impedance of the network Z_{in} are *independent* of s, the e.m.f. $\hat{\mathscr{E}}$ is also *independent* of s, i.e., it can be included into *any* cross-section of the branch. If, in addition, the network has everywhere the same temperature ($\Theta = \text{const}$), then $\hat{\mathscr{E}}$ can be identified with the *local Nyquist* e.m.f. e of the two-terminal network that emerges when the branch under consideration is disconnected. Under these conditions (quasi-stationary and equilibrium), (3.78) becomes the Nyquist equation

$$\langle |\tilde{e}(\omega)|^2 \rangle = \frac{2}{\pi} \Theta \frac{Q_0}{|I_0|^2} = \frac{1}{\pi} \Theta R_{\text{en}} \ , \tag{3.79}$$

where R_{en} is the energy-related resistance of the two-terminal network

$$Q_0 = \tfrac{1}{2} R_{\text{en}} |I_0|^2 \ .$$

Let us return to the antenna formula (3.78), which we illustrate using two examples:

1) Antenna in an Equilibrium Radiation Field. Consider an antenna in a free space filled with a transparent medium with constant temperature throughout ($\Theta = \text{const}$).

Across a section s of the antenna, an auxiliary e.m.f. $\mathscr{E}_0$ is connected that produces a diffraction field of the antenna that is by no means localized only in the antenna wires or their immediate vicinity. On the contrary, the field also contains the waves beamed by the antenna, so that the Joulean losses Q_0 occur both in the antenna material, and in any conducting bodies that happen to be in the path of the radiated waves. We will assume for the moment that there are no such bodies (free space), and, accordingly, the power given off by $\mathscr{E}_0$ goes only into the heating of the antenna itself and into the radiation, the latter accounting for most of the power in an antenna in normal operation.

Assuming that the medium has an infinitesimal conductivity and neglecting any thermal losses in the antenna itself, we can consider that the total losses Q_0 of the diffraction field are simply equal to the *radiation power* P_0. The latter is usually represented in the form $P_0 = R_\Sigma(s)\,|I_0(s)|^2/2$, where R_Σ is the so-called *radiation resistance* of the antenna, which clearly depends on the section s across which the e.m.f. $\mathscr{E}_0$ is connected. Thus

$$Q_0 = \tfrac{1}{2}R_\Sigma(s)|I_0(s)|^2 \ . \tag{3.80}$$

At $\Theta = \text{const}$, (3.78) has the form $\langle|\hat{\mathscr{E}}(s)|^2\rangle = 2\Theta Q_0/\pi|I_0(s)|^2$. Substituting (3.80) gives

$$\langle|\hat{\mathscr{E}}(s)|^2\rangle = \frac{1}{\pi}\Theta R_\Sigma(s) \ ,$$

but the spectral density of the fluctuational e.m.f. for positive frequencies is twice as great

$$\hat{\mathscr{E}}^2_\omega(s) = 2\langle|\hat{\mathscr{E}}(s)|^2\rangle = \frac{2}{\pi}\Theta R_\Sigma(s) \ . \tag{3.81}$$

This treatment suggests that this result is valid for any antenna of whatever complexity in free space, if we only refer R_Σ and $\hat{\mathscr{E}}^2_\omega$ to the same antenna section.

2) Thermal Noise Induced by Remote Bodies. If a body is at a sufficiently large distance R from the antenna, or rather within its Fraunhofer zone, and the body size is also small as compared to the Fresnel zone, then in the spatial region occupied by the body we can regard the antenna field as a superposition of a plane wave and a secondary field diffracted by the body. If, as in Sect. 3.6, we denote the effective absorption cross-section of the body for an appropriately polarized incident wave by σ_{eff}, we can express the power absorbed by the body in the form $Q_0 = \sigma_{\text{eff}}\mathscr{S}_0$, where $\mathscr{S}_0$ is the Poynting vector of the primary wave. The relationship of $\mathscr{S}_0$ to the total power $P_0 = R_\Sigma|I_0|^2/2$ radiated by the antenna is given by the well-known equation

$$\mathscr{S}_0 = P_0\frac{G}{4\pi R^2} \ ,$$

where G is a function of direction, called the *reduced directive gain* of the antenna, which describes the angular distribution of the radiated power P_0 (the mean of G over the unit sphere is $\oint G\,do/4\pi = 1$). Thus

$$Q_0 = \sigma_{\text{eff}}\mathscr{S}_0 = P_0\frac{G\sigma_{\text{eff}}}{4\pi R^2} = \frac{R_\Sigma|I_0|^2 G\sigma_{\text{eff}}}{8\pi R^2} \ .$$

According to (3.78), the mean square of the equivalent thermal e.m.f. caused at s by the body in question is (for $\omega > 0$)

$$\hat{\mathscr{E}}^2_\omega(s) = 2\langle|\hat{\mathscr{E}}(s)|^2\rangle = \frac{\Theta R_\Sigma(s)G\sigma_{\text{eff}}}{2\pi^2 R^2} \ . \tag{3.82}$$

We assume here that the body has the same temperature throughout ($\Theta = \text{const}$). If there are several bodies with different temperatures and separations from the antenna in the antenna radiation field, then

$$\hat{\mathscr{E}}^2_\omega(s) = \frac{R_\Sigma(s)}{2\pi^2}\sum_\alpha \frac{\Theta_\alpha G_\alpha \sigma_{\text{eff}\alpha}}{R_\alpha^2} \ . \tag{3.83}$$

For completely black bodies, the effective cross-section coincides with the geometric one ($\sigma_{\text{eff}} = \sigma$; the linear dimensions σ are assumed to be very much larger than the wavelength λ). Accordingly, $\sigma_{\text{eff}}/R^2 = \sigma/R^2 = o$, where o is the solid angle from which the body is seen at the antenna location. Equation (3.83) then becomes

$$\hat{\mathscr{E}}^2_\omega(s) = \frac{|2R_\Sigma(s)}{\pi}\sum_\alpha \Theta_\alpha G_\alpha \frac{o_\alpha}{4\pi} \ ,$$

or even

$$\hat{\mathscr{E}}^2_\omega(s) = \frac{2R_\Sigma(s)}{\pi}\int \Theta G \frac{do}{4\pi}$$

(provided that $do \gg \lambda^2/R^2$). If the temperatures of all the bodies are the same, we will get

$$\hat{\mathscr{E}}^2_\omega(s) = \frac{2}{\pi}\Theta R_\Sigma(s)\frac{1}{4\pi}\int G\, do \ .$$

If the periphery is filled with black bodies having the same temperatures (black envelope), we return to (3.81).

3.9 Equilibrium Thermal Fields. Equilibrium Form of the FDT

Up until now, we have been concerned mainly with *nonequilibrium* problems—the field produced by heated bodies in an environment[16] that was taken to be either transparent (and hence making no contribution to the thermal losses of the auxiliary diffraction field), or so cold that its own fluctuational field could be neglected. In this section we will examine the case where *the temperature is the same throughout the volume of the complete field*, i.e., we will consider the *equilibrium* fluctuational field. If, in particular, we deal with bodies placed in a transparent medium that has no ideally reflecting envelope, we will suppose that, at sufficiently large distances, the field is bounded by a completely absorbing envelope having the same temperature as the heated bodies.

[16] Recall that the heated bodies themsèlves must be assumed to be in a quasi-equilibrium state.

The general expressions (3.37–39) for the spatial covariances of the spectral amplitudes of the fluctuational field caused by uniformly heated bodies can be combined into a single formula, if we denote by A and B any two of the six components of $\boldsymbol{E}$ and $\boldsymbol{H}$. Instead of (3.37–39), we can then write

$$\langle A(\boldsymbol{r}_1)B^*(\boldsymbol{r}_2)\rangle = \pm\frac{2}{\pi}\Theta(\omega,T)\,Q_{0AB^*}\ , \tag{3.84}$$

The positive sign is for the case where both components are either electric or magnetic, the negative where one component is electric and the other magnetic. Q_{0AB^*} is the mixed losses of the diffraction field produced by unit point sources of an appropriate type placed at points $\boldsymbol{r}_1$ and $\boldsymbol{r}_2$.

The Kirchhoff form of FDT (3.84) is, of course, also valid for the equilibrium field, but by Q_{0AB^*} we now understand the mixed losses of the diffraction field *throughout* the space occupied by it. This enables the expression for Q_{0AB^*} to be greatly simplified, using the so-called *complex Lorentz lemma.* This lemma is in a sense analogous to the reciprocity theorem, differing from the latter only in that it relates "crosswise" the field of one system of sources to the *complex conjugate* field of the other system of sources.

As in the derivation of the reciprocity theorem (3.28), we now write two sets of Maxwell's equations (3.23)—one for the field and sources 1, the other for the field and sources 2. We multiply the equations of the first set by—$\boldsymbol{H}_2^*$ and $\boldsymbol{E}_2^*$ respectively, the complex conjugate equations of the second set by—$\boldsymbol{H}_1$ and $\boldsymbol{E}_1$, and add up the results. We thus arrive (with the usual convention $\boldsymbol{E}\times\boldsymbol{B}=[\boldsymbol{E}\boldsymbol{B}]$) at

$$\begin{aligned}-\frac{c}{4\pi}&\operatorname{div}\{[\boldsymbol{E}_1\boldsymbol{H}_2^*]+[\boldsymbol{E}_2^*\boldsymbol{H}_1]\}\\ &=\boldsymbol{j}_{e1}\cdot\boldsymbol{E}_2^*+\boldsymbol{j}_{m1}\cdot\boldsymbol{H}_2^*+\boldsymbol{j}_{e2}^*\cdot\boldsymbol{E}_1+\boldsymbol{j}_{m2}^*\cdot\boldsymbol{H}_1\\ &\quad-\frac{i\omega}{4\pi}(\boldsymbol{D}_1\cdot\boldsymbol{E}_2^*-\boldsymbol{E}_1\cdot\boldsymbol{D}_2^*+\boldsymbol{B}_1\cdot\boldsymbol{H}_2^*-\boldsymbol{H}_1\cdot\boldsymbol{B}_2^*)\ .\end{aligned}$$

In integrating this relation over V of the *total* field, the integral of div vanishes and we obtain the Lorentz lemma

$$\begin{aligned}-\frac{i\omega}{16\pi}&\int_V(\boldsymbol{D}_1\cdot\boldsymbol{E}_2^*-\boldsymbol{E}_1\cdot\boldsymbol{D}_2^*+\boldsymbol{B}_1\cdot\boldsymbol{H}_2^*-\boldsymbol{H}_1\cdot\boldsymbol{B}_2^*)d^3r\\ &=-\frac{1}{4}\int_V(\boldsymbol{j}_{e1}\cdot\boldsymbol{E}_2^*+\boldsymbol{j}_{m1}\cdot\boldsymbol{H}_2^*+\boldsymbol{j}_{e2}^*\cdot\boldsymbol{E}_1+\boldsymbol{j}_{m2}^*\cdot\boldsymbol{H}_1)d^3r\ .\end{aligned} \tag{3.85}$$

Considering that $D_{1\alpha}=\varepsilon_{\alpha\beta}E_{1\beta}$, $D_{2\beta}^*=\varepsilon_{\beta\alpha}^*E_{2\alpha}^*$ (and similarly for $\boldsymbol{B}_1$ and $\boldsymbol{B}_2$), the expression on the left-hand side of (3.85), is nothing other than the mixed thermal losses Q_{12^*} of the fields 1 and 2 throughout the space [see (3.35, 36)], and hence

$$Q_{12^*}=-\tfrac{1}{4}\int_V(\boldsymbol{j}_{e1}\cdot\boldsymbol{E}_2^*+\boldsymbol{j}_{m1}\cdot\boldsymbol{H}_2^*+\boldsymbol{j}_{e2}^*\cdot\boldsymbol{E}_1+\boldsymbol{j}_{m2}^*\cdot\boldsymbol{H}_1)d^3r\ . \tag{3.86}$$

We will now apply this result of the Lorentz lemma to diffractional fields produced by unit point sources.

Let us take the case where A and B in (3.84) are the components of the *electric* strength of the thermal field, $\boldsymbol{E}$. Then

$$\boldsymbol{j}_1 \equiv \boldsymbol{j}_{e1} = \boldsymbol{l}_1 \delta(\boldsymbol{r}-\boldsymbol{r}_1), \quad \boldsymbol{j}_2 \equiv \boldsymbol{j}_{e2} = \boldsymbol{l}_2 \delta(\boldsymbol{r}-\boldsymbol{r}_2), \quad \boldsymbol{j}_{m1} = \boldsymbol{j}_{m2} = 0 \ ,$$

and (3.86) gives

$$Q_{ee^*}(\boldsymbol{r}_1,\boldsymbol{l}_1;\boldsymbol{r}_2,\boldsymbol{l}_2) = \tfrac{1}{4}\{\boldsymbol{l}_1\cdot\boldsymbol{E}^*_{02}(\boldsymbol{r}_1) + \boldsymbol{l}_2\cdot\boldsymbol{E}_{01}(\boldsymbol{r}_2)\} \ . \tag{3.87}$$

But it follows from the conventional reciprocity theorem (3.28) that in this case also

$$\boldsymbol{l}_1\cdot\boldsymbol{E}_{02}(\boldsymbol{r}_1) = \boldsymbol{l}_2\cdot\boldsymbol{E}_{01}(\boldsymbol{r}_2) \ .$$

Using this expression, we eliminate from (3.87) either $\boldsymbol{E}_{01}$ or $\boldsymbol{E}^*_{02}$. We can thus write the losses Q_{ee^*} in the following two forms:

$$Q_{ee^*}(\boldsymbol{r}_1,\boldsymbol{l}_1;\boldsymbol{r}_2,\boldsymbol{l}_2) = -\tfrac{1}{2}\mathrm{Re}\{\boldsymbol{l}_1\cdot\boldsymbol{E}_{02}(\boldsymbol{r}_1)\} = -\tfrac{1}{2}\mathrm{Re}\{\boldsymbol{l}_2\cdot\boldsymbol{E}_{01}(\boldsymbol{r}_2)\} \ .$$

Substituting these expressions into (3.38) gives

$$\begin{aligned}\langle E_{l_1}(\boldsymbol{r}_1)E^*_{l_2}(\boldsymbol{r}_2)\rangle &= -\frac{1}{\pi}\Theta(\omega,T)\mathrm{Re}\{\boldsymbol{l}_1\cdot\boldsymbol{E}_{02}(\boldsymbol{r}_1)\}\\ &= -\frac{1}{\pi}\Theta(\omega,T)\mathrm{Re}\{\boldsymbol{l}_2\cdot\boldsymbol{E}_{01}(\boldsymbol{r}_2)\} \ .\end{aligned} \tag{3.88}$$

For two components of the magnetic strength of the thermal field, $\boldsymbol{H}$, (3.86) yields a similar expression for Q_{mm^*} and, by (3.39),

$$\begin{aligned}\langle H_{l_1}(\boldsymbol{r}_1)H^*_{l_2}(\boldsymbol{r}_2)\rangle &= -\frac{1}{\pi}\Theta(\omega,T)\mathrm{Re}\{\boldsymbol{l}_1\cdot\boldsymbol{H}_{02}(\boldsymbol{r}_1)\}\\ &= -\frac{1}{\pi}\Theta(\omega,T)\mathrm{Re}\{\boldsymbol{l}_2\cdot\boldsymbol{H}_{01}(\boldsymbol{r}_2)\} \ .\end{aligned} \tag{3.89}$$

Lastly, when A is a component of $\boldsymbol{E}$ and B a component of $\boldsymbol{H}$, we derive Q_{em^*}. It follows from (3.37) that

$$\begin{aligned}\langle E_{l_1}(\boldsymbol{r}_1)H^*_{l_2}(\boldsymbol{r}_2)\rangle &= -\frac{\mathrm{i}}{\pi}\Theta(\omega,T)\mathrm{Im}\{\boldsymbol{l}_1\cdot\boldsymbol{E}_{02}(\boldsymbol{r}_1)\}\\ &= \frac{\mathrm{i}}{\pi}\Theta(\omega,T)\mathrm{Im}\{\boldsymbol{l}_2\cdot\boldsymbol{H}_{01}(\boldsymbol{r}_2)\} \ ,\end{aligned} \tag{3.90}$$

i.e., the covariances for electric and magnetic strengths are purely imaginary.

Equations (3.88–90) indicate that, up to factor $-\Theta/\pi$, *the covariances of the equilibrium field coincide with the real or imaginary parts of the appropriate Green's functions.* These equations may be termed the *equilibrium form of the* FDT. Once we have found this, we no longer need to calculate the thermal losses of the auxiliary diffraction field. It is sufficient to take the real or imaginary part of the strengths of the field. At the same time, the *linear* relationship between the second moments of the thermal field and the strengths of the diffraction field enables us to use the tools of the theory of analytic functions of complex variables when we seek some integral (in spectrum) effects in an equilibrium field, i.e., when we integrate the covariances of the thermal field with respect to ω.

Given the covariances of the spectral components of a fluctuational field and, specifically, the means of the products of those components at the same point, we are able to find not only the means of the energy densities and fluxes, but other bilinear variables as well. We are, for instance, able to compute the means of the components of the Maxwellian stress tensor, i.e., to find the mechanical (ponderomotive) forces exerted on bodies by an equilibrium field. This applies both to quasi-equilibrium forces, when we have the generalized Kirchhoff law (3.84) at our disposal, and to equilibrium ones, when we can use the equilibrium form of the FDT (3.88, 89). These fascinating physical applications of the above theorems [3.6] are, however, of no direct concern to radiophysics. We will, therefore, confine ourselves to a few remarks.

On the face of it, it might appear unusual that even in the *equilibrium* case, the fluctuational electromagnetic field governs the action of forces on bodies (including ideal conductors). But there is nothing remarkable here simply because, when speaking about "equilibrium", we at all times refer only to *thermal* equilibrium, i.e., a uniform temperature T throughout the system at hand (in particular, $T = 0$). This by no means excludes the action of electrodynamic forces that are either compensated for by *external connections*, or make the bodies shift until *they achieve an arrangement that is in mechanical equilibrium.* It is also not surprising, therefore, that, for example, an equilibrium field in the half-space above an ideally plane mirror exerts on the latter a pressure whose spectral density, for a nondispersive medium, is

$$p_\omega = -\frac{1}{4\pi}\{\varepsilon\langle|E_n|^2\rangle - \mu\langle|H_t|^2\rangle\} = \frac{\Theta k^2}{3\pi^2\omega} = \frac{u_\omega}{3} .$$

As always, the term $\hbar\omega/2$ in $\Theta(\omega, T)$ (i.e., zero oscillations) gives an expression which diverges when we integrate with respect to ω. This part of the pressure is, however, compensated for by the "zero" pressure of the field on the other side of the mirror, where the equilibrium radiation may, generally speaking, have a different temperature. The resultant force on the mirror is determined by the pressure difference on either side, and in this case contains no contribution from the zero oscillations.

This does not suggest that zero oscillations cause no mechanical forces at all. When the mirror is plane, the geometry is symmetric on either side of the mirror. If we take, say, a dihedral angle formed by two ideal mirrors (Fig. 3.7), then the "zero" standing waves within the angle and beyond will have different structures. As a result, even at $T = 0$ the mirrors are subjected to a "closing" integral (in ω) moment. In that case, we should integrate with respect to ω not up to $\omega = \infty$, but only

Fig. 3.7. Dihedral angle formed by two mirrors subject to a "closing" moment

up to a certain frequency $\omega_m \ll \sigma$ (σ is the conductivity; the inequality is the condition that with normal skin-effect, a metal can still be considered to be ideally conducting).

Of course, the fluctuational electromagnetic field gives rise to forces which also act on absorbing bodies. Furthermore, because the surfaces of such bodies are "lined" with a layer of near (quasi-stationary) thermal field, which builds up steeply in the direction of the surface, the effect of mechanical interaction (cohesion forces) becomes especially strong for small gaps between the body surfaces. This was pointed out by *Lifshits* [3.8] (see also [3.12, 13] and [Ref. 3.6, Sect. 18]).

The theory of macroscopic cohesion forces was developed earlier based on the elementary law of Van der Waals forces interacting pairwise between atoms or molecules, thereby *a priori* confining the result to the case of rarefied media. A purely phenomenological theory based on fluctuational electrodynamics removes this constraint. However, starting with an expression for the macroscopic cohesion force, for rarefied media we can draw a contrary conclusion about the law of pairwise interaction of individual neutral atoms and molecules. However paradoxical at first sight, this path appears to be simpler than a direct quantum-mechanical treatment for two neutral particles, in which the interaction law is only derived for higher orders by using the perturbation method.

3.10 Thermal Fields in Gyrotropic Bodies

Many media (e.g., ferrites, plasma) become gyrotropic in a sufficiently strong, constant (in time) external magnetic field $\boldsymbol{B}_0$ [17]. Specifically, this means that for a homogeneous medium in a homogeneous field $\boldsymbol{B}_0$, the proper waves are monochromatic with *circular* (right or left) polarization; only these waves propagate without changing their polarization type. If, say, a linearly polarized wave is passed through the medium, its polarization plane will, generally speaking, turn about the direction of propagation, due to the difference in the phase velocities of waves with right and left circular polarization into which the linearly polarized wave may be decomposed.

[17] By "sufficiently strong" we mean a field whose magnitude B_0 is significantly larger than the strengths of the fluctuational field. Therefore, in linearizing Maxwell's equations in these fluctuational fields, it still makes sense to account for the variation of medium variables with $\boldsymbol{B}_0$.

In an anisotropic medium the permittivity and permeability tensors are symmetrical

$$\varepsilon_{\alpha\beta}(\omega,\boldsymbol{r}) = \varepsilon_{\beta\alpha}(\omega,\boldsymbol{r}) \ ,$$

$$\mu_{\alpha\beta}(\omega,\boldsymbol{r}) = \mu_{\beta\alpha}(\omega,\boldsymbol{r}) \ .$$

In a gyrotropic medium, being functions of $\boldsymbol{B}_0$, they are also symmetrical, but now $\boldsymbol{B}_0$ *changes its sign*

$$\varepsilon_{\alpha\beta}(\omega,\boldsymbol{r},\boldsymbol{B}_0) = \varepsilon_{\beta\alpha}(\omega,\boldsymbol{r},-\boldsymbol{B}_0) \ ,$$

$$\mu_{\alpha\beta}(\omega,\boldsymbol{r},\boldsymbol{B}_0) = \mu_{\beta\alpha}(\omega,\boldsymbol{r},-\boldsymbol{B}_0) \ .$$

It follows that for gyrotropic media, the conventional reciprocity theorem (3.28) does not hold. Instead, an equation is used that relates "crosswise" the sources and strengths of field 1 in a medium exposed to the magnetic field $\boldsymbol{B}_0(\boldsymbol{j}_{e1}, \boldsymbol{j}_{m1}, \boldsymbol{E}_1, \boldsymbol{H}_1)$ to the sources and strengths of field 2 in the same medium, but with the *reversed external magnetic field* $(-\boldsymbol{B}_0)$. We will mark the electromagnetic quantities relating to field 2 in such a "reverse medium" by a tilde $(\tilde{\boldsymbol{j}}_{e2}, \tilde{\boldsymbol{j}}_{m2}, \tilde{\boldsymbol{E}}_2, \tilde{\boldsymbol{H}}_2)$. The gyrotropic medium obeys the following theorem instead of (3.28):

$$\int_V (\tilde{\boldsymbol{E}}_2 \cdot \boldsymbol{j}_{e1} - \tilde{\boldsymbol{H}}_2 \cdot \boldsymbol{j}_{m1}) d^3r = \int_V (\boldsymbol{E}_1 \cdot \tilde{\boldsymbol{j}}_{e2} - \boldsymbol{H}_1 \cdot \tilde{\boldsymbol{j}}_{m2}) d^3r \ . \tag{3.91}$$

Clearly, equations (3.26) for the covariances of *external currents* also hold for the gyrotropic medium, but the generalized Kirchhoff law (3.84) providing the correlation of *fluctuational field strengths* is no longer valid, because it was derived using the reciprocity theorem (3.28).

Using the same treatment, from (3.91) we will arrive at the following result that generalizes (3.84) to gyrotropic media:

$$\langle A(\boldsymbol{r}_1) B^*(\boldsymbol{r}_2) \rangle = \pm \frac{2}{\pi} \Theta(\omega, T) \tilde{Q}_{0AB^*} \ , \tag{3.92}$$

and with the temperature varying sufficiently smoothly in space:

$$\langle A(\boldsymbol{r}_1) B^*(\boldsymbol{r}_2) \rangle = \pm \frac{2}{\pi} \int \Theta(\omega, T, \boldsymbol{r}) d\tilde{Q}_{0AB^*} \ . \tag{3.93}$$

The correlation of the spectral amplitudes of the fluctuational field in the gyrotropic medium exposed to the external magnetic field $\boldsymbol{B}_0(\boldsymbol{r})$ is thus controlled by the thermal losses of the diffraction field in the reverse medium, i.e., the *losses should be calculated with the sign of the external magnetic field changed* $[\boldsymbol{B}_0(\boldsymbol{r}) \to -\boldsymbol{B}_0(\boldsymbol{r})]$[18]. Equations (3.92, 93) are universal in the sense that they

[18] Not only the generalized, but also the classical Kirchhoff law for gyrotropic media remained unknown for a long time. It was not until the late 1950s that the law (3.93) was derived by M. L. Levin and later published in [3.6].

impose constraints neither on the size and shape of bodies, nor on their electric and magnetic properties, including gyrotropy, which also depends on the intensity and configuration of the constant (in time) bias field $\boldsymbol{B}_0$.

Equations (3.92, 93) simplify fluctuational problems for gyrotropic bodies to no lesser degree than do (3.84) for nongyrotropic ones. Of course, the solution of the auxiliary diffraction problem, necessary to work out the thermal losses, generally presents greater difficulties than that for nongyrotropic bodies and media. Suffice it to say that even finding the Fresnel coefficients of the reflection of a plane wave from a homogeneous, gyrotropic half-space is a rather unwieldy problem for an arbitrary orientation of the homogeneous field $\boldsymbol{B}_0$. The treatment is therefore usually confined to special cases (with $\boldsymbol{B}_0$ normal or parallel to the boundary and the primary wave normally incident). The number of exactly solvable diffraction problems, i.e., ones permitting the finding of exact Green's functions, is limited for gyrotropic bodies still more strictly than for isotropic bodies, but this is due to the physical nature of the problem, but not the method of solving the fluctuational problems. The statistical characteristics of thermal field are also easily determined.

Naturally, the equations derived with the help of the generalized Kirchhoff law undergo the same transformation in dealing with gyrotropic bodies, as is contained in (3.92). So, for example, instead of (3.55), the energy flux of the thermal radiation within the solid angle *do* in the wave zone of a *gyrotropic* body will be

$$dP_\omega = \frac{\Theta(\omega, T)}{\pi\lambda^2}\tilde{\sigma}_{\text{eff}}\, do \ ,$$

where $\tilde{\sigma}_{\text{eff}}$ is the effective absorption cross-section of a given body for the *reverse* field $-\boldsymbol{B}_0$. The waveguide from of Kirchhoff's law (3.61) is also replaced by

$$P_{\omega n}^{\mp} = \frac{\Theta}{2\pi}\tilde{A}_n^{\mp} \ . \tag{3.95}$$

Examples of the uses of (3.92–95) are given in [Ref. 3.6, Sects. 21–23].

As for the equilibrium form of the FDT, it is affected by gyrotropy as follows. Equations (3.88–90) were derived using the generalized Kirchhoff law (3.84) and the complex Lorentz lemma (3.87). The Lorentz lemma holds for any media with *linear* material equations, but for gyrotropic media (3.84) is replaced by (3.92). Therefore, for example, instead of [see (3.88)]

$$\langle E_{l_1}(\boldsymbol{r}_1)E_{l_2}^*(\boldsymbol{r}_2)\rangle = -\frac{1}{2\pi}\Theta(\omega, T)\{\boldsymbol{l}_1\cdot\boldsymbol{E}_{02}^*(\boldsymbol{r}_1) + \boldsymbol{l}_2\cdot\boldsymbol{E}_{01}(\boldsymbol{r}_2)\} \ ,$$

we now have

$$\langle E_{l_1}(\boldsymbol{r}_1)E_{l_2}^*(\boldsymbol{r}_2)\rangle = -\frac{1}{2\pi}\Theta(\omega, T)\{\boldsymbol{l}_1\cdot\tilde{\boldsymbol{E}}_{02}^*(\boldsymbol{r}_1) + \boldsymbol{l}_2\cdot\tilde{\boldsymbol{E}}_{01}(\boldsymbol{r}_2)\} \ , \tag{3.96}$$

where Green's functions on the right-hand side appear in a reverse medium.

But it follows from the theorem (3.91), i.e., the reciprocity theorem generalized to gyrotropic media, that for the point sources at hand

$$\boldsymbol{l}_1 \cdot \tilde{\boldsymbol{E}}_{02}(\boldsymbol{r}_1) = \boldsymbol{l}_2 \cdot \boldsymbol{E}_{01}(\boldsymbol{r}_2) \ . \tag{3.97}$$

Using this relation, we can eliminate from (3.96), say, $\tilde{\boldsymbol{E}}^*_{02}(\boldsymbol{r}_1)$, which gives

$$\langle E_{l_1}(\boldsymbol{r}_1) E^*_{l_2}(\boldsymbol{r}_2)\rangle = -\frac{1}{2\pi}\Theta(\omega,T)\boldsymbol{l}_2 \cdot \{\boldsymbol{E}^*_{01}(\boldsymbol{r}_2) + \tilde{\boldsymbol{E}}_{01}(\boldsymbol{r}_2)\} \ . \tag{3.98}$$

This is the mixed form of the covariance containing Green's functions both in the initial medium, and in the medium with a reverse bias field. At $\boldsymbol{B}_0 = 0$, equations of the mixed type [in particular, (3.98)] transform directly into the earlier formulas for nongyrotropic bodies [in particular (3.88)]. Using (3.97), we can also substitute in (3.96) the strengths in the initial medium (without reversing $\boldsymbol{B}_0$) for *both* strengths with tildes, and then

$$\langle E_{l_1}(\boldsymbol{r}_1) E^*_{l_2}(\boldsymbol{r}_2)\rangle = -\frac{1}{2\pi}\Theta(\omega,T)\{\boldsymbol{l}_2 \cdot \boldsymbol{E}^*_{01}(\boldsymbol{r}_2) + \boldsymbol{l}_1 \cdot \boldsymbol{E}_{02}(\boldsymbol{r}_1)\} \ . \tag{3.99}$$

As in (3.98, 99), we can also write the covariances for the components of $\boldsymbol{H}$ and the mutual covariances for the components of $\boldsymbol{E}$ and $\boldsymbol{H}$.

For equilibrium fields, gyrotropy causes no problem in principle. As with nongyrotropic bodies, it is only necessary to know Green's functions, although their actual derivation is, of course, much more complicated than for isotropic bodies.

Clearly, gyrotropy should not affect the universal relationship between the *integral* radiation and the absorption of a body, which follows from the second law of thermodynamics. If, for example, the body is surrounded by a sufficiently remote completely black envelope with the same temperature as that of the body, then a thermal equilibrium between the body and the envelope will be established, regardless of whether or not the body is gyrotropic. This implies that the integral (over the solid angle) total (over both polarizations) intensity of the body's radiation should in both cases be the same ($P_\omega = \tilde{P}_\omega$). But in the absence of gyrotropy, according to (3.55),

$$P_\omega = \frac{\Theta}{\pi\lambda^2}\oint \sigma_{\text{eff}}\, do$$

and with it, according to (3.94),

$$\tilde{P}_\omega = \frac{\Theta}{\pi\lambda^2}\oint \tilde{\sigma}_{\text{eff}}\, do \ .$$

It follows from $P_\omega = \tilde{P}_\omega$ that

$$\oint \sigma_{\text{eff}}\, do = \oint \tilde{\sigma}_{\text{eff}}\, do \ . \tag{3.100}$$

In exactly the same way, the total (i.e., to the right and left) radiation of a body disposed in a waveguide between two black plugs

$$P_\omega = P_\omega^+ + P_\omega^- = \sum_n{}' (P_{\omega n}^+ + P_{\omega n}^-)$$

must be independent of gyrotropicity, i.e., it must be $P_\omega = \tilde{P}_\omega$. Summing (3.60) over all the numbers n of travelling waves gives

$$P_\omega^+ = \frac{\Theta}{2\pi}\left\{\sum_n{}' A_n^- + \sum_{m,n}{}' (D_{nm}^- - D_{mn}^+)\right\} ,$$

$$P_\omega^- = \frac{\Theta}{2\pi}\left\{\sum_n{}' A_n^+ + \sum_{m,n}{}' (D_{nm}^+ - D_{mn}^-)\right\} .$$

Clearly, in the double sums over m and n these subscripts may be interchanged. Therefore

$$P_\omega = P_\omega^+ + P_\omega^- = \frac{\Theta}{2\pi}\sum_n{}' (A_n^- + A_n^+) .$$

Since $P_\omega = \tilde{P}_\omega$, we thus arrive at the relationship, similar to (3.100),

$$\sum_n{}' (A_n^- + A_n^+) = \sum_n{}' (\tilde{A}_n^- + \tilde{A}_n^+) . \tag{3.101}$$

3.11 Thermal Fields in Media with Spatial Dispersion

Up until now, we have confined our discussion to the material equations (3.2) describing a medium (generally speaking, inhomogeneous) in the absence of spatial dispersion: the inductions $\boldsymbol{D}$ and $\boldsymbol{B}$ at a point $\boldsymbol{r}$ were dependent on the behavior and time history of the medium (ε and μ), and on the field ($\boldsymbol{E}$ and $\boldsymbol{H}$) at $\boldsymbol{r}$. If the state of the medium at $\boldsymbol{r}$ depends on the time history (which predetermines the *temporal* nonlocality) not only at this point, but also in its neighborhood (*spatial* nonlocality), then the material equations should be of a more general kind

$$\begin{aligned} \boldsymbol{D}(t,\boldsymbol{r}) &= \int_{-\infty}^{+\infty} dt' \int_{-\infty}^{+\infty} \varepsilon(t-t',\boldsymbol{r},\boldsymbol{r}')\boldsymbol{E}(t',\boldsymbol{r}')d^3r' , \\ \boldsymbol{B}(t,\boldsymbol{r}) &= \int_{-\infty}^{+\infty} dt' \int_{-\infty}^{+\infty} \mu(t-t',\boldsymbol{r},\boldsymbol{r}')\boldsymbol{H}(t',\boldsymbol{r}')d^3r' . \end{aligned} \tag{3.102}$$

The variation of the kernels ε and μ with the *difference* $t-t'$ suggests that the processes of interest are stationary, i.e., presupposes that the properties of the medium remain unchanged in time (if the medium changed in time, ε and μ would depend on t and t' separately). The dependence of ε and μ on $\boldsymbol{r}$ and $\boldsymbol{r}'$ means that the medium is spatially *inhomogeneous*. In a homogeneous medium, ε and μ vary only with the difference $\boldsymbol{\varrho}=\boldsymbol{r}-\boldsymbol{r}'$. If the region of spatial nonlocality contracts into a point, then

$$\varepsilon(t-t',\boldsymbol{r},\boldsymbol{r}')\rightarrow\varepsilon(t-t',\boldsymbol{r})\,\delta(\boldsymbol{r}-\boldsymbol{r}')\ ,$$

and similarly for μ. The material equations (3.102) immediately transform into the local equations (3.2)[19].

Equations (3.102), like (3.2), apply to isotropic and nongyrotropic media. For anisotropic or gyrotropic medium the relationship between $\boldsymbol{D}$ and $\boldsymbol{E}$ is given by the electric permittivity tensor $\varepsilon_{\alpha\beta}$ $(\alpha,\beta=1,2,3)$, so that the first of the equations (3.102) is replaced by

$$D_\alpha(t,\boldsymbol{r})=\int_{-\infty}^{+\infty}dt'\int_{-\infty}^{+\infty}\varepsilon_{\alpha\beta}(t-t',\boldsymbol{r},\boldsymbol{r}')E_\beta(t',\boldsymbol{r}')d^3r'\ . \tag{3.103a}$$

As for the second of these equations, it follows from general considerations that one additional tensor $\mu_{\alpha\beta}$ would be superfluous. The point is that by introducing $\boldsymbol{D}=\boldsymbol{E}+4\pi\boldsymbol{P}$ and $\boldsymbol{B}=\boldsymbol{H}+4\pi\boldsymbol{M}$, we describe the medium by the polarization $\boldsymbol{P}$ and magnetization $\boldsymbol{M}$, classifying for this purpose the induced currents and charges in the medium into different "types" (bound and free charges with their polarization current and conduction current, and, in addition, molecular Amperian currents). But for a nonlocal medium it is simpler and more natural to include into $\boldsymbol{D}$ *all* types of induced currents, so that

$$\boldsymbol{B}=\boldsymbol{H}\ , \tag{3.103b}$$

and $\boldsymbol{D}$ obeys the material equation (3.103a). Clearly, $\boldsymbol{D}$ and $\boldsymbol{B}$ (i.e., their relation to microfields and microcurrents) are defined in (3.103a, b) in a different way than in (3.102), so that, without spatial dispersion, (3.103a, b) *do not transform* into (3.102).

We can thus completely describe a medium by the material equations (3.103a, b), which indicates that *one* tensor $\varepsilon_{\alpha\beta}$ at $\mu_{\alpha\beta}=\delta_{\alpha\beta}$ would be enough (for further details see [3.9,15]). Note that in the *bulk* of a nonlocal medium, even one that is homogeneous and isotropic, the tensor $\varepsilon_{\alpha\beta}(t-t',\boldsymbol{r},\boldsymbol{r}')$ in (3.103a) does not degenerate into a diagonal, but has a form that is quite general for this case

$$\varepsilon_{\alpha\beta}(\tau,\boldsymbol{\varrho})=\left(\delta_{\alpha\beta}-\frac{\varrho_\alpha\varrho_\beta}{\varrho^2}\right)\varepsilon_t(\tau,\varrho)+\frac{\varrho_\alpha\varrho_\beta}{\varrho^2}\varepsilon_l(\tau,\varrho)\ , \tag{3.104}$$

where $\tau=t-t'$, $\boldsymbol{\varrho}=\boldsymbol{r}-\boldsymbol{r}'$, i.e., it contains *two* independent scalar functions ε_t and ε_l—the transverse and longitudinal (relative to $\boldsymbol{\varrho}$) electric permittivities. This

---

[19] The state of the medium at a time t and at a point $\boldsymbol{r}$ can only depend on events inside a light cone directed into the past, whose vertex is at the "point" $(t,\boldsymbol{r})$. In other words, these events occur before the instant t and produce fields that travel no faster than light in a vacuum (the causality principle). In consequence, the integration with respect to t' and $\boldsymbol{r}'$ in (3.102) and in (3.103a) with infinite limits supposes that ε, μ, and $\varepsilon_{\alpha\beta}$ vanish beyond the cone. This imposes certain constraints on their $\omega\kappa$-amplitudes, too; namely, it establishes a relationship between the real and imaginary parts of these functions. In the absence of spatial dispersion, this relationship is expressed by the Cramers–Kroning equations [Ref. 3.3, Sect. 62], and when this dispersion is present, by more general equations [3.14].

strongly suggests that for a *homogeneous*, *isotropic*, nonlocal medium, material equations of the form (3.102) are also permissible, since these also contain only the two independent scalar functions $\varepsilon(\tau, \varrho)$ and $\mu(\tau, \varrho)$. We will look at the connection between ε_t, ε_l and ε, μ later in the chapter.

It was with good reason that we mentioned earlier that the expression (3.104) refers to the bulk of the medium, i.e., to regions lying sufficiently far from its boundaries. Near the boundary $\varepsilon_{\alpha\beta}$ cannot remain a function of $\boldsymbol{\varrho} = \boldsymbol{r} - \boldsymbol{r}'$ only, as the direction normal to the boundary becomes distinguishable. In other words, near the boundary, the layer with thickness of the order of the nonlocality radius is inhomogeneous and anisotropic in its electrodynamic properties. This complicates the electrodynamic part of the problem, especially when finding auxiliary diffraction fields (Green's functions) or the densities of their Joule losses, if it is also necessary to make allowance for the boundary layer. As to the fluctuational part proper, the situation is the same as with local media, since both the generalized Kirchhoff law (3.37–39), and the equilibrium form of the FDT (3.88–90) hold for *any* form of material equations.

In the following we will only take the simplest case of the *homogeneous*, *isotropic*, nonlocal medium that is also *infinite*. Now over the entire space we can use both $\varepsilon_{\alpha\beta}(\tau, \boldsymbol{\varrho})$ in the form (3.104), and the scalars $\varepsilon(\tau, \varrho)$, $\mu(\tau, \varrho)$. It is expedient here to pass over to Fourier transforms, both in τ and in $\boldsymbol{\varrho}$

$$\varepsilon(\tau, \boldsymbol{\varrho}) = \frac{1}{(2\pi)^4} \iint_{-\infty}^{+\infty} \varepsilon(\omega, \boldsymbol{\kappa}) \exp[-\mathrm{i}(\omega\tau - \boldsymbol{\kappa} \cdot \boldsymbol{\varrho})]\, d\omega\, d^3\kappa \ , \tag{3.105}$$

and similarly for $\mu(\tau, \boldsymbol{\varrho})$. The same Fourier transform for $\varepsilon_{\alpha\beta}(\tau, \boldsymbol{\varrho})$ contains in the integrand the following transform

$$\varepsilon_{\alpha\beta}(\omega, \boldsymbol{\kappa}) = \left(\delta_{\alpha\beta} - \frac{\kappa_\alpha \kappa_\beta}{\kappa^2}\right) \varepsilon_t(\omega, \kappa) + \frac{\kappa_\alpha \kappa_\beta}{\kappa^2} \varepsilon_l(\omega, \kappa) \ , \tag{3.106}$$

where $\varepsilon_t(\omega, \kappa)$ and $\varepsilon_l(\omega, \kappa)$ are the transverse and longitudinal (in relation to $\boldsymbol{\kappa}$) permittivities, which, of course, *are not* Fourier transforms of $\varepsilon_t(\tau, \varrho)$ and $\varepsilon_t(\tau, \varrho)$ in (3.104). The scalar functions $\varepsilon(\omega, \kappa)$ and $\mu(\omega, \kappa)$ are related to $\varepsilon_t(\omega, \kappa)$ and $\varepsilon_l(\omega, \kappa)$ [3.9] in a very simple way

$$\begin{aligned} &\varepsilon(\omega, \kappa) = \varepsilon_l(\omega, \kappa) \ , \\ &1 - \frac{1}{\mu(\omega, \kappa)} = \frac{k_0^2}{\kappa^2} [\varepsilon_t(\omega, \kappa) - \varepsilon_l(\omega, \kappa)], \quad k_0 = \frac{\omega}{c} \ . \end{aligned} \tag{3.107}$$

The second of these equations indicates that $\mu(\omega, \kappa)$ cannot vanish.

Under these conditions, it is also expedient to apply space-time Fourier transforms ($\omega\kappa$-amplitudes) or spatial Fourier transforms of spectral amplitudes, since in this case all the linear equations for the fields become *algebraic* linear equations of their $\omega\kappa$-amplitude. Specifically, Maxwell's equations (3.6), if we assume

$$\boldsymbol{E}(t, \boldsymbol{r}) = \iint_{-\infty}^{+\infty} \boldsymbol{E}(\omega, \boldsymbol{\kappa}) \exp[-\mathrm{i}(\omega t - \boldsymbol{\kappa} \cdot \boldsymbol{r})]\, d\omega\, d^3\kappa \ , \tag{3.108}$$

and similarly for $\boldsymbol{H}$, $\boldsymbol{D}$, $\boldsymbol{B}$ and the external currents $\boldsymbol{j}_0, \boldsymbol{j}_\mathrm{m}$, for $\omega\boldsymbol{\kappa}$-amplitudes take the following form. (For simplicity we have omitted ω and $\boldsymbol{\kappa}$):

$$\begin{aligned} \mathrm{i}[\boldsymbol{\kappa}\boldsymbol{H}] &= -\mathrm{i}k_0\boldsymbol{D} + \frac{4\pi}{c}\boldsymbol{j}_\mathrm{e} \ , \\ \mathrm{i}[\boldsymbol{\kappa}\boldsymbol{E}] &= \mathrm{i}k_0\boldsymbol{B} - \frac{4\pi}{c}\boldsymbol{j}_\mathrm{m} \ . \end{aligned} \tag{3.109}$$

The material equations for $\omega\boldsymbol{\kappa}$-amplitudes, derived from (3.102, 105, 108) (the medium is homogeneous!), are reduced to the relationships

$$\boldsymbol{D}(\omega,\boldsymbol{\kappa}) = \varepsilon(\omega,\boldsymbol{\kappa})\boldsymbol{E}(\omega,\boldsymbol{\kappa}), \qquad \boldsymbol{B}(\omega,\boldsymbol{\kappa}) = \mu(\omega,\boldsymbol{\kappa})\boldsymbol{H}(\omega,\boldsymbol{\kappa}) \ . \tag{3.110}$$

The presence of spatial dispersion manifests itself in that ε and μ vary with $\boldsymbol{\kappa}$ (for an isotropic medium these are only dependent on $\kappa = |\boldsymbol{\kappa}|$).

The FDT also becomes an algebraic expression. In the general case of the multicomponent fluctuational field $\xi^{(j)}(t,\boldsymbol{r})$ and the Lagrangian conjugate field $f^{(j)}(t,\boldsymbol{r})$, the operator equations (3.20) for spectral amplitudes, as a result of the spatial Fourier transformation, become

$$\begin{aligned} \xi^{(j)}(\omega,\boldsymbol{\kappa}) &= \sum_k A_{jk}(-\mathrm{i}\omega,\ \mathrm{i}\boldsymbol{\kappa}) f^{(k)}(\omega,\boldsymbol{\kappa}) \ , \\ f^{(j)}(\omega,\boldsymbol{\kappa}) &= \sum_k A_{jk}^{-1}(-\mathrm{i}\omega,\ \mathrm{i}\boldsymbol{\kappa})\, \xi^{(k)}(\omega,\boldsymbol{\kappa}) \ , \end{aligned} \tag{3.111}$$

and the expressions for the correlation matrices of the $\omega\boldsymbol{\kappa}$-amplitudes $\xi^{(j)}$ and $f^{(j)}$, which follow from the FDT, become (see exercise 6)

$$\begin{aligned} &\langle \xi^{(j)}(\omega,\boldsymbol{\kappa})\xi^{(k)*}(\omega,\boldsymbol{\kappa}')\rangle \equiv G_{jk}^{(\xi)}(\omega,\boldsymbol{\kappa})\delta(\boldsymbol{\kappa}-\boldsymbol{\kappa}') \\ &\quad = -\frac{\mathrm{i}\Theta(\omega,T)}{(2\pi)^4\omega}\{A_{jk}(-\mathrm{i}\omega,\mathrm{i}\boldsymbol{\kappa}) - A_{kj}^*(-\mathrm{i}\omega,\ \mathrm{i}\boldsymbol{\kappa})\}\,\delta(\boldsymbol{\kappa}-\boldsymbol{\kappa}') \ , \end{aligned} \tag{3.112}$$

$$\begin{aligned} &\langle f^{(j)}(\omega,\boldsymbol{\kappa})f^{(k)*}(\omega,\boldsymbol{\kappa}')\rangle \equiv G_{jk}^{(f)}(\omega,\boldsymbol{\kappa})\delta(\boldsymbol{\kappa}-\boldsymbol{\kappa}') \\ &\quad = -\frac{\mathrm{i}\Theta(\omega,T)}{(2\pi)^4\omega}\{A_{kj}^{-1*}(-\mathrm{i}\omega,\ \mathrm{i}\boldsymbol{\kappa}) - A_{jk}^{-1}(-\mathrm{i}\omega,\ \mathrm{i}\boldsymbol{\kappa})\}\,\delta(\boldsymbol{\kappa}-\boldsymbol{\kappa}') \ , \end{aligned} \tag{3.113}$$

where $G_{jk}(\omega,\boldsymbol{\kappa})$ are the $\omega\boldsymbol{\kappa}$-densities. Fourier transformation of these densities in $\boldsymbol{\kappa}$, clearly, gives the spectral densities. For example,

$$\begin{aligned} g_{jk}^{(\xi)}(\omega,\boldsymbol{\varrho}) &\equiv \langle \xi^{(j)}(\omega,\boldsymbol{r})\xi^{(k)*}(\omega,\boldsymbol{r}')\rangle \\ &= \iint_{-\infty}^{+\infty} \langle \xi^{(j)}(\omega,\boldsymbol{\kappa})\xi^{(k)*}(\omega,\boldsymbol{\kappa}')\rangle \exp[\mathrm{i}(\boldsymbol{\kappa}\cdot\boldsymbol{r}-\boldsymbol{\kappa}'\cdot\boldsymbol{r}')]\,d^3\kappa\, d^3\kappa' \\ &= \iint_{-\infty}^{+\infty} G_{jk}^{(\xi)}(\omega,\boldsymbol{\kappa})\delta(\boldsymbol{\kappa}-\boldsymbol{\kappa}')\exp[\mathrm{i}(\boldsymbol{\kappa}\cdot\boldsymbol{r}-\boldsymbol{\kappa}\cdot\boldsymbol{r}')]\,d^3\kappa\, d^3\kappa' \\ &= \int_{-\infty}^{+\infty} G_{jk}^{(\xi)}(\omega,\boldsymbol{\kappa})\exp(\mathrm{i}\boldsymbol{\kappa}\cdot\boldsymbol{\varrho})\,d^3\kappa \quad (\boldsymbol{\varrho}=\boldsymbol{r}-\boldsymbol{r}') \ . \end{aligned} \tag{3.114}$$

We now apply (3.112) to Maxwell's equations (3.109) having eliminated the inductions $\boldsymbol{D}$ and $\boldsymbol{B}$ using (3.110), we obtain

$$\mathrm{i}k_0\varepsilon\boldsymbol{E}+\mathrm{i}[\boldsymbol{\kappa}\boldsymbol{H}]=\frac{4\pi}{c}\boldsymbol{j}_{\mathrm{e}},\quad \mathrm{i}k_0\mu\boldsymbol{H}-\mathrm{i}[\boldsymbol{\kappa}\boldsymbol{E}]=\frac{4\pi}{c}\boldsymbol{j}_{\mathrm{m}}\ . \tag{3.115}$$

These equations should be solved for $\boldsymbol{E}$ and $\boldsymbol{H}$ if the formula (3.112) is to be used. The determinant of (3.115) is $\mathscr{D}=k_0^2\varepsilon\mu\Delta^2$, where

$$\Delta=\kappa^2-k_0^2\varepsilon\mu\ . \tag{3.116}$$

Let the medium be homogeneous. Since $\mu(\omega,\kappa)$ cannot vanish, the dispersion equation $\mathscr{D}=0$ splits into two: either $\varepsilon(\omega,\kappa)=0$ (longitudinal waves), or $\Delta(\omega,\kappa)=0$ (transverse waves). The solution to (3.115) is

$$\boldsymbol{E}=\frac{4\pi}{\mathrm{i}\omega\Delta}\left\{-k_0^2\mu\boldsymbol{j}_{\mathrm{e}}+\frac{\boldsymbol{\kappa}}{\varepsilon}(\boldsymbol{\kappa}\cdot\boldsymbol{j}_{\mathrm{e}})+k_0[\boldsymbol{\kappa}\cdot\boldsymbol{j}_{\mathrm{m}}]\right\},$$

$$\boldsymbol{H}=\frac{4\pi}{\mathrm{i}\omega\Delta}\left\{-k_0^2\varepsilon\boldsymbol{j}_{\mathrm{m}}+\frac{\boldsymbol{\kappa}}{\mu}(\boldsymbol{\kappa}\cdot\boldsymbol{j}_{\mathrm{m}})-k_0[\boldsymbol{\kappa}\cdot\boldsymbol{j}_{\mathrm{e}}]\right\},$$

or in components

$$\begin{aligned}E_\alpha&=-\frac{1}{\mathrm{i}\omega}\left(A_{\alpha\beta}^{(E,\mathrm{e})}j_{\mathrm{e}\beta}+A_{\alpha\beta}^{(E,\mathrm{m})}j_{\mathrm{m}\beta}\right),\\ H_\alpha&=-\frac{1}{\mathrm{i}\omega}\left(A_{\alpha\beta}^{(H,\mathrm{e})}j_{\mathrm{e}\beta}+A_{\alpha\beta}^{(H,\mathrm{m})}j_{\mathrm{m}\beta}\right),\end{aligned} \tag{3.117}$$

where

$$\left[\begin{array}{c|c}A_{\alpha\beta}^{(E,\mathrm{e})} & A_{\alpha\beta}^{(E,\mathrm{m})}\\ \hline A_{\alpha\beta}^{(H,\mathrm{e})} & A_{\alpha\beta}^{(H,\mathrm{m})}\end{array}\right]=\frac{4\pi}{\Delta}\left[\begin{array}{c|c}k_0^2\mu\delta_{\alpha\beta}-\dfrac{\kappa_\alpha\kappa_\beta}{\varepsilon} & -k_0\varepsilon_{\alpha\gamma\beta}\kappa_\gamma\\ \hline k_0\varepsilon_{\alpha\gamma\beta}\kappa_\gamma & k_0^2\varepsilon\delta_{\alpha\beta}-\dfrac{\kappa_\alpha\kappa_\beta}{\mu}\end{array}\right]. \tag{3.118}$$

Here $\varepsilon_{\alpha\gamma\beta}$ is a completely antisymmetric unit tensor of the third rank ($\varepsilon_{123}=1$), and the summation is over subscript β that enters into (3.117) twice. Note that, by (3.116),

$$\frac{1}{\Delta\varepsilon}=\frac{1}{\kappa^2\varepsilon}+\frac{k_0^2\mu}{\Delta\kappa^2},\quad \frac{1}{\Delta\mu}=\frac{1}{\kappa^2\mu}+\frac{k_0^2\varepsilon}{\Delta\kappa^2}\ .$$

Therefore, the elements of the diagonal squares of the matrix (3.118) may be represented as

$$A_{\alpha\beta}^{(E,\mathrm{e})}=4\pi\left[\frac{k_0^2\mu}{\Delta}\left(\delta_{\alpha\beta}-\frac{\kappa_\alpha\kappa_\beta}{\kappa^2}\right)-\frac{\kappa_\alpha\kappa_\beta}{\kappa^2\varepsilon}\right],$$

$$A_{\alpha\beta}^{(H,\mathrm{m})}=\left[\frac{k_0^2\varepsilon}{\Delta}\left(\delta_{\alpha\beta}-\frac{\kappa_\alpha\kappa_\beta}{\kappa^2}\right)-\frac{\kappa_\alpha\kappa_\beta}{\kappa^2\mu}\right].$$

As shown in Sect. 3.4, the quantities $\mathfrak{D}/4\pi = -\boldsymbol{j}_e/\mathrm{i}\omega$ and $\mathfrak{B}/4\pi = -\boldsymbol{j}_m/\mathrm{i}\omega$ for $\boldsymbol{E}$ and $\boldsymbol{H}$ are generalized forces, so that the coefficients A introduced in (3.117) are exactly those coefficients A_{jk} that appear in (3.112). Using this formula, we obtain for the $\omega\kappa$-densities $G^{(E)}_{\alpha\beta}(\omega,\kappa) \equiv \langle E_\alpha(\omega,\kappa)E^*_\beta(\omega,\kappa)\rangle$, etc., the following expressions:

$$G^{(E)}_{\alpha\beta}(\omega,\kappa) = -\frac{\mathrm{i}\Theta(\omega,T)}{4\pi^3\omega}\left\{\frac{k_0^2\mu}{\Delta}\left(\delta_{\alpha\beta}-\frac{\kappa_\alpha\kappa_\beta}{\kappa^2}\right)-\frac{\kappa_\alpha\kappa_\beta}{\varepsilon\kappa^2}-\text{complex conjugate}\right\}, \tag{3.119a}$$

$$G^{(H)}_{\alpha\beta}(\omega,\kappa) = -\frac{\mathrm{i}\Theta(\omega,T)}{4\pi^3\omega}\left\{\frac{k_0^2\varepsilon}{\Delta}\left(\delta_{\alpha\beta}-\frac{\kappa_\alpha\kappa_\beta}{\kappa^2}\right)-\frac{\kappa_\alpha\kappa_\beta}{\mu\kappa^2}-\text{complex conjugate}\right\}, \tag{3.119b}$$

$$G^{(EH)}_{\alpha\beta}(\omega,\kappa) = -G^{(HE)}_{\alpha\beta}(\omega,\kappa) = \frac{\mathrm{i}\Theta(\omega,T)}{4\pi^3 c}\varepsilon_{\alpha\gamma\beta}\kappa_\gamma\left(\frac{1}{\Delta}-\frac{1}{\Delta^*}\right). \tag{3.119c}$$

It is seen from (3.119c) that, unlike the external currents $\boldsymbol{j}_e$ and $\boldsymbol{j}_m$, only the orthogonal components of $\boldsymbol{E}$ and $\boldsymbol{H}$ are mutually correlated, for $\varepsilon_{\alpha\gamma\beta} = 0$ at $\alpha = \beta$. The first terms in (3.119a, b), containing Δ in denominators, determine the correlation of the transverse (in relation to $\boldsymbol{\kappa}$) field, i.e., of the strengths $\boldsymbol{E}_\perp = \boldsymbol{E} - \boldsymbol{E}_\parallel$ and $\boldsymbol{H}_\perp = \boldsymbol{H} - \boldsymbol{H}_\parallel$; and the second terms, the correlation of the longitudinal field $\boldsymbol{E}_\parallel = \boldsymbol{\kappa}(\boldsymbol{\kappa}\boldsymbol{E})/\kappa^2$, and similarly for $\boldsymbol{H}_\parallel$. The transverse and longitudinal strengths are not intercorrelated, whereby $G = G_\perp + G_\parallel$.

To conclude this section, we will look at the fluctuations of the *total* current in a medium, which is taken to mean the current consisting of all "types" of electric current, with the exception of the vacuum displacement current. In terms of the microscopic theory, the total current is the sum of the convection currents caused by the motion of any (not subdivided into "types") microcharges averaged over physically infinitesimal volumes of the medium. This averaging of the Lorentz equations for microfields $\boldsymbol{e}$ and $\boldsymbol{h}$ gives for $\boldsymbol{E} = \tilde{\boldsymbol{e}}$ and $\boldsymbol{B} = \tilde{\boldsymbol{h}}$ (the tilde denotes averaging over a physically infinitesimal volume) the macroscopic equations of the following form:

$$\operatorname{rot}\boldsymbol{B}' = \frac{1}{c}\frac{\partial\boldsymbol{E}}{\partial t} + \frac{4\pi}{c}\boldsymbol{j}_t, \quad \operatorname{rot}\boldsymbol{E} = -\frac{1}{c}\frac{\partial\boldsymbol{B}'}{\partial t}. \tag{3.120}$$

Here $\boldsymbol{j}_t$ denotes the volume density of the total current and the prime at $\boldsymbol{B}$ is used here in order to distinguish this magnetic induction, which also includes the *external* magnetic induction, from the induction $\boldsymbol{B}$ in (3.3), where the external magnetic current has been separated out. The volume densities of the total charge and total current are, of course, related by the continuity equation

$$\frac{\partial\varrho_t}{\partial t} + \operatorname{div}\boldsymbol{j}_t = 0. \tag{3.121}$$

It is these total densities that are of interest in a number of problems, specifically in problems concerned with plasma fluctuations.

We will first of all establish the relationship between $\boldsymbol{B}'$ and $\boldsymbol{j}_t$ and the quantities appearing in the standard symmetrical form of Maxwell's equations. We write (3.120, 121) for the $\omega\boldsymbol{\kappa}$-amplitudes

$$\mathrm{i}[\boldsymbol{\kappa}\boldsymbol{B}'] = -\mathrm{i}k_0\boldsymbol{E} + \frac{4\pi}{c}\boldsymbol{j}_t, \quad \mathrm{i}[\boldsymbol{\kappa}\boldsymbol{E}] = \mathrm{i}k_0\boldsymbol{B}' \ , \tag{3.122}$$

$$-\omega\varrho_t + \boldsymbol{\kappa}\cdot\boldsymbol{j}_t = 0 \ . \tag{3.123}$$

A comparison of the second equation in (3.122) and (3.109) indicates that $\boldsymbol{B}' = \boldsymbol{B} - 4\pi\boldsymbol{j}_m/\mathrm{i}k_0 c$, whereby the first of (3.122) takes the form

$$\mathrm{i}[\boldsymbol{\kappa}\boldsymbol{B}] - \frac{4\pi}{k_0 c}[\boldsymbol{\kappa}\boldsymbol{j}_m] = -\mathrm{i}k_0\boldsymbol{E} + \frac{4\pi}{c}\boldsymbol{j}_t \ .$$

Subtracting from this the first of (3.109) gives

$$\boldsymbol{j}_t = \boldsymbol{j}_e - \frac{\mathrm{i}k_0 c}{4\pi}(\boldsymbol{D} - \boldsymbol{E}) - \frac{1}{k_0}[\boldsymbol{\kappa}\boldsymbol{j}_m] + \frac{\mathrm{i}c}{4\pi}[\boldsymbol{\kappa}(\boldsymbol{B} - \boldsymbol{H})] \ . \tag{3.124}$$

It follows that the total current is the sum of the external electric current (the first term), the induced polarization current (the second term that includes the conduction current, of course), and the electric currents caused by the external magnetic current (the third term) and the induced magnetization (the last term that also includes the magnetic conductivity).

We can easily write the total current as a function of $\boldsymbol{E}$ only. To this end, it is sufficient to substitute into (3.124) the expressions for $\boldsymbol{j}_e$ and $\boldsymbol{j}_m$ that follow from (3.103). This gives

$$\boldsymbol{j}_t = \frac{\mathrm{i}\omega}{4\pi}\left\{\boldsymbol{E} + \frac{1}{k_0^2}[\boldsymbol{\kappa}[\boldsymbol{\kappa}\boldsymbol{E}]]\right\} ,$$

or in components

$$j_{t\alpha} = \frac{\mathrm{i}\omega}{4\pi}\left\{(1-\eta^2)\delta_{\alpha\beta} + \eta^2\frac{\kappa_\alpha\kappa_\beta}{\kappa^2}\right\}E_\beta \quad \left(\eta = \frac{\kappa}{k_0}\right) .$$

Consequently, the $\omega\boldsymbol{\kappa}$-densities of the total current $\boldsymbol{j}_t$ and the electric field strength $\boldsymbol{E}$ are related by

$$G_{\alpha\beta}^{(j_t)}(\omega,\boldsymbol{\kappa}) = \frac{\omega^2}{16\pi^2}\left[(1-\eta^2)\delta_{\alpha\gamma} + \eta^2\frac{\kappa_\alpha\kappa_\gamma}{\kappa^2}\right]\left[(1-\eta^2)\delta_{\beta\delta} + \eta^2\frac{\kappa_\beta\kappa_\delta}{\kappa^2}\right] \times G_{\gamma\delta}^{(E)}(\omega,\boldsymbol{\kappa}) \ .$$

We next substitute the expression (3.119a) for $G_{\gamma\delta}^{(E)}$ and sum over the subscripts γ

and δ that appear twice on the right

$$G^{(j_t)}_{\alpha\beta}(\omega,\boldsymbol{\kappa}) = -\frac{i\omega\Theta(\omega,T)}{64\pi^5}\left\{\frac{k_0^2\mu}{\Delta}(1-\eta^2)^2\left(\delta_{\alpha\beta}-\frac{\kappa_\alpha\kappa_\beta}{\kappa^2}\right)-\frac{\kappa_\alpha\kappa_\beta}{\varepsilon\kappa^2}\right.$$

$$\left. - \text{complex conjugate}\right\} . \tag{3.125}$$

According to (3.123, 125), the $\omega\boldsymbol{\kappa}$-density of fluctuations of the total charge ϱ_t for a unit volume of the medium will be

$$G^{(\varrho_t)}(\omega,\boldsymbol{\kappa}) = \frac{\kappa_\alpha\kappa_\beta}{\omega^2}G^{(j_t)}_{\alpha\beta}(\omega,\boldsymbol{\kappa}) = \frac{i\Theta(\omega,T)\kappa^2}{64\pi^5\omega}\left(\frac{1}{\varepsilon}-\frac{1}{\varepsilon^*}\right) . \tag{3.126}$$

Both the transverse and longitudinal waves contribute to the total current (3.125), but fluctuations of the volume density of the total charge (3.126) are only related to the longitudinal waves: the denominator of the expression for $G^{(\varrho_t)}$ contains only $\varepsilon(\omega,\kappa)$, and not $\Delta(\omega,\kappa)$. The factor κ^2 in (3.126) testifies to the larger contributions of the small-scale fluctuations. Since, in a homogeneous medium, ε is a function of $\kappa=|\boldsymbol{\kappa}|$ only, the Fourier transform (3.114) which gives the *spectral* density of the fluctuations, for the charge ϱ_t takes the form

$$g^{(\varrho_t)}(\omega,\varrho) = \frac{i\Theta(\omega,T)}{64\pi^5\omega}\int_{-\infty}^{+\infty}\exp(i\boldsymbol{\kappa}\cdot\boldsymbol{\varrho})\left(\frac{1}{\varepsilon}-\frac{1}{\varepsilon^*}\right)\kappa^2 d^3\kappa$$

$$= \frac{i\Theta(\omega,T)}{16\pi^4\omega\varrho}\int_0^\infty\left(\frac{1}{\varepsilon}-\frac{1}{\varepsilon^*}\right)\kappa^3\sin\kappa\varrho\, d\kappa \quad (\varrho=|\boldsymbol{r}-\boldsymbol{r}'|) . \tag{3.127}$$

In exercise 7, (3.125, 126) are applied to a simple model of a medium showing spatial dispersion, which is of interest for describing fluctuations in ionospheric plasma.

Before concluding this chapter, we will make some additional remarks. We noted in Sect. 3.8 that the theory of random fields due to equilibrium thermal fluctuations refers to statistical scheme (1), i.e., it deals with problems in which the statistics of random field sources is known. Nevertheless, we devoted a separate chapter to thermal fields, thus stressing not only their importance but also their specificity. The latter resides precisely in the fact that for this group of problems, the statistics governing the sources is not given from the outset, but is *defined by the dynamical equations of the problem.* To be more exact, the covariances of the external (Langevin) forces are governed by the skew-Hermitian parts of those deterministic linear operators, in terms of which the dynamical equations are written. This is what forms the contents of the fluctuation–dissipation theorem (Sect. 3.5).

But in the case of substantially nonequilibrium thermal fluctuations or, in general, fields of nonthermal origin, the statistics of the sources is not normally defined by the form of the equations themselves, but should be specified—either formally, or based on a statistical analysis of microscopic models.

The universal character of the various forms of the FDT is also worthy of comment. In this chapter we considered *electromagnetic* thermal fields, although we relied on the FDT that is applicable to equilibrium thermal fluctuations of *any* physical nature (Sect. 3.3). For electromagnetic fields, we obtained other forms of the FDT–Kirchhoffian (Sects. 3.5 and 3.6), waveguide (Sect. 3.7), equilibrium (Sect. 3.9)—which are physically more lucid, more "economical" (in solving specific problems) and more promising as regards the scope of solvable problems. One might well ask whether all these forms of the FDT are as physically versatile as their initial, "canonical" form (3.22).

All we needed to take account of in electrodynamics in order to arrive at those forms of the FDT comes down to the electrodynamic reciprocity theorem (3.28) or (3.91), and the complex Lorentz lemma (3.85). We may expect, therefore, that for any equilibrium (and quasi-equilibrium) fluctuational fields coming under the umbrella of the reciprocity theorem and the analog of the Lorentz lemma, the above-mentioned forms of the FDT will also hold. An examination of this issue indicates that this is in actual fact the case [3.16], thus suggesting a number of interesting consequences.

For example, the Kirchhoffian form of the FDT, specifically (3.55), is fully applicable to acoustical waves. Thermal losses here stem from the viscosity and thermal conductivity of a medium. Imagine a body with such losses immersed in a "transparent" liquid, i.e., one with small absorption of longitudinal waves. On exposing the body to a plane wave of frequency ω (say, in the sound or ultrasound range), we can find the effective absorption cross-section of the body (for a given orientation). Equation (3.55), in which we can clearly put $\Theta = kT$, then governs the energy flux of longitudinal waves of frequency ω given off by the body within the solid angle do in the opposite direction to the incident auxiliary plane wave. Although the physical mechanism of thermal fluctuations in a body (fluctuations of density, temperature, velocities of macroparticles) is absolutely different from that in electrodynamics, the sound radiation of a heated body into the surrounding "transparent" medium follows the same law as the radiation of electromagnetic waves, namely Kirchhoff's law.

3.12 Exercises

3.12.1 For small velocities $|\dot{\boldsymbol{r}}| \ll c$ of an elastically bound electron, its equation of motion (including the reaction of the radiation) has the form

$$m\ddot{\boldsymbol{r}} - \frac{2e^2}{3c^3}\dddot{\boldsymbol{r}} + m\omega_0^2\boldsymbol{r} = \boldsymbol{f}(t) = e\boldsymbol{E}(t) \;, \tag{3.128}$$

where e and m are the electron charge and mass, respectively; c is the velocity of light in a vacuum; ω_0 is the natural frequency of the oscillator. Let $\boldsymbol{E}(t)$ be the electric field of an equilibrium radiation. Applying the FDT to (3.128), find the spectral density of the energy of this radiation (in positive frequencies ω), i.e., the quantity (the electron is in a vacuum)

$$u_{0\omega} = \frac{1}{8\pi}\langle |\boldsymbol{E}|^2 + |\boldsymbol{H}|^2 \rangle \ .$$

Solution. The equation of motion (3.128) in spectral form is

$$\boldsymbol{r}(\omega) = \alpha(\omega)\boldsymbol{f}(\omega) \ , \quad \text{where}$$

$$\frac{1}{\alpha(\omega)} = m(\omega_0^2 - \omega^2) - \mathrm{i}\,\frac{2e^2\omega^3}{3c^3} \ . \tag{3.129}$$

The spectral density of the component $E_x(t) = f_x(t)/e$ of the fluctuational field for $\omega > 0$ is

$$g^{+}_{E_x}(\omega) = \frac{1}{e^2} g^{+}_{f_x}(\omega) \equiv \frac{2}{e^2}\langle |f_x(\omega)|^2 \rangle \ , \tag{3.130}$$

where $\langle |f_x(\omega)|^2 \rangle$ is the spectral density of the force $f_x(t)$ in ω from $-\infty$ to $+\infty$. According to the discrete FDT, $\langle |f_x(\omega)|^2 \rangle$ in terms of generalized susceptibility $\alpha(\omega)$ will be

$$\langle |f_x(\omega)|^2 \rangle = -\frac{\mathrm{i}\,\Theta(\omega, T)}{2\pi\omega}\left[\frac{1}{\alpha^*(\omega)} - \frac{1}{\alpha(\omega)}\right] \tag{3.131}$$

[this is equation (3.17b) in the special case of a single variable]. It follows from (3.129–131) that

$$g^{+}_{E_x}(\omega) = \frac{4\omega^2}{3\pi c^3}\Theta(\omega, T) \ .$$

Note that $\mathrm{Re}\{1/\alpha(\omega)\}$ is insignificant, therefore the electron may just as well be free ($\omega_0 = 0$).

To derive the spectral energy density of the electromagnetic field, it is sufficient to remember that equilibrium radiation is isotropic in an infinite space, whereby

$$\langle |E_x|^2 \rangle = \langle |E_y|^2 \rangle = \langle |E_z|^2 \rangle \ ,$$

and that its magnetic and electric energies are equal

$$\langle |\boldsymbol{E}|^2 \rangle = \langle |\boldsymbol{H}|^2 \rangle \ .$$

Therefore,

$$u_{0\omega} = 6\frac{1}{8\pi}\, g^{+}_{E_x}(\omega) = \frac{\omega^2}{\pi^2 c^3}\Theta(\omega, T) \ ,$$

i.e., we have arrived at the Planck formula.

3.12.2 Equations (3.26) for the covariances of external currents may also be obtained in a more "physical" way than in Sect. 3.4, i.e., from the classical Kirchhoff law and the *detailed balancing principle.* The latter requires that for two similarly heated bodies 1 and 2, the power Q_{12} absorbed by body 1 from the radiation coming from body 2 be at each frequency ω equal to the power Q_{21} absorbed by body 2 from the radiation coming from body 1. Bearing this in mind derive (3.26) by considering the energy exchange between the body and a plate (Fig. 3.8) separated by such a large distance R that their dimensions are much smaller than the radius $\sqrt{\lambda R}$ of the first Fresnel zone (i.e., these lie within the Fraunhoffer zone of each other). At the same time, the plate is so much larger than the wavelength λ that the plate radiation can be calculated using the classical Kirchhoff law.

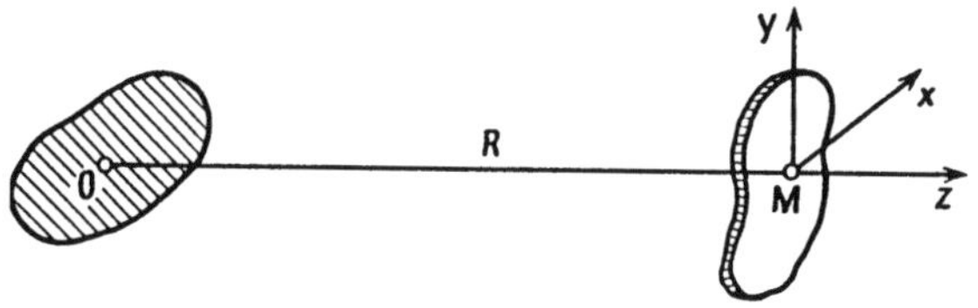

Fig. 3.8. Thermal exchange between a body and a plate

Solution. Under these conditions we compute the powers Q_b and Q_p absorbed by the body and plate, respectively, at frequency ω.

Let A be the absorption coefficient of the plate and Σ its area ($\lambda R \gg \Sigma \gg \lambda^2$). The power absorbed by the plate is

$$Q_p = \mathcal{S}_z A \Sigma \ , \tag{3.132}$$

where $\mathcal{S}_z$ is the z-component of the Poynting vector of the thermal radiation of the body. For $\mathcal{S}_z$ at one polarization we have the expression (3.53), so that, considering both independent polarizations, $\mathcal{S}_z = c\langle |E_1|^2\rangle/\pi$. We next substitute E_1 from (3.52) and write the scalar products of the vectors in components. This, by (3.132), gives

$$\begin{aligned} Q_p = \frac{cA\Sigma}{\pi} \iint_V \{ & E_{0\alpha}(\boldsymbol{r}) E^*_{0\beta}(\boldsymbol{r}') \langle j_{e\alpha}(\boldsymbol{r}) j^*_{e\beta}(\boldsymbol{r}')\rangle \\ & + H_{0\alpha}(\boldsymbol{r}) H^*_{0\beta}(\boldsymbol{r}') \langle j_{m\alpha}(\boldsymbol{r}) j^*_{m\beta}(\boldsymbol{r}')\rangle - E_{0\alpha}(\boldsymbol{r}) H^*_{0\beta}(\boldsymbol{r}') \langle j_{e\alpha}(\boldsymbol{r}) j^*_{m\beta}(\boldsymbol{r}')\rangle \\ & - H_{0\alpha}(\boldsymbol{r}) E^*_{0\beta}(\boldsymbol{r}') \langle j_{m\alpha}(\boldsymbol{r}) j^*_{e\beta}(\boldsymbol{r}')\rangle \} \, d^3r \, d^3r' \ . \end{aligned} \tag{3.133}$$

Let us now compute Q_b—the power absorbed by the body from the radiation of the plate. The size of the latter being large as compared with λ, the intensity of plate radiation in a solid angle do with the axis directed at the body will be given by (3.56).

We compare (3.56) with the power beamed within the same solid angle do by a unit point source (an electric dipole at point M with the moment $p = 1/i\omega$ that lies in the plane of the plate)

$$dP_{0\omega} = \frac{\omega^2}{8\pi c^3} do \ .$$

We see that the plate radiates

$$\frac{dP_\omega}{dP_{0\omega}} = \frac{2c\Theta}{\pi^2} A\Sigma$$

times more than the point source. Hence, by exactly the same factor, Q_b is greater than Q_0 which is absorbed by the body from the radiation by a point source

$$Q_b = \frac{2c\Theta}{\pi^2} A\Sigma Q_0 \ .$$

But the losses Q_0 of the diffraction field produced by the electric dipole are

$$Q_0 = -\frac{i\omega}{16\pi}\int_V \{E_{0\alpha}(\boldsymbol{r})E^*_{0\beta}(\boldsymbol{r})(\varepsilon_{\alpha\beta}-\varepsilon^*_{\beta\alpha}) + H_{0\alpha}(\boldsymbol{r})H^*_{0\beta}(\boldsymbol{r})(\mu_{\alpha\beta}-\mu^*_{\beta\alpha})\}d^3r \ ,$$

so that

$$Q_b = -\frac{i\omega c A\Sigma\Theta(\omega, T)}{8\pi^3}\int_V \{E_{0\alpha}(\boldsymbol{r})E^*_{0\beta}(\boldsymbol{r})(\varepsilon_{\alpha\beta}-\varepsilon^*_{\beta\alpha}) + H_{0\alpha}(\boldsymbol{r})H^*_{0\beta}(\boldsymbol{r})(\mu_{\alpha\beta}-\mu^*_{\beta\alpha})\}d^3r \ . \quad (3.134)$$

Expressions (3.133, 134) must be equal. A term-by-term comparison of the integrands gives (3.26). The uniqueness of the results follows from the fact that the integrals must be equal for any form, size and arrangement of the body and plate (of course the conditions $\lambda R \gg \Sigma \gg \lambda^2$ must not be violated).

It should be observed that in this derivation of (3.26), one should understand by $\Theta(\omega, T)$ only that part of the mean energy of the oscillator that varies with temperature. The energy of the zero oscillations $\hbar\omega/2$ should not be taken into account because we consider here the energy *exchange* through the agency of *radiation* (Sect. 3.5).

3.12.3 In a waveguide there are two black-body radiators with temperature T_0 and, in between them, a body with temperature $T \neq T_0$. Find the energy flux of the fluctuational field at the nth natural wave of the waveguide, and show that at any T and T_0 the zero oscillations are excluded from this flux.

Solution. The energy flux, say, on the right-hand side of the body equals its self-radiation ($P^+_{\omega n}$) plus the reflected radiation (by the body) of the right radiator $\Theta_0\sum'_m R^-_{mn}/2\pi$ plus the transmitted radiation of the left radiator $\Theta_0\sum'_m D^+_{mn}/2\pi$ minus the back radiation flux on the right-hand side $\Theta_0/2\pi$

$$P_{\omega n} = P^+_{\omega n} + \frac{\Theta_0}{2\pi}\sum_m{}' (R^-_{mn} + D^+_{mn}) - \frac{\Theta_0}{2\pi} \ .$$

According to (3.59),

$$P^+_{\omega n} = \frac{\Theta}{2\pi}\left[1 - \sum_m{}' (R^-_{mn} + D^+_{mn})\right] ,$$

so that

$$P_{\omega n} = \frac{\Theta - \Theta_0}{2\pi}\left[\sum_m{}' (R_{mn}^- + D_{mn}^+) - 1\right].$$

But now the zero energy $\hbar\omega/2$ does not enter into the difference $\Theta - \Theta_0$, i.e., the opposite energy fluxes of the zero oscillations cancel out at any T and T_0. Because these oscillations never contribute to the energy *transport*, the term $\hbar\omega/2$ in $\Theta(\omega, T)$ may be ignored in all cases where we deal with the (mean) energy flux.

3.12.4 Let a radiator in an ideal waveguide consist of a transverse ideally reflecting partition and a transverse semi-transparent plate at a distance l from the partition, see Fig. 3.9. The thickness of the plate is taken to be small as compared to the wavelength in its material, so that we can ignore the interference within the plate and characterize the plate by *real* amplitude coefficients of reflection r and transmission d. So $r^2 + d^2 + a = 1$, where a is the energy absorption coefficient of the plate. Find the power this radiator emits for a wave with constant wavenumber h.

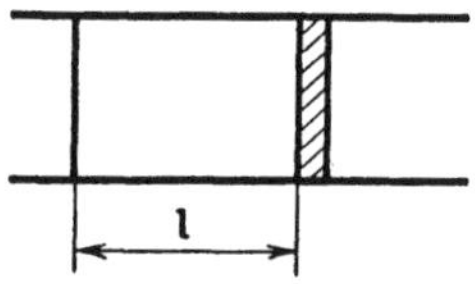

Fig. 3.9. A semi-transparent plate in a waveguide

Solution. The radiated power is $P = \Theta A/2\pi$, where $A = 1 - \mathscr{R}$ (the radiator as a whole is opaque), $\mathscr{R}$ being the energy reflection coefficient of the radiator. If the wave incident on the right is $\exp(-ihz)$, then, as a result of a first reflection from the plate and subsequent multiple reflections of the transmitted wave between the plate and the ideally reflecting partition, the amplitude of the wave reflected by the radiator will be

$$r + d^2 e^{-2ihl} + rd^2 e^{-4ihl} + r^2 d^2 e^{-6ihl} + \ldots = r + \frac{d^2}{e^{2ihl} - r}.$$

Hence,

$$A = 1 - \mathscr{R} = 1 - \left| r + \frac{d^2}{e^{2ihl} - r}\right|^2 = a\left[1 + \frac{d^2}{1 + r^2 - 2r\cos 2hl}\right].$$

The absorption depends on l in an oscillatory manner with a period equal to half the wavelength $\Lambda = 2\pi/h$, and with maxima at resonance lengths of the resonator formed by partition and plate. As the plate transparency is reduced ($d \to 0$), we eventually arrive at $A = a$. If $r^2 \to 1$ (in that case, of necessity, $d \to 0$, $a \to 0$), then $A \to 0$.

3.12.5 The *Nyquist formula* for the spectral density of the thermal e.m.f. $e(t)$ in the two-terminal network with impedance $Z(\omega)$ has the form

$$\langle|\tilde{e}(\omega)|^2\rangle = \frac{1}{\pi}\Theta(\omega, T)\mathrm{Re}\{Z(\omega)\} \ .$$

For good conductors, $\mathrm{Re}\{Z\}$ is often replaced by R. When is this change legitimate? After all, as one moves from conductor to ideal dielectrics (conductivity $\sigma \to 0$, and hence $R \to \infty$), the e.m.f. grows indefinitely. As was noted in Sect. II.3.5, this apparent paradox only occurs if we forget that the substitution of R for $\mathrm{Re}\{Z\}$ is legitimate solely for conductors that are so good that within them we can neglect the displacement current as compared to the conduction current. It is quite obvious that as $\sigma \to 0$, this condition is sooner or later violated. Derive an equation for $\langle|\tilde{e}(\omega)|^2\rangle$ in a thin conductor, taking into consideration the displacement current.
Solution. If in deriving the equation (3.74) for $\langle|I(s)|^2\rangle$, and hence the expression (3.78) for $\langle|\hat{\mathscr{E}}(s)|^2\rangle$, we do not ignore the displacement current in the conductor, i.e., we use the general expression (3.70) for the correlation of the thermal field components, then we will again arrive at (3.74) and (3.78), but, as opposed to (3.75), we now have

$$dQ_0 = \frac{|\boldsymbol{j}_0|^2}{2\sigma}\frac{d^3r}{1+(\varepsilon\omega/4\pi\sigma)^2} \ .$$

Accordingly, for the quasi-stationary (I_0 is independent of s), equilibrium ($\Theta = \mathrm{const}$) network, the "energy" resistance, defined by an equation similar to (3.80), is

$$R_{\mathrm{en}} = \frac{1}{|I_0|^2}\int\frac{|\boldsymbol{j}_0|^2 d^3r}{\sigma[1+(\varepsilon'\omega/4\pi\sigma)^2]} \ . \tag{3.135}$$

This expression is most conveniently clarified by passing to the limiting case of the weak skin-effect in a conductor made of a homogeneous material (ε' and σ are constant). Then $\boldsymbol{j}_0 = I_0/\Sigma$, and for a conductor of length l, (3.135) becomes

$$R_{\mathrm{en}} = \frac{R}{1+(RC\omega)^2} \ ,$$

where $R = 1/\sigma\Sigma$ is the *resistance*, $C = \varepsilon'\Sigma/4\pi l$ is the *internal* capacitance of the conductor. Thus, R_{en} is here the resistance of an RC-cell (Fig. 3.10), and as $R \to \infty$

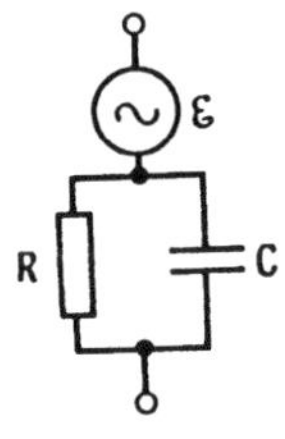

Fig. 3.10. The equivalent diagram of a section of a hot body

we get $R_{en} = 0$. The substitution of R for R_{en} is only legitimate on condition that $(RC\omega)^2 \ll 1$.

3.12.6 Use (3.19a, b) to obtain the covariances of $\omega\kappa$-amplitudes of the fields $\xi(t, \boldsymbol{r})$ and $f(t, \boldsymbol{r})$ in an infinite homogeneous medium, i.e., of the amplitudes in the spatial Fourier transforms

$$\xi(\omega, \boldsymbol{r}) = \int_{-\infty}^{+\infty} \xi(\omega, \boldsymbol{\kappa})\exp(\mathrm{i}\boldsymbol{\kappa}\cdot\boldsymbol{r})\, d^3\kappa, \quad f(\omega, \boldsymbol{r}) = \int_{\infty}^{+\infty} f(\omega, \boldsymbol{\kappa})\exp(\mathrm{i}\boldsymbol{\kappa}\cdot\boldsymbol{r})\, d^3\kappa \ .$$

Solution. First, we work out the covariance for the $\omega\kappa$-amplitude of $f(t, \boldsymbol{r})$

$$f(\omega, \boldsymbol{\kappa}) = \frac{1}{(2\pi)^3} \int_{-\infty}^{+\infty} f(\omega, \boldsymbol{r})\exp(-\mathrm{i}\boldsymbol{\kappa}\cdot\boldsymbol{r})\, d^3r \ .$$

From (3.19a) we have

$$\begin{aligned}
&\langle f(\omega, \boldsymbol{\kappa}) f^*(\omega, \boldsymbol{\kappa}')\rangle \\
&= \frac{1}{(2\pi)^6} \iint_{-\infty}^{+\infty} \langle f(\omega, \boldsymbol{r}) f^*(\omega, \boldsymbol{r}')\rangle \exp(-\mathrm{i}\boldsymbol{\kappa}\cdot\boldsymbol{r} + \mathrm{i}\boldsymbol{\kappa}'\cdot\boldsymbol{r}')\, d^3r\, d^3r' \\
&= -\frac{\mathrm{i}\Theta(\omega, T)}{(2\pi)^7\omega} \iint_{-\infty}^{+\infty} (\hat{A}^{-1*} - \hat{A}^{-1})\delta(\boldsymbol{r}-\boldsymbol{r}')\exp(-\mathrm{i}\boldsymbol{\kappa}\cdot\boldsymbol{r} + \mathrm{i}\boldsymbol{\kappa}'\cdot\boldsymbol{r}')\, d^3r\, d^3r' \ .
\end{aligned} \tag{3.136}$$

Now we represent the delta-function as a Fourier integral

$$\delta(\boldsymbol{r}-\boldsymbol{r}') = \frac{1}{(2\pi)^3} \int_{-\infty}^{+\infty} \mathrm{e}^{\mathrm{i}\boldsymbol{p}\cdot(\boldsymbol{r}-\boldsymbol{r}')}\, d^3p \ .$$

Since $\nabla\exp(\mathrm{i}\boldsymbol{p}\boldsymbol{r}) = \mathrm{i}\boldsymbol{p}\exp(\mathrm{i}\boldsymbol{p}\cdot\boldsymbol{r})$, operating with $\hat{A}^{-1}(\nabla)$ on $\exp(\mathrm{i}\boldsymbol{p}\cdot\boldsymbol{r})$ gives

$$\hat{A}^{-1}(\nabla)\delta(\boldsymbol{r}-\boldsymbol{r}') = \frac{1}{(2\pi)^3} \int_{-\infty}^{+\infty} A^{-1}(\mathrm{i}\boldsymbol{p})\mathrm{e}^{\mathrm{i}\boldsymbol{p}\cdot(\boldsymbol{r}-\boldsymbol{r}')}\, d^3p \ ,$$

where $A^{-1}(\mathrm{i}\boldsymbol{p})$ is a function of $\boldsymbol{p}$ resulting from the change $\nabla \to \mathrm{i}\boldsymbol{p}$ in the operator $\hat{A}^{-1}(\nabla)$. Similarly[20],

$$\hat{A}^{-1*}(\nabla')\delta(\boldsymbol{r}-\boldsymbol{r}') = \frac{1}{(2\pi)^3} \int_{-\infty}^{+\infty} A^{-1*}(\mathrm{i}\boldsymbol{p})\mathrm{e}^{\mathrm{i}\boldsymbol{p}\cdot(\boldsymbol{r}-\boldsymbol{r}')}\, d^3p \ ,$$

[20] In connection with this and subsequent formulas a clarification is in order. The complex conjugation asterisk is hereinafter understood in its conventional sense, i.e., as a substitution of $-\mathrm{i}$ for i in *all* the arguments of a real function of $-\mathrm{i}\omega$, $\mathrm{i}\boldsymbol{\kappa}$, $\mathrm{i}\boldsymbol{p}$, and so on. There is another (less convenient) convention [3.6]: the asterisk denotes complex conjugation in the *spectral amplitude* alone, i.e., in the real function of $\mathrm{i}\omega$ and $\boldsymbol{r}$, and thus reduces to the change $-\mathrm{i}\omega \to +\mathrm{i}\omega$. Therefore, in [3.6] the integrand in the expression for $\hat{A}^{-1*}(\nabla')\delta(\boldsymbol{r}-\boldsymbol{r}')$ contains the function $\hat{A}^{-1*}(-\mathrm{i}\boldsymbol{p})$. The asterisk denotes a change of sign at $\mathrm{i}\omega$, and in the argument $-\mathrm{i}\boldsymbol{p}$ the minus sign is written explicitly. Clearly, the conventional notation used in this book is simpler.

so that (3.136) takes the form

$$\langle f(\omega,\boldsymbol{\kappa})f^*(\omega,\boldsymbol{\kappa}')\rangle = -\frac{\mathrm{i}\Theta(\omega,T)}{(2\pi)^{10}\omega}\int\limits_{-\infty}^{+\infty}\{A^{-1*}(\mathrm{i}\boldsymbol{p})-A^{-1}(\mathrm{i}\boldsymbol{p})\}\,d^3p$$
$$\times\iint\limits_{-\infty}^{+\infty}\exp\{\mathrm{i}[(\boldsymbol{\varrho}-\boldsymbol{\kappa})\cdot\boldsymbol{r}-(\boldsymbol{\varrho}-\boldsymbol{\kappa}')\cdot\boldsymbol{r}']\}\,d^3r\,d^3r' \ .$$

The integrals with respect to $\boldsymbol{r}$ and $\boldsymbol{r}'$ give $(2\pi)^6\delta(\boldsymbol{p}-\boldsymbol{\kappa})\delta(\boldsymbol{p}-\boldsymbol{\kappa}')$, whereupon integration with respect to $\boldsymbol{p}$ gives

$$\langle f(\omega,\boldsymbol{\kappa})f^*(\omega,\boldsymbol{\kappa}')\rangle = -\frac{\mathrm{i}\Theta(\omega,T)}{(2\pi)^4\omega}\{A^{-1*}(\mathrm{i}\boldsymbol{\kappa})-A^{-1}(\mathrm{i}\boldsymbol{\kappa})\}\,\delta(\boldsymbol{\kappa}-\boldsymbol{\kappa}') \ . \tag{3.137}$$

We may obtain the covariance for $\xi(\omega,\boldsymbol{\kappa})$ in an analogous way [using (3.19a)], or by a different method based on the fact that $\xi(\omega,\boldsymbol{\kappa})=A(-\mathrm{i}\omega,\mathrm{i}\boldsymbol{\kappa})f(\omega,\boldsymbol{\kappa})$. Putting $\boldsymbol{\kappa}'=\boldsymbol{\kappa}$ in the factor at $\delta(\boldsymbol{\kappa}-\boldsymbol{\kappa}')$ gives

$$\begin{aligned}\langle\xi(\omega,\boldsymbol{\kappa})\xi^*(\omega,\boldsymbol{\kappa}')\rangle &= A(\mathrm{i}\boldsymbol{\kappa})A^*(\mathrm{i}\boldsymbol{\kappa}')\langle f(\omega,\boldsymbol{\kappa})f^*(\omega,\boldsymbol{\kappa}')\rangle\\ &= -\frac{\mathrm{i}\Theta(\omega,T)}{(2\pi)^4\omega}A(\mathrm{i}\boldsymbol{\kappa})A^*(\mathrm{i}\boldsymbol{\kappa})\{A^{-1*}(\mathrm{i}\boldsymbol{\kappa})-A^{-1}(\mathrm{i}\boldsymbol{\kappa})\}\,\delta(\boldsymbol{\kappa}-\boldsymbol{\kappa}')\\ &= -\frac{\mathrm{i}\Theta(\omega,T)}{(2\pi)^4\omega}\{A(\mathrm{i}\boldsymbol{\kappa})-A^*(\mathrm{i}\boldsymbol{\kappa})\}\,\delta(\boldsymbol{\kappa}-\boldsymbol{\kappa}') \ .\end{aligned} \tag{3.138}$$

Equations (3.9) (if the variation of $\hat{A}$ and $\hat{A}^{-1}$ with $-\mathrm{i}\omega$ is written explicitly) for the $\omega\boldsymbol{\kappa}$-amplitudes take the form

$$\xi(\omega,\boldsymbol{\kappa})=A(-\mathrm{i}\omega,\mathrm{i}\boldsymbol{\kappa})f(\omega,\boldsymbol{\kappa}),\quad f(\omega,\boldsymbol{\kappa})=A^{-1}(-\mathrm{i}\omega,\mathrm{i}\boldsymbol{\kappa})\xi(\omega,\boldsymbol{\kappa}) \ .$$

The first of these has just been used to derive (3.138), and the second suggests that the equation

$$A^{-1}(-\mathrm{i}\omega,\mathrm{i}\boldsymbol{\kappa})=0 \tag{3.139}$$

is the dispersion equation of the problem. It is seen from (3.138) that any integral with respect to $\boldsymbol{\kappa}$ and $\boldsymbol{\kappa}'$, containing $\langle\xi(\omega,\boldsymbol{\kappa})\xi^*(\omega,\boldsymbol{\kappa}')\rangle$, is reduced to an integral with respect to $\boldsymbol{\kappa}$ only [due to $\delta(\boldsymbol{\kappa}-\boldsymbol{\kappa}')$], and the latter can often be taken by residues at the poles of the function $A(-\mathrm{i}\omega,\mathrm{i}\boldsymbol{\kappa})$, i.e., at the zeros of $A^{-1}(\mathrm{i}\omega,\mathrm{i}\boldsymbol{\kappa})$ or, in other words, at the values of $\boldsymbol{\kappa}$ that are the solutions to the dispersion equation (3.139).

It is a straightforward exercise to reproduce the calculations performed in deriving (3.137, 138) for the case of the *multicomponent* homogeneous fields $\xi^{(j)}(t,\boldsymbol{r})$ and $f^{(j)}(t,\boldsymbol{r})$ using (3.22a, b). This leads to the correlation *matrices* (3.112, 113) for the $\omega\boldsymbol{\kappa}$-amplitudes.

3.12.7 For longitudinal waves, long as compared to the nonlocality radius a ($ka \ll 1$), a not very rarefied plasma consisting of electrons, various species of ions, and neutral atoms and molecules may adequately be described by the so-called *quasi-hydrodynamic* approximation. The equations of hydrodynamics have to be written here for all the species of particles, but we will confine ourselves to the simplest model of the *single-liquid* (electron) plasma, i.e., we will take the ions to be stationary[21].

The linearized equation of motion for the electron liquid (gas) will then become

$$m\frac{\partial \boldsymbol{v}}{\partial t} + m\nu\boldsymbol{v} = -\frac{1}{N}\nabla\tilde{p} - e\boldsymbol{E} \ , \tag{3.140}$$

where m is the electron mass, e the electron charge, N the *mean* electron concentration, and $\tilde{p}$ the variable part of the pressure

$$\tilde{p} = \gamma\Theta\tilde{N} \ . \tag{3.141}$$

(γ is the Poisson ratio, $\Theta = kT$ the energy-related temperature of the electron gas, and $\tilde{N}$ the variable part of the concentration). Equation (3.140) also includes the "friction"—through the effective number of collisions per unit time (ν). The electron current density (in this model this is also the *total* current) in a linear approximation is $\boldsymbol{j} = -Ne\boldsymbol{v}$, and the variable part of the charge density is $\tilde{\varrho} = -\tilde{N}e$ (the average charge densities of electrons and ions compensate each other so that the plasma is quasi-neutral). Therefore, the linearized continuity equation will be

$$\frac{\partial \tilde{N}}{\partial t} + N\mathrm{div}\,\boldsymbol{v} = 0 \ . \tag{3.142}$$

Find the spectral density of the electron concentration fluctuations for this plasma model.

Solution. According to (3.126), in order to find the $\omega\boldsymbol{\kappa}$-density of the fluctuations $\tilde{\varrho}$, it is sufficient to know the longitudinal electric permittivity $\varepsilon(\omega, \boldsymbol{\kappa})$. For the $\omega\boldsymbol{\kappa}$-amplitudes ($\tilde{N}, \tilde{p}, v, E \sim \exp[-\mathrm{i}(\omega t - \boldsymbol{\kappa r})]$), by eliminating from (3.140–142) all the variables save for $\boldsymbol{v} = -\boldsymbol{j}/Ne$, we obtain the following equation for the total current $\boldsymbol{j}$ (in components):

$$\left(1 + \mathrm{i}\frac{\nu}{\omega}\right) j_\alpha - \frac{\gamma\Theta}{m\omega^2}\kappa_\alpha\kappa_\beta j_\beta = \frac{\mathrm{i}\omega_\mathrm{e}^2}{4\pi\omega}E_\alpha \ ,$$

where ω_e is the plasma (electron) frequency

$$\omega_\mathrm{e}^2 = 4\pi Ne^2/m \ .$$

It is convenient to introduce the notation

$$X = \omega_\mathrm{e}^2/\omega^2, \quad Z = \nu/\omega_\mathrm{e}, \quad a = \sqrt{\gamma}d \ , \quad \text{where}$$

$$d = \frac{1}{\omega_\mathrm{e}}\sqrt{\frac{\Theta}{m}}$$

---

[21] For more general cases and, in particular, the *two-liquid* nonequilibrium model (electrons and one species of ions, the temperature of electrons and ions being different) see [3.15, 17].

is the so-called *Debye distance* (or the screening distance of the electric field, see [3.16]). In this notation the equation for $\boldsymbol{j}$ takes the form

$$(1+\mathrm{i}Z\sqrt{X})j_\alpha - a^2 X\kappa_\alpha\kappa_\beta j_\beta = \frac{\mathrm{i}\omega}{4\pi}XE_\alpha\ , \quad \text{whence}$$

$$j_\alpha = \frac{\mathrm{i}\omega X}{4\pi(1+\mathrm{i}Z\sqrt{X})}\left(E_\alpha + \frac{a^2X\kappa_\alpha\kappa_\beta E_\beta}{1+\mathrm{i}Z\sqrt{X}-\kappa^2a^2X}\right) .$$

The electric induction $D_\alpha = \varepsilon_{\alpha\beta}E_\beta$ is related to the total current $\boldsymbol{j}$ by the relationship $\boldsymbol{D} = \boldsymbol{E} - 4\pi\,\boldsymbol{j}/\mathrm{i}\omega$[22], so that

$$\begin{aligned} D_\alpha &= E_\alpha - \frac{4\pi}{\mathrm{i}\omega}j_\alpha \\ &= \left\{\left(1-\frac{X}{1+\mathrm{i}Z\sqrt{X}}\right)\delta_{\alpha\beta} - \frac{a^2X^2\kappa_\alpha\kappa_\beta}{(1+\mathrm{i}Z\sqrt{X})[1+\mathrm{i}Z\sqrt{X}-\kappa^2a^2X]}\right\}E_\beta\ . \end{aligned}$$

The expression in braces is the electric permittivity $\varepsilon_{\alpha\beta}(\omega, \boldsymbol{\kappa})$. It is easily transformed to the form

$$\begin{aligned} \varepsilon_{\alpha\beta}(\omega, \boldsymbol{\kappa}) = &\left(1-\frac{X}{1+\mathrm{i}Z\sqrt{X}}\right)\left(\delta_{\alpha\beta} - \frac{\kappa_\alpha\kappa_\beta}{\kappa^2}\right) \\ &+ \left(1-\frac{X}{1+\mathrm{i}Z\sqrt{X}-\kappa^2a^2X}\right)\frac{\kappa_\alpha\kappa_\beta}{\kappa^2}\ , \end{aligned}$$

whence the expressions for the transverse and longitudinal permittivities will be

$$\begin{aligned} &\varepsilon_\mathrm{t}(\omega, \kappa) = 1 - \frac{X}{1+\mathrm{i}Z\sqrt{X}}\ , \\ &\varepsilon_\mathrm{l}(\omega, \kappa) \equiv \varepsilon(\omega, \kappa) = 1 - \frac{X}{1+\mathrm{i}Z\sqrt{X}-\kappa^2a^2X}\ . \end{aligned} \tag{3.143}$$

Thus, in the model considered, ε_t does not possess spatial dispersion (it is independent of κ). But the spatial dispersion of $\varepsilon_\mathrm{l} = \varepsilon$ is due to the term with $\nabla\tilde{p}$ in (3.140), i.e., due to the *elasticity* of the electron gas.

According to expression (3.143), for ε we have

$$\frac{1}{\varepsilon} - \frac{1}{\varepsilon^*} = -\frac{2\mathrm{i}ZX^{3/2}}{(1-X-\kappa^2a^2X)^2+Z^2X}\ .$$

Substituting this into (3.126, 127), we find that the $\omega\boldsymbol{\kappa}$-density of the fluctuations of the electron concentration $\tilde{N} = -\tilde{\varrho}/e$ is

$$G^{(\tilde{N})}(\omega, \kappa) = \frac{1}{e^2}G^{(\tilde{\varrho})}(\omega, \kappa) = \frac{\Theta ZX^2}{32\pi^5\omega_\mathrm{e}e^2}\,\frac{\kappa^2}{(1-X-\kappa^2a^2X)^2+Z^2X}\ , \tag{3.144}$$

[22] This follows from the first of (3.122) if written in the form $\mathrm{i}[\boldsymbol{\kappa}\tilde{\boldsymbol{B}}] = -\mathrm{i}k_0\boldsymbol{D}$.

and its spectral density is

$$g^{(\tilde{N})}(\omega, \varrho) \equiv \frac{\Theta Z X^2}{8\pi^4 \omega_e e^2 \varrho} \int_0^\infty \frac{\kappa^3 \sin \kappa\varrho}{(1 - X - \kappa^2 a^2 X)^2 + Z^2 X} \, d\kappa \,. \tag{3.145}$$

The integral in (3.145) can easily be taken. The integrand with respect to κ being even, we can take half the value of the integral from $-\infty$ to $+\infty$. Further, we write $\sin \kappa\varrho$ in the form $[\exp(\mathrm{i}\kappa\varrho) - \exp(-\mathrm{i}\kappa\varrho)]/2\mathrm{i}$ and change the sign of κ in the integral with $\exp(-\mathrm{i}\kappa\varrho)$. This simply doubles the first integral [with $\exp(\mathrm{i}\kappa\varrho)$] and gives

$$g^{(\tilde{N})}(\omega, \varrho) \equiv \frac{\Theta Z X^2}{16\pi^4 \mathrm{i} \omega_e e^2 \varrho} \int_{-\infty}^{+\infty} \frac{\mathrm{e}^{\mathrm{i}\kappa\varrho} \kappa^3 d\kappa}{(1 - X - \kappa^2 a^2 X)^2 + Z^2 X} \,.$$

The closing of the integration path in the upper half-plane of the complex κ reduces the integral to the residues at the poles $\kappa = k$ and $\kappa = -k^*$; where k is the solution to the dispersion equation $\varepsilon(\omega, \kappa) = 0$ that lies in the first quadrant of the κ-plane

$$k = k' + \mathrm{i}k'' = \frac{1}{a\sqrt{2X}} \left\{ \sqrt{\sqrt{(1-X)^2 + Z^2 X} + (1 - X)} \right.$$

$$\left. + \mathrm{i} \sqrt{\sqrt{(1-X)^2 + Z^2 X} - (1 - X)} \right\} \,. \tag{3.146}$$

The final result is

$$g^{(\tilde{N})}(\omega, \varrho) = \frac{N m v}{4\pi^2 \gamma^2 \Theta} \frac{\mathrm{e}^{-k''\varrho}}{\varrho} \left[\cos k'\varrho + \frac{1 - X}{Z\sqrt{X}} \sin k'\varrho \right] \,. \tag{3.147}$$

As is seen from (3.145), $g^{(\tilde{N})}(\omega, 0) = \infty$. This singularity at $\varrho = 0$, which is quite obvious in (3.147), stems from the fact that $1/\varepsilon - 1/\varepsilon^*$ vanishes insufficiently fast with increasing κ. But the hydrodynamic description itself is only valid for $\kappa a \ll 1$, i.e., for spatial harmonics of fluctuations of $\tilde{N}$ with wavelengths $2\pi/\kappa \equiv \Lambda \gg a$. It serves no purpose, therefore, to take values $\varrho \lesssim a$. In this region of κ, the phase velocity of longitudinal waves becomes of the same order as the thermal velocity of electrons and, as is shown by the kinetic analysis, the longitudinal waves are strongly damped. The hydrodynamic model does not include this damping caused by the thermal motion of charges [Ref. 3.15, Sect. 2]. In this connection we note that, say, in the F layer of the ionosphere, if we put $N = 5 \cdot 10^5\ \mathrm{cm}^{-3}$, $T = 300\ \mathrm{K}$, and $\gamma = 3$[23], then

$$\omega_e = 4 \cdot 10^7\ \mathrm{Hz}, \quad a \approx 0.3\ \mathrm{cm} \,.$$

In the frequency range $\omega > \omega_e$ ($X < 1$) and for $Z \ll 1 - X$, by (3.146), we have approximately

[23] This choice of γ is based on a kinetic treatment [Ref. 3.15, Sect. 2].

$$k' \approx \frac{1}{a}\sqrt{\frac{1-X}{X}}\,, \quad k'' \approx \frac{Z}{2a\sqrt{1-X}}\,,$$

i.e., $k' \gg k''$, and (3.147) is dominated by the term with $\sin k'\varrho$. The spectral density (3.147) is a slowly damped oscillation as a function of ϱ. On the contrary, for $\omega < \omega_e$ $(X > 1)$ and $Z \ll X-1$, the values of k' and k'' are interchanged

$$k' \approx \frac{Z}{2a\sqrt{X-1}}\,, \quad k'' \approx \frac{1}{a}\sqrt{\frac{X-1}{X}}\,,$$

so that the exponential function decays faster than during a period of oscillation $2\pi/k'$. In the vicinity of ω_e, i.e., at $X \approx 1$ and $|1-X| \ll Z$, we have

$$k' \approx k'' \approx \frac{1}{a}\sqrt{\frac{Z}{2}}\,.$$

Note in conclusion that, in confining ourselves to a description of the electron gas in terms of $\boldsymbol{v}$, $\tilde{p}$, and $\tilde{N}$ within the framework of the hydrodynamic model, we have thereby excluded from consideration fluctuations of temperature $\tilde{T} = T - \bar{T}$.

4. Single Scattering Theory

This chapter discusses the simplest case of wave scattering in a randomly inhomogeneous medium, or rather the situation where fluctuations of the properties of the medium are so small that the scattered wave field is fairly small as compared to the field of the primary wave. Of the approximate techniques for handling wave problems, it is natural in this situation to give preference to the *method of small perturbations*, small enough so that one can confine oneself in first approximation to fluctuations of the wave, i.e., to the theory of the so-called single scattering (Sect. 4.1).

Given this solution, we can then easily derive the covariance and the mean intensity of the scattered field, and also find the effective scattering cross-section and estimate the applicability of the above approximation (Sects. 4.2, 3). In Sect. 4.4 we take a closer look at the spatial correlation of the scattered field.

For simplicity, the above issues are considered for *scalar* wave fields, *monochromatic* primary waves and spatially quasi-nonhomogeneous, *time-independent* media. This is sufficient to identify some typical wave features in scattering.

Of considerable interest, however, are also those problems in physics in which the above constraints are partially lifted. Therefore, later sections discuss more general cases: scattering from time-dependent inhomogeneities (Sect. 4.5) and the scattering of the quasi-monochromatic primary wave, pulsed or modulated (Sect. 4.6). The scattering of vector (electromagnetic) waves and inherent polarization effects are only studied for the case of an isotropic nondispersive random medium (Sect. 4.7).

Under certain constraints the approximation of single scattering is also applicable to the case of discrete scatterers, noncontinuous inhomogeneities of a continuous medium (Sect. 4.8).

4.1 Method of Small Perturbations

In this chapter and later on in Volume IV we will be concerned with the fluctuations of waves propagating through randomly inhomogeneous media, i.e., with problems of type (2) according to the classification given in Sect. 1.8.

Random inhomogeneities of real media affect the characteristics of waves propagating in these media, thus giving rise to an astonishing variety of phenomena. Stellar scintillations and extraterrestrial radiation fluctuations, the fading of radio waves and the Rayleigh scattering of light, laser beam broadening in the troposphere and sound scattering in the sea, to name but a few examples of the effects observed. These effects are the concern of the statistical theory of wave propagation and scattering.

The problems of wave propagation in media with fluctuating parameters are generally handled using *approximate* methods. The point is that the appropriate differential equations have coefficients that include random functions of the coordinates (and perhaps time, too) that describe the random medium. An *exact* solution of such a *parametric* problem would suggest that we are able to write, say, Green's function for any realizations of the random functions involved, which was never the case in practice[1]. It is for this reason that approximate methods have to be utilized here. Clearly, the choice of the approximation depends on the formulation of the problem, e.g., on the amount of fluctuations of the medium parameters, on the ratio of wavelength to inhomogeneity size, on the geometry of the problem (path length, beam width), and so on. For all the diversity of specific conditions, most problems of type (2) can be solved by a few techniques available so far.

If the relative fluctuations of the medium parameters are fairly weak, and the scattered field is small in comparison with the field of a primary wave, then the *method of small perturbations* is used. The fields derived using *first*-order perturbation theory are the subject matter of *single scattering theory* discussed in this chapter.

Should the assumptions of single scattering theory be violated (the fluctuations are not weak and/or the scattered field is not small), it will be necessary to consider the double, triple, etc., scattering of the field, i.e., to establish a theory of multiple scattering. In the case of weak but large (as compared to wavelength) inhomogeneities, multiply scattered waves deviate only slightly from the direction of propagation of the primary wave. Multiple scattering is then effectively described by the *geometrical optics method* (GOM) and related, more general, short-wave asymptotic methods of diffraction theory — the *method of smooth perturbation* (MSP) and the *parabolic equation method* (PEM). The last three methods will be considered in Volume IV.

Multiple scattering can also be taken into account by approximate summation of perturbative series (mainly using techniques that originated in quantum electrodynamics). Incidentally, this approach makes it possible to handle not only weak, but also strong, fluctuations of the medium. However, it requires that the inhomogeneities be small-scale. The elements of multiple scattering theory are also outlined in Volume IV.

Let us begin with the simplest case: the scalar monochromatic wave field $u(t, \boldsymbol{r})$ $[u(t, \boldsymbol{r}) = u(\boldsymbol{r})\exp(-\mathrm{i}\omega t)]$, and the *time-constant* inhomogeneities[2]. Although this scalar picture ignores polarization, it is still adequate to describe a number of general phenomena of wave theory, such as interference and diffraction. We will look at polarization effects in the scattering of electromagnetic waves in Sect. 4.7.

---

[1] With the exception of those rather specific cases where the medium is described by special forms of deterministic functions varying with a finite number of random parameters.

[2] In this chapter we will follow for the most part the treatment given in [4.1]. A bibliography on single scattering theory is also to be found in [4.2]. Among later works, *Ishimaru* [4.18] is worth noting.

Under the above conditions, the propagation of a wave in an inhomogeneous medium is described by the Helmholtz equation

$$\Delta u(\boldsymbol{r}) + k_0^2 \varepsilon(\boldsymbol{r}) u(\boldsymbol{r}) = 0 \ , \tag{4.1}$$

where $k_0 = \omega/c$ is the wave number in an unperturbed medium or (in the case of an electric field) in a vacuum. We will refer to $\varepsilon(\boldsymbol{r})$, a function describing the inhomogeneity of a medium, as (electric) permittivity, mainly of the electromagnetic field. For a randomly inhomogeneous medium, $\varepsilon(\boldsymbol{r})$ can be represented as

$$\varepsilon(\boldsymbol{r}) = \bar{\varepsilon}(\boldsymbol{r}) + \tilde{\varepsilon}(\boldsymbol{r}) \ , \tag{4.2}$$

where $\bar{\varepsilon}$ is the ensemble-averaged value of ε (over the ensemble of realizations of the medium), and $\tilde{\varepsilon}$ is the permittivity fluctuations. The Helmholtz equation then becomes[3]

$$\Delta u(\boldsymbol{r}) + k_0^2 [\bar{\varepsilon}(\boldsymbol{r}) + \tilde{\varepsilon}(\boldsymbol{r})] u(\boldsymbol{r}) = 0 \ . \tag{4.3}$$

There are no general techniques for solving even as simple a wave equation as this. Of the approximate methods, the most common is the *perturbation method*, which assumes the fluctuations $\tilde{\varepsilon}$ to be sufficiently weak, and seeks the wave field $u(\boldsymbol{r})$ as a series in $\tilde{\varepsilon}(\boldsymbol{r})$, or (which amounts to the same) in $\sigma_\varepsilon \ll \bar{\varepsilon}$. To construct the series it is convenient to go over from the differential equation (4.3) to an equivalent integral equation.

Let $u_0(\boldsymbol{r})$ be a primary wave field obeying the unperturbed Helmholtz equation,

$$\Delta u_0(\boldsymbol{r}) + k_0^2 \bar{\varepsilon}(\boldsymbol{r}) u_0(\boldsymbol{r}) = 0 \ . \tag{4.4}$$

We denote by $G(\boldsymbol{r}, \boldsymbol{r}')$ an unperturbed Green's function that obeys the equation for a point source

$$\Delta G(\boldsymbol{r}, \boldsymbol{r}') + k_0^2 \bar{\varepsilon}(\boldsymbol{r}) G(\boldsymbol{r}, \boldsymbol{r}') = \delta(\boldsymbol{r}, \boldsymbol{r}') \ . \tag{4.5}$$

Of course, the primary field u_0 and Green's function G meet the appropriate boundary conditions. The solution to the inhomogeneous equation

$$\Delta u + k_0^2 \bar{\varepsilon} u = F(\boldsymbol{r})$$

is expressed in terms of Green's function in the following way:

$$u(\boldsymbol{r}) = u_0(\boldsymbol{r}) + \int G(\boldsymbol{r}, \boldsymbol{r}') F(\boldsymbol{r}') d^3 r' \ . \tag{4.6}$$

[3] The scattering of sound waves in a turbulent medium is described by a more complicated scalar equation than (4.3), because as sound waves propagate through these media they interact nonlinearly with hydrodynamic turbulences. The treatment is simplified by linearizing the initial nonlinear equations of hydrodynamics in sound perturbations, if the latter are sufficiently weak. The relevant distinctive features of sound scattering are discussed in [4.1].

Writing the original equation (4.3) in the form

$$\Delta u + k_0^2 \bar{\varepsilon} u = -k_0^2 \tilde{\varepsilon} u \equiv F(\boldsymbol{r}) \tag{4.7}$$

and using (4.6), we obtain for the wave field

$$u(\boldsymbol{r}) = u_0(\boldsymbol{r}) - k_0^2 \int G(\boldsymbol{r}, \boldsymbol{r}') \tilde{\varepsilon}(\boldsymbol{r}') u(\boldsymbol{r}') d^3 r' , \tag{4.8}$$

where we clearly integrate over a region V occupied by the inhomogeneities $\tilde{\varepsilon}(\boldsymbol{r})$. Equation (4.8) is equivalent to the original differential equation (4.3), but it also meets (through G) all the boundary conditions.

The expansion in perturbations is derived by iterating the integral equation (4.8). To obtain the first iteration we write the value of the field at $\boldsymbol{r} = \boldsymbol{r}'$

$$u(\boldsymbol{r}') = u_0(\boldsymbol{r}') - k_0^2 \int G(\boldsymbol{r}', \boldsymbol{r}'') \tilde{\varepsilon}(\boldsymbol{r}'') u(\boldsymbol{r}'') d^3 r''$$

and substitute it into the right-hand side of (4.8). We thus have

$$\begin{aligned} u(\boldsymbol{r}) = {} & u_0(\boldsymbol{r}) - k_0^2 \int G(\boldsymbol{r}, \boldsymbol{r}') \tilde{\varepsilon}(\boldsymbol{r}') u_0(\boldsymbol{r}') d^3 r' \\ & + k_0^4 \int G(\boldsymbol{r}, \boldsymbol{r}') \tilde{\varepsilon}(\boldsymbol{r}') d^3 r' \int G(\boldsymbol{r}', \boldsymbol{r}'') \tilde{\varepsilon}(\boldsymbol{r}'') u(\boldsymbol{r}'') d^3 r'' . \end{aligned} \tag{4.9}$$

We now write the value of $u(\boldsymbol{r})$ at $\boldsymbol{r} = \boldsymbol{r}''$ and substitute it into the right-hand side of (4.9) to obtain the second iteration. By performing the operation repeatedly, we obtain the infinite perturbation series

$$\begin{aligned} u(\boldsymbol{r}) = {} & u_0(\boldsymbol{r}) - k_0^2 \int G(\boldsymbol{r}, \boldsymbol{r}') \tilde{\varepsilon}(\boldsymbol{r}') u_0(\boldsymbol{r}') d^3 r' \\ & + k_0^4 \iint G(\boldsymbol{r}, \boldsymbol{r}') G(\boldsymbol{r}', \boldsymbol{r}'') \tilde{\varepsilon}(\boldsymbol{r}') \tilde{\varepsilon}(\boldsymbol{r}'') u_0(\boldsymbol{r}'') d^3 r' \, d^3 r'' \\ & - k_0^6 \iiint G(\boldsymbol{r}, \boldsymbol{r}') G(\boldsymbol{r}', \boldsymbol{r}'') G(\boldsymbol{r}'', \boldsymbol{r}''') \tilde{\varepsilon}(\boldsymbol{r}') \tilde{\varepsilon}(\boldsymbol{r}'') \tilde{\varepsilon}(\boldsymbol{r}''') \\ & \times u_0(\boldsymbol{r}''') d^3 r' \, d^3 r'' \, d^3 r''' + \dots . \end{aligned} \tag{4.10}$$

In mathematics this series is called the Neumann series for the integral equation (4.8), whereas in physics it is known as the Born expansion.[4]

The first term of the Born series (4.10) is the primary field $u_0(\boldsymbol{r})$. The second,

$$u_1(\boldsymbol{r}) = -k_0^2 \int G(\boldsymbol{r}, \boldsymbol{r}') \tilde{\varepsilon}(\boldsymbol{r}') u_0(\boldsymbol{r}') d^3 r' , \tag{4.11}$$

describes the single-scattered field. It is caused directly by $u_0(\boldsymbol{r})$ and is linear in the perturbation $\tilde{\varepsilon}(\boldsymbol{r})$. The third term can be represented in the form (4.11)

$$u_2(\boldsymbol{r}) = -k_0^2 \int G(\boldsymbol{r}, \boldsymbol{r}') \tilde{\varepsilon}(\boldsymbol{r}') u_1(\boldsymbol{r}') d^3 r' . \tag{4.12}$$

This is the double-scattered field, which is now caused by the single-scattered field, not by the primary one. The double-scattered field in turn excites the tripple-scattered waves u_3, and so on. The series (4.10) is an expansion of the *scattered field*

---

[4] Born first applied perturbation theory to the scattering problem in quantum mechanics. It is true that, long before Born, a similar form of perturbation method was used by Rayleigh to treat light scattering by transparent bodies. In optics, even today, one speaks about *Rayleigh*, and not Born, light scattering.

$u_s \equiv u - u_0$ in the *scattering multiplicity*

$$u_s \equiv u - u_0 = u_1 + u_2 + u_3 + \dots \quad . \tag{4.13}$$

It is apparent from the method of constructing the series that the nth term describing the nth scattering contains the product $\tilde{\varepsilon}(\boldsymbol{r}_1) \dots \tilde{\varepsilon}(\boldsymbol{r}_n)$ in the integrand. It follows that even for the mean $\bar{u}$ to be computed, we have to know the moments $\langle \tilde{\varepsilon}(\boldsymbol{r}_1) \dots \tilde{\varepsilon}(\boldsymbol{r}_n) \rangle$ of any order. With arbitrary statistics of $\tilde{\varepsilon}$, to find these moments is a difficult problem, but even if it is solvable (as is, for example, the case with normal distribution), we still have to cope with the question of the *methods of summation of averaged expansions* in perturbations. We will discuss the general approach to the question in Volume IV. As already stated, we will confine ourselves here to the simpler problem of finding the statistical characteristics of the field in the *single* scattering approximation (the so-called *first Born* approximation, or simply the *Born* approximation).

This approximation assumes the fluctuations $\tilde{\varepsilon}$ to be so small that in (4.13) we can keep only the first term u_1. We already dealt with expressions of this type in Sect. 2.5, when we considered the excitation of fields by random sources. In this case, the role of the sources $q(\boldsymbol{r})$ is played by the right-hand side of (4.7), with u_0 substituted for u: $q(\boldsymbol{r}) = -k_0^2 \tilde{\varepsilon}(\boldsymbol{r}) u_0(\boldsymbol{r})$. Thus, in the single scattering approximation the problem of wave propagation through random media [a problem of type (2)] is reduced to a problem of type (1), i.e., the excitation of fields by given random sources.

According to (4.11), the scattered field $u_s \approx u_1$ is a *linear* functional of $\tilde{\varepsilon}$. Therefore, all the moments of u_1 are also expressed *linearly* in terms of the moments of $\tilde{\varepsilon}$ of the same order. In particular, the mean of the single-scattered field u_1 is zero, since $\langle \tilde{\varepsilon}(\boldsymbol{r}) \rangle = 0$, and the covariance $\psi_u(\boldsymbol{r}_1, \boldsymbol{r}_2) = \langle u_1(\boldsymbol{r}_1) u_1^*(\boldsymbol{r}_2) \rangle$ is expressed linearly through the covariance of the inhomogeneities $\psi_\varepsilon(\boldsymbol{r}', \boldsymbol{r}'')$

$$\psi_u(\boldsymbol{r}_1, \boldsymbol{r}_2) = k_0^4 \iint G(\boldsymbol{r}_1, \boldsymbol{r}') G^*(\boldsymbol{r}_2, \boldsymbol{r}'') u_0(\boldsymbol{r}') u_0^*(\boldsymbol{r}'') \psi_\varepsilon(\boldsymbol{r}', \boldsymbol{r}'') \, d^3r' \, d^3r'' \quad . \tag{4.14}$$

Expression (4.14) and similar quadratures for the higher order moments of u_1 in principle give the *complete* statistical solution of the problem in the Born approximation. We need, however, to attach a physical interpretation to this mathematical result.

4.2 Mean Intensity of Scattered Fields

In order to elucidate the behavior of scattering we will make a number of assumptions that simplify the treatment but, at the same time, are still sufficiently general for the many applications of the theory. The assumptions are the following:

a) The medium is on average homogeneous, i.e., $\bar{\varepsilon}_r = \text{const}$.
b) The primary field $u_0(\boldsymbol{r})$ is an omnidirectional *spherical* wave with center at $\boldsymbol{r}_0$

$$u_0(\boldsymbol{r}) = \frac{A}{|\boldsymbol{r} - \boldsymbol{r}_0|} e^{ik|\boldsymbol{r} - \boldsymbol{r}_0|} \quad , \tag{4.15}$$

where k is the wave number in the homogeneous medium

$$k = k_0\sqrt{\bar{\varepsilon}} = \frac{\omega}{c}\sqrt{\bar{\varepsilon}} \ .$$

c) The Green's function $G(\boldsymbol{r}, \boldsymbol{r}')$ describes the field of a point source in an infinite homogeneous medium

$$G(\boldsymbol{r}, \boldsymbol{r}') = -\frac{1}{4\pi}\frac{e^{ik|\boldsymbol{r}-\boldsymbol{r}'|}}{|\boldsymbol{r}-\boldsymbol{r}'|} \ . \tag{4.16}$$

d) The random field $\tilde{\varepsilon}$ is *statistically quasi-homogeneous*, i.e., the covariance of ε has the form (Sect. 1.5)

$$\psi_\varepsilon(\boldsymbol{r}', \boldsymbol{r}'') \equiv \langle\tilde{\varepsilon}(\boldsymbol{r}')\tilde{\varepsilon}(\boldsymbol{r}'')\rangle = \psi_\varepsilon(\boldsymbol{\varrho}, \boldsymbol{R}) \ , \tag{4.17}$$

where $\boldsymbol{\varrho} = \boldsymbol{r}' - \boldsymbol{r}''$, $\boldsymbol{R} = (\boldsymbol{r}' + \boldsymbol{r}'')/2$ and $\psi_\varepsilon(\boldsymbol{\varrho}, \boldsymbol{R})$ varies slowly with $\boldsymbol{R}$, i.e., the space scale L_ε for the variation of ψ_ε with $\boldsymbol{R}$ is substantially larger than the characteristic scale l_ε (the correlation radius for $\tilde{\varepsilon}$) for $\boldsymbol{\varrho}$. The spectral density of such fluctuations given by

$$\Phi_\varepsilon(\boldsymbol{\kappa}, \boldsymbol{R}) = \frac{1}{8\pi^3}\int \psi_\varepsilon(\boldsymbol{\varrho}, \boldsymbol{R})\exp(-i\boldsymbol{\kappa}\cdot\boldsymbol{\varrho})\,d^3\varrho \tag{4.18}$$

is a slow function of $\boldsymbol{R}$, i.e., it varies only slightly over distances of order l_ε.

e) Lastly, we will consider that random inhomogeneities fill the *bounded* volume V, and that there are *many* inhomogeneities in this volume. The last condition can be represented symbolically by

$$V \sim L^3 \gg l_\varepsilon^3 \quad \text{or} \quad L \sim V^{1/3} \gg l_\varepsilon \ , \tag{4.19}$$

where L is the transverse dimension of the region occupied by the inhomogeneities. The assumption that the scattering volume is finite is required so that the single-scattered field will be small, whereas (4.19) is only used to simplify the calculations.

The finite character of the scattering volume is conveniently described by using the step function

$$M(\boldsymbol{r}) = \begin{cases} 1 & \text{inside } V \ , \\ 0 & \text{elsewhere} \ . \end{cases} \tag{4.20}$$

Bringing it under the integration sign in (4.11) enables us to extend the integration to the whole of the space. For the primary spherical wave (4.1) and the Green's function of free space (4.2), the single-scattered field (4.11) will be

$$u_1(\boldsymbol{r}) = \frac{k_0^2 A}{4\pi}\int_{-\infty}^{\infty}\frac{\exp[ik(|\boldsymbol{r}-\boldsymbol{r}'| + |\boldsymbol{r}'-\boldsymbol{r}_0|)]}{|\boldsymbol{r}-\boldsymbol{r}'||\boldsymbol{r}'-\boldsymbol{r}_0|}M(\boldsymbol{r}')\tilde{\varepsilon}(\boldsymbol{r}')\,d^3r' \ . \tag{4.21}$$

In later sections we shall be looking at some of the more general formulations of the problem (variation of $\tilde{\varepsilon}$ with time, calculations of the polarization characteristics of an electromagnetic field, etc.). At the same time, we will sometimes make specific assumptions (plane wave, statistically homogeneous fluctuations, etc.). However, here we will derive expressions for the covariance and, in particular, for the mean intensity of the scattered field from (4.21).

According to (4.21), the spatial covariance of the field is

$$\psi_u(\boldsymbol{r}_1, \boldsymbol{r}_2) = \frac{k_0^4|A|^2}{(4\pi)^2} \iint\limits_{-\infty}^{\infty} M(\boldsymbol{r}')M(\boldsymbol{r}'')\psi_\varepsilon(\boldsymbol{\varrho}, \boldsymbol{R}) \times \frac{\exp\{ik(|\boldsymbol{r}_1-\boldsymbol{r}'|-|\boldsymbol{r}_2-\boldsymbol{r}''|+|\boldsymbol{r}'-\boldsymbol{r}_0|-|\boldsymbol{r}''-\boldsymbol{r}_0|)\}}{|\boldsymbol{r}_1-\boldsymbol{r}'||\boldsymbol{r}_2-\boldsymbol{r}''||\boldsymbol{r}'-\boldsymbol{r}_0||\boldsymbol{r}''-\boldsymbol{r}_0|} d^3r'\,d^3r'' . \tag{4.22}$$

This equation differs from (4.14) only in that the forms of the primary field and the Green's function are explicit here.

In (4.22) we will pass over to the integration variables $\boldsymbol{\varrho}=\boldsymbol{r}'-\boldsymbol{r}''$, $\boldsymbol{R}=(\boldsymbol{r}'+\boldsymbol{r}'')/2$, whence $\boldsymbol{r}'=\boldsymbol{R}+\boldsymbol{\varrho}/2$, $\boldsymbol{r}''=\boldsymbol{R}-\boldsymbol{\varrho}/2$, and $d^3r'\,d^3r''=d^3\varrho\,d^3R$.

Equation (4.22) takes the form

$$\psi_u(\boldsymbol{r}_1, \boldsymbol{r}_2) = \frac{k_0^4|A|^2}{(4\pi)^2} \iint\limits_{-\infty}^{\infty} \frac{e^{i(\Psi_1+\Psi_2)} M\left(\boldsymbol{R}+\frac{\boldsymbol{\varrho}}{2}\right) M\left(\boldsymbol{R}-\frac{\boldsymbol{\varrho}}{2}\right)\psi_\varepsilon(\boldsymbol{\varrho}, \boldsymbol{R})\, d^3\varrho\, d^3R}{\left|\boldsymbol{r}_1-\boldsymbol{R}-\frac{\boldsymbol{\varrho}}{2}\right|\left|\boldsymbol{r}_2-\boldsymbol{R}+\frac{\boldsymbol{\varrho}}{2}\right|\left|\boldsymbol{R}+\frac{\boldsymbol{\varrho}}{2}-\boldsymbol{r}_0\right|\left|\boldsymbol{R}-\frac{\boldsymbol{\varrho}}{2}-\boldsymbol{r}_0\right|} , \tag{4.23}$$

where

$$\begin{aligned} \Psi_1 &= k(|\boldsymbol{r}_1-\boldsymbol{r}'|-|\boldsymbol{r}_2-\boldsymbol{r}''|) = k\left(\left|\boldsymbol{r}_1-\boldsymbol{R}-\frac{\boldsymbol{\varrho}}{2}\right|-\left|\boldsymbol{r}_2-\boldsymbol{R}+\frac{\boldsymbol{\varrho}}{2}\right|\right) , \\ \Psi_2 &= k(|\boldsymbol{r}'-\boldsymbol{r}_0|-|\boldsymbol{r}''-\boldsymbol{r}_0|) = k\left(\left|\boldsymbol{R}-\boldsymbol{r}_0+\frac{\boldsymbol{\varrho}}{2}\right|-\left|\boldsymbol{R}-\boldsymbol{r}_0-\frac{\boldsymbol{\varrho}}{2}\right|\right) . \end{aligned} \tag{4.24}$$

All further simplifications of (4.23) are based on the fact that the covariance of the inhomogeneous medium, $\psi_\varepsilon(\boldsymbol{\varrho}, \boldsymbol{R})$, becomes extremely small when $\varrho \gg l_\varepsilon$. We can, therefore, make use of the expansion of $|\boldsymbol{r}-\boldsymbol{\varrho}|$ into a Taylor series in the powers of the small ratio ϱ/r

$$|\boldsymbol{r}-\boldsymbol{\varrho}| = \sqrt{(\boldsymbol{r}-\boldsymbol{\varrho})^2} \approx r - \boldsymbol{n}\cdot\boldsymbol{\varrho} + \frac{1}{2r}\varrho_\perp^2 + O\left(\frac{\varrho^3}{r^2}\right) , \tag{4.25}$$

where $\boldsymbol{n}=\boldsymbol{r}/r$, and $\boldsymbol{\varrho}_\perp$ is the component of $\boldsymbol{\varrho}$ which is normal to $\boldsymbol{n}: \boldsymbol{\varrho}_\perp = \boldsymbol{\varrho}-\boldsymbol{n}(\boldsymbol{n}\cdot\boldsymbol{\varrho})$.

We apply (4.25) to each of the terms in Ψ_1 and Ψ_2. For Ψ_1 we get

$$\Psi_1 = k\left[s-\frac{1}{2}(\boldsymbol{n}_{s1}+\boldsymbol{n}_{s2})\cdot\boldsymbol{\varrho}+\frac{\varrho_{\perp 1}^2}{8|\boldsymbol{r}_1-\boldsymbol{R}|}-\frac{\varrho_{\perp 2}^2}{8|\boldsymbol{r}_2-\boldsymbol{R}|}+O\left(\frac{\varrho^3}{|\boldsymbol{r}_{1,2}-\boldsymbol{R}|^2}\right)\right] , \tag{4.26}$$

where

$$s = |\boldsymbol{r}_1-\boldsymbol{R}|-|\boldsymbol{r}_2-\boldsymbol{R}| \tag{4.27}$$

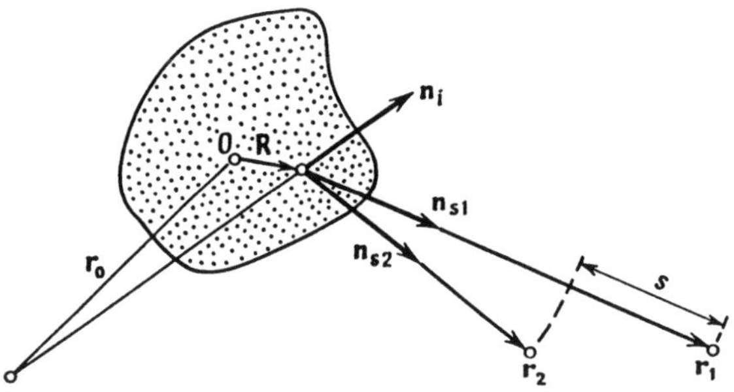

Fig. 4.1. The arrangement of the observation points $\boldsymbol{r}_1$ and $\boldsymbol{r}_2$ and the scattering volume

is the path difference from the "scattering point"[5] $\boldsymbol{R}$ to the observation points $\boldsymbol{r}_1$ and $\boldsymbol{r}_2$; $\boldsymbol{n}_{s1}$ and $\boldsymbol{n}_{s2}$ are unit vectors directed at the points $\boldsymbol{r}_1$ and $\boldsymbol{r}_2$ (Fig. 4.1)

$$\boldsymbol{n}_{s1}=\frac{\boldsymbol{r}_1-\boldsymbol{R}}{|\boldsymbol{r}_1-\boldsymbol{R}|}\,,\quad \boldsymbol{n}_{s2}=\frac{\boldsymbol{r}_2-\boldsymbol{R}}{|\boldsymbol{r}_2-\boldsymbol{R}|}\,, \tag{4.28}$$

and $\boldsymbol{\varrho}_{\perp 1}$ and $\boldsymbol{\varrho}_{\perp 2}$ are the components of $\boldsymbol{\varrho}$ normal to $\boldsymbol{n}_{s1}$ and $\boldsymbol{n}_{s2}$, respectively.

Let us now identify the conditions under which we could ignore in (4.26) the square and cubic terms in $\boldsymbol{\varrho}$. The actual integration region with respect to $\boldsymbol{\varrho}$ is confined by the inequality $\varrho \lesssim l_\varepsilon$. If we require that the quadratic terms in (4.26) be small as compared to π at $\varrho \sim l_\varepsilon$, we will get the inequality

$$\frac{k}{8}\left(\frac{\varrho_{\perp 1}^2}{|\boldsymbol{r}_1-\boldsymbol{R}|}-\frac{\varrho_{\perp 2}^2}{|\boldsymbol{r}_2-\boldsymbol{R}|}\right)\bigg|_{\varrho\sim l_\varepsilon} \ll \pi\,, \tag{4.29}$$

which will be met *a fortiori*, if it holds for each of the terms

$$\frac{k\varrho_{\perp 1,2}^2}{8|\boldsymbol{r}_{1,2}-\boldsymbol{R}|}\bigg|_{\varrho\sim l_\varepsilon} \ll \pi\,.$$

But the implication of these inequalities is that the observation points $\boldsymbol{r}_1$ and $\boldsymbol{r}_2$, with respect to inhomogeneities of size l_ε, must lie within the *Fraunhofer zone*

$$|\boldsymbol{r}_{1,2}-\boldsymbol{R}| \gg \frac{kl_\varepsilon^2}{8\pi} \sim \frac{l_\varepsilon^2}{\lambda}\,. \tag{4.30}$$

It is, however, clear, that (4.30) should be regarded as a *sufficient* condition for discarding the higher powers of $\boldsymbol{\varrho}$, since (4.29) contains the *difference* of two terms that are almost equal in magnitude. If $\boldsymbol{r}_1=\boldsymbol{r}_2$, then the left-hand side of (4.29) vanishes and in (4.26) quadratic terms are absent. Instead of (4.29), we should require here that the cubic, and not the quadratic, terms be small (as

[5] Or, rather, from the "center of mass", i.e., from the point lying midway between the scattering points $\boldsymbol{r}'$ and $\boldsymbol{r}''$ separated by the small distance $\varrho \lesssim l_\varepsilon$.

compared to π)

$$\left.\frac{k\varrho^3}{|\boldsymbol{r}_{1,2}-\boldsymbol{R}|^2}\right|_{\varrho\sim l_\varepsilon} \ll \pi \ .$$

This inequality imposes on $|\boldsymbol{r}_{1,2}-\boldsymbol{R}|$ the limitation

$$|\boldsymbol{r}_{1,2}-\boldsymbol{R}| \gg \sqrt{kl_\varepsilon^3} \approx \sqrt{l_\varepsilon^3/\lambda} \ . \tag{4.31}$$

For large inhomogeneities ($l_\varepsilon \gg \lambda$), the condition (4.31) is much weaker than the condition (4.30), which requires that observation points lie in the Fraunhofer zone of an individual inhomogeneity (of size $\sim l_\varepsilon$). On the contrary, for $l_\varepsilon \lesssim \lambda$ the inequality (4.31) is stronger than (4.16). For inhomogeneities whose sizes are comparable to the wavelength, (4.30) and (4.31) are equivalent.

As for Ψ_2, its expansion only contains odd powers of $\boldsymbol{\varrho}$. The condition

$$|\boldsymbol{R}-\boldsymbol{r}_0| \gg \sqrt{l_\varepsilon^3/\lambda} \tag{4.32}$$

is analogous to (4.31) and enables us to discard the cubic terms. Therefore, if it is met, we have

$$\Psi_2 \approx k\boldsymbol{n}_\mathrm{i}\cdot\boldsymbol{\varrho} \ , \qquad \text{where}$$

$$\boldsymbol{n}_\mathrm{i}(\boldsymbol{R}) = \frac{\boldsymbol{R}-\boldsymbol{r}_0}{|\boldsymbol{R}-\boldsymbol{r}_0|} \tag{4.33}$$

is the unit vector in the direction of propagation of the primary wave at $\boldsymbol{R}$ (Fig. 4.1). We thus arrive at

$$\Psi = \Psi_1 + \Psi_2 \approx k[s - \tfrac{1}{2}(\boldsymbol{n}_{s1}+\boldsymbol{n}_{s2})\cdot\boldsymbol{\varrho} + \boldsymbol{n}_\mathrm{i}\cdot\boldsymbol{\varrho}] \ . \tag{4.34}$$

In the denominator of the integrand in (4.23), and in the product $M(\boldsymbol{R}+\boldsymbol{\varrho}/2)\times M(\boldsymbol{R}-\boldsymbol{\varrho}/2)$, we simply neglect $\boldsymbol{\varrho}$. For the denominator to be simplified in this way, it is sufficient that both observation points $\boldsymbol{r}_1$ and $\boldsymbol{r}_2$ and the radiation point $\boldsymbol{r}_0$ be separated from the scattering point $\boldsymbol{R}$ by no less than l_ε. At the same time, in the entire volume V, except for the boundary layer of thickness l_ε, we can replace $M(\boldsymbol{R}+\boldsymbol{\varrho}/2)M(\boldsymbol{R}-\boldsymbol{\varrho}/2)$ by $M^2(\boldsymbol{R})$, or, equivalently, by $M(\boldsymbol{R})$. The relative error involved in this change is of the order of the ratio of the boundary layer volume $L^2 l_\varepsilon$ to the complete volume $V\sim L^3$, i.e., of the order of $l_\varepsilon/L \ll 1$ [see (4.19)].

Given these assumptions, (4.23) takes the form

$$\psi_u(\boldsymbol{r}_1,\boldsymbol{r}_2) = \frac{k_0^4|A|^2}{(4\pi)^2}\iint\limits_{-\infty}^{\infty}\frac{M(\boldsymbol{R})\exp(iks - i\boldsymbol{Q}\cdot\boldsymbol{\varrho})\psi_\varepsilon(\boldsymbol{\varrho},\boldsymbol{R})}{|\boldsymbol{r}_1-\boldsymbol{R}||\boldsymbol{r}_2-\boldsymbol{R}||\boldsymbol{R}-\boldsymbol{r}_0|^2}d^3\varrho\, d^3R \ , \tag{4.35}$$

where we have introduced

$$\boldsymbol{Q}(\boldsymbol{R}) = k\{\tfrac{1}{2}[\boldsymbol{n}_{s1}(\boldsymbol{R})+\boldsymbol{n}_{s2}(\boldsymbol{R})] - \boldsymbol{n}_\mathrm{i}(\boldsymbol{R})\} \ . \tag{4.36}$$

According to (4.18), the integral with respect to $\boldsymbol{\varrho}$ can be expressed in terms of the spectral density of the quasi-homogeneous fluctuations

$$\int_{-\infty}^{\infty} \psi_\varepsilon(\varrho, \boldsymbol{R}) \exp(-\mathrm{i}\boldsymbol{Q}\cdot\boldsymbol{\varrho})\, d^3\varrho = 8\pi^3 \Phi_\varepsilon(\boldsymbol{Q}, \boldsymbol{R}) \ , \tag{4.37}$$

so that we finally obtain

$$\psi_u(\boldsymbol{r}_1, \boldsymbol{r}_2) = \frac{\pi k_0^4 |A|^2}{2} \int_{-\infty}^{\infty} \frac{M(\boldsymbol{R})\Phi_\varepsilon(\boldsymbol{Q}, \boldsymbol{R})\mathrm{e}^{\mathrm{i}ks}}{|\boldsymbol{r}_1 - \boldsymbol{R}||\boldsymbol{r}_2 - \boldsymbol{R}||\boldsymbol{R} - \boldsymbol{r}_0|^2} d^3R \ . \tag{4.38}$$

We will defer the examination of this covariance to Sect. 4.4, and here we will only look at the spatial distribution of the *mean intensity*

$$\bar{I}_1(\boldsymbol{r}) = \langle |u_1(\boldsymbol{r})|^2 \rangle = \psi_u(\boldsymbol{r}, \boldsymbol{r}) \ . \tag{4.39}$$

When the observation points coincide ($\boldsymbol{r}_1 = \boldsymbol{r}_2 = \boldsymbol{r}$, so that $s = 0$), Q in (4.36) becomes a so-called *scattering vector*

$$\boldsymbol{q}(\boldsymbol{R}) = k[\boldsymbol{n}_\mathrm{s}(\boldsymbol{R}) - \boldsymbol{n}_\mathrm{i}(\boldsymbol{R})] \ . \tag{4.40}$$

As a result, the mean intensity will be

$$\bar{I}_1(\boldsymbol{r}) = \frac{\pi k_0^4 |A|^2}{2} \int_{-\infty}^{\infty} \frac{M(\boldsymbol{R})\Phi_\varepsilon(\boldsymbol{q}, \boldsymbol{R})}{|\boldsymbol{r} - \boldsymbol{R}|^2 |\boldsymbol{R} - \boldsymbol{r}_0|^2} d^3R \ . \tag{4.41}$$

It is seen that waves scattered by individual volume elements d^3R add up in an *incoherent* way, i.e., their intensities are added up. Of course, the integrand [with the coefficient that in (4.41) is taken outside the integration sign] is the mean intensity $\overline{dI_1}$ of the field scattered by d^3R.

Equation (4.41) derived for the spherical primary wave (4.15) can readily be extended to waves of a different shape. The quantity $I_0(\boldsymbol{R}) = |A|^2/|\boldsymbol{R} - \boldsymbol{r}_0|^2$ is the intensity of the spherical wave at $\boldsymbol{R}$. Writing (4.41) in the form

$$\bar{I}_1(\boldsymbol{r}) = \frac{\pi k_0^4}{2} \int_{-\infty}^{\infty} \frac{M(\boldsymbol{R}) I_0(\boldsymbol{R}) \Phi_\varepsilon(\boldsymbol{q}, \boldsymbol{R})}{|\boldsymbol{r} - \boldsymbol{R}|^2} d^3R \tag{4.42}$$

gives an expression that holds for *any* primary field if, for all practical purposes, its structure within the confines of one inhomogeneity of size l_ε is similar to that of the spherical wave. This means that, within the limits of an inhomogeneity, the field amplitude and the curvature radius R_curv of the wave front can be considered to be constant. In addition, R_curv must satisfy inequalities of the form (4.32), i.e.,

$$R_\mathrm{curv} \gg \sqrt{l_\varepsilon^3/\lambda} \ .$$

We may thus apply (4.42) both to the *plain* primary wave

$$u_\mathrm{i} = A_0 \mathrm{e}^{\mathrm{i}k\boldsymbol{n}_\mathrm{i}\cdot\boldsymbol{r}} \tag{4.43}$$

for which $I_0 = |A_0|^2 = \mathrm{const}$ [unlike (4.33), the unit vector $\boldsymbol{n}_\mathrm{i}$ is constant here throughout the entire space] and to the *directed* spherical wave

$$u_i = Af_0(\boldsymbol{n}_i)\frac{e^{ik|\boldsymbol{r}-\boldsymbol{r}_0|}}{|\boldsymbol{r}-\boldsymbol{r}_0|} \tag{4.44}$$

[$f_0(\boldsymbol{n}_i)$ is the radiation pattern of the radiator], for which

$$I_0(\boldsymbol{R}) = \frac{|A|^2|f_0(\boldsymbol{n}_i)|^2}{|\boldsymbol{R}-\boldsymbol{r}_0|^2} \ . \tag{4.45}$$

The key point about equations (4.41, 42) is that they reflect the *selective* nature of scattering. Let θ be the angle, called the *scattering angle*, between the wave vectors of the incident ($\boldsymbol{k}_i$) and scattered ($\boldsymbol{k}_s$) waves. Clearly,

$$q = k|\boldsymbol{n}_s - \boldsymbol{n}_i| = 2k\sin(\theta/2) \ . \tag{4.46}$$

Corresponding to this value of q are the spatial harmonic perturbations $\exp(i\boldsymbol{qr})$ with wavelength

$$\Lambda_q = \frac{2\pi}{q} = \frac{\lambda}{2\sin(\theta/2)} \ , \tag{4.47}$$

with the normal to the wave parallel to $\boldsymbol{q}$, i.e., to $(\boldsymbol{n}_s - \boldsymbol{n}_i)$ (Fig. 4.2). Consequently, (4.47) is the Bragg equation determining the spatial period of the harmonic for which the wave diffracts at the angle θ.

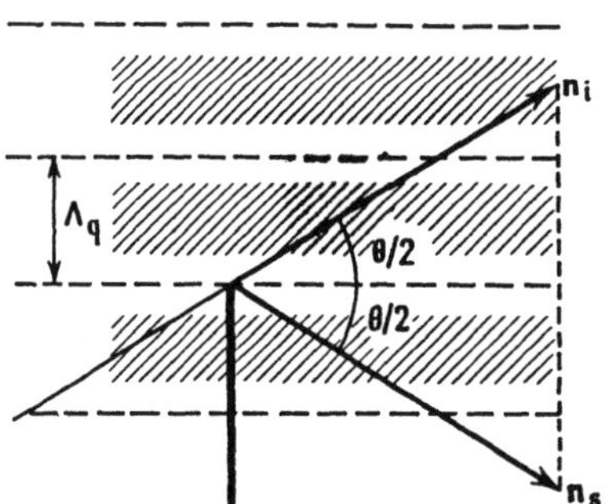

Fig. 4.2. The scattering vector $\boldsymbol{q}$ and the unit vectors of the incident ($\boldsymbol{n}_i$) and scattered ($\boldsymbol{n}_s$) waves

Of course, the spatial spectrum of the fluctuations $\tilde{\varepsilon}(\boldsymbol{r})$ contains an infinite set of such harmonics (volume "diffraction gratings") with a wide variety of periods and orientations. The scattering is selective in that in a given direction $\boldsymbol{n}_s$, the only contributions come from waves scattered from a definite grating—the one with spatial period (4.47) and orientation described by the law of reflection (at $\boldsymbol{R}$) of the primary wave coming from $\boldsymbol{r}_0$ to $\boldsymbol{r}$. The intensity of this harmonic is proportional to the spectral density $\Phi_\varepsilon(\boldsymbol{q})$ in (4.41, 42).

With *back*scattering, where $\boldsymbol{k}_s = -\boldsymbol{k}_i$ and the scattering vector is $\boldsymbol{q} = -2\boldsymbol{k}_i$ and $\theta = \pi$, the scattering is dominated by the harmonic with period $\Lambda_q = \lambda/2$, i.e., equal to the half wave-length of the incident radiation. Backscattering is observed in a variety of cases, e.g., in radar and sound ranging, where the source and receiver of the radiation lie at the same point. Scattering at an arbitrary angle is often used in optics when investigating inhomogeneities in transparent materials.

The period Λ_q increases with decreasing scattering angle θ. The values of Λ_q and λ coincide at $\theta = \pi/3$. As θ decreases further, the spatial period of the scattering grating Λ_q exceeds λ, and for *forward* scattering ($\theta \to 0$) q vanishes and Λ_q becomes infinite. This implies that the forward scattering is due to the largest volume perturbations with space scales $\Lambda_q \gg \lambda$.

Equation (4.42) is simplified for a small scattering volume such that the quantities $\Phi_\varepsilon(\boldsymbol{q}, \boldsymbol{R})$, $I_0(\boldsymbol{R})$ and $1/|\boldsymbol{r}-\boldsymbol{R}|^2$ are virtually constant. Bringing them outside the integration sign with values corresponding to, say, the center 0 of the scattering volume ($\boldsymbol{R} = 0$), we obtain

$$\bar{I}_1(\boldsymbol{r}) = \frac{\pi k_0^4}{2} \frac{I_0(0)\Phi_\varepsilon(\boldsymbol{q}_0, 0)}{r^2} V \, , \tag{4.48}$$

where $\boldsymbol{q}_0 = \boldsymbol{q}(0) = k(\boldsymbol{n}_{s0} - \boldsymbol{n}_{i0})$ is the scattering vector corresponding to the center of the scattering volume, and $\boldsymbol{n}_{s0} = \boldsymbol{r}/r$ and $\boldsymbol{n}_{i0} = -\boldsymbol{r}_0/r_0$ are unit vectors directed from 0 (Fig. 4.3). We may substitute $1/r^2$ for $1/|\boldsymbol{r} - \boldsymbol{R}|^2$ under the obvious condition that the observation point lies at a distance significantly larger than the transverse size L of the scatterer, i.e.,

$$r \gg L \, . \tag{4.49}$$

If the incident wave is spherical, then instead of the exact expression for the intensity $I_0(R) = |A|^2/|\boldsymbol{R} - \boldsymbol{r}_0|^2$, we may use the approximate one $I_0(0) = |A|^2/r_0^2$, if in addition $r_0 \gg L$.

$\Phi_\varepsilon(\boldsymbol{q}, \boldsymbol{R})$ can be replaced by the constant $\Phi_\varepsilon(\boldsymbol{q}_0, 0)$ provided that, first, fluctuations within V are statistically *homogeneous* (the scale L_ε must be larger than L); and, second, that the variation (within V) of $\boldsymbol{q}$ is small as compared to the

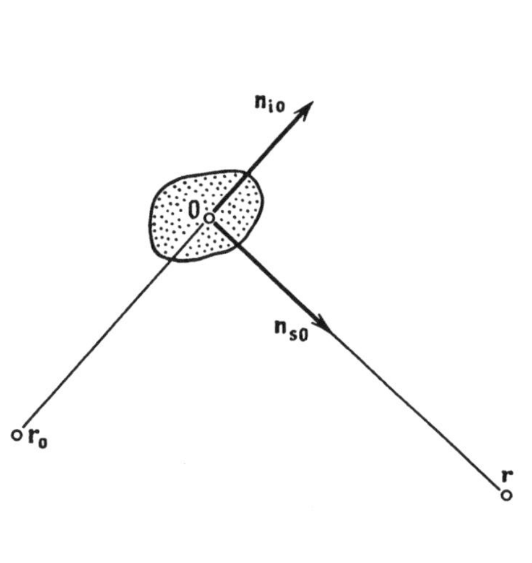

Fig. 4.3. Scattering at a small volume

characteristic space scale of the spectrum $\Phi_\varepsilon(\boldsymbol{q}, 0)$ that is equal to $1/l_\varepsilon$

$$\Delta q \ll 1/l_\varepsilon \ . \tag{4.50}$$

The variation of q is of order of magnitude $\Delta q \sim k\Delta n_s$ or $k\Delta n_i$ (where $\Delta \boldsymbol{n}_s$ and $\Delta \boldsymbol{n}_i$ are the variations of direction of the incident and scattered waves within V). Obviously, $\Delta n_s \sim L/r$, so that (4.50) is reduced to the condition

$$r \gg kLl_\varepsilon \sim Ll_\varepsilon/\lambda \ . \tag{4.51}$$

Since the restrictions on the distance r_0 to the source of the spherical wave are derived in a similar manner, we will in the following only discuss those inequalities that refer to the distance r from the observation point.

In the case of small-scale inhomogeneities ($l_\varepsilon \ll \lambda$), of the two inequalities (4.49) and (4.51) the first is clearly the more rigid, because for large-scale inhomogeneities ($l_\varepsilon \gg \lambda$) the minimal distance to the observation point is limited by the condition (4.51). Denoting by $r_{\min}$ the smaller of L and Ll_ε/λ, we may write (4.40) and (4.51) in the form

$$r \gg r_{\min} = \min\{L, Ll_\varepsilon/\lambda\} \ . \tag{4.52}$$

The simplified expression (4.48) is generally derived on the assumption that the observation point lies in the Fraunhofer zone of the entire volume V. For this purpose, in (4.21) (or a similar expression for the plane primary wave) we expand $|\boldsymbol{r} - \boldsymbol{r}'|$ into a series in $\boldsymbol{r}'$ and ignore the quadratic term $kr_\perp^2/2r$ in the Green's function (4.16). This leads to the Fraunhofer approximation

$$u_1(\boldsymbol{r}) = \frac{k_0^2 A e^{ikr}}{4\pi r r_0} \int_{-\infty}^{\infty} e^{-i\boldsymbol{q}_0\cdot \boldsymbol{r}'} M(\boldsymbol{r}')\tilde{\varepsilon}(\boldsymbol{r}')\, d^3r' \ . \tag{4.53}$$

The expression for the mean intensity derived using (4.53) is much simpler than in the general case, namely it coincides with (4.48). However, the condition

$$\frac{kr'^2}{2r} \lesssim \frac{kL^2}{2r} \ll \pi \quad \text{or} \quad r \gg \frac{L^2}{\lambda} \ , \tag{4.54}$$

necessary for the Fraunhofer approximation (4.53), is far more stringent than the condition (4.52), which is sufficient for (4.48) to be applied.

In many applications we measure not the scattered field $u_1(\boldsymbol{r})$ itself, but the signal at the receiving antenna output. If the center of the receiving aperture is at $\boldsymbol{r}$, then the output signal v is the field $u_1(\boldsymbol{r} + \boldsymbol{\varrho})$ integrated over the aperture Σ with the weight function $b(\boldsymbol{\varrho})$ that describes the current distribution in the antenna operating in transmit conditions

$$v = \widetilde{u_1(\boldsymbol{r})} \equiv \int_\Sigma u_1(\boldsymbol{r} + \boldsymbol{\varrho}) b(\boldsymbol{\varrho})\, d^2\varrho \ . \tag{4.55}$$

Integrating over the aperture (wavy line) is, of course, not equivalent to ensemble averaging.

Let us now substitute the scattered field (4.21) into (4.55), leaving aside for the moment the issue of the concrete form of the primary wave $u_0(\boldsymbol{r})$,

$$v = \int_{\Sigma} b(\boldsymbol{\varrho})\, d^2\varrho \int \frac{k_0^2}{4\pi} \frac{\exp(ik|\boldsymbol{r}+\boldsymbol{\varrho}-\boldsymbol{r}'|)}{|\boldsymbol{r}+\boldsymbol{\varrho}-\boldsymbol{r}'|} M(\boldsymbol{r}')\tilde{\varepsilon}(\boldsymbol{r}')u_0(\boldsymbol{r}')\, d^3r' \tag{4.56}$$

Next we expand $|\boldsymbol{r}+\boldsymbol{\varrho}-\boldsymbol{r}'|$ into a Taylor series in

$$|\boldsymbol{r}+\boldsymbol{\varrho}-\boldsymbol{r}'| = |\boldsymbol{r}-\boldsymbol{r}'| + \boldsymbol{n}_s'\cdot\boldsymbol{\varrho} + \frac{1}{2|\boldsymbol{r}-\boldsymbol{r}'|}[\varrho^2 - (\boldsymbol{n}_s'\cdot\boldsymbol{\varrho})^2] + \dots, \tag{4.57}$$

where $\boldsymbol{n}_s' \equiv (\boldsymbol{r}-\boldsymbol{r}')/|\boldsymbol{r}-\boldsymbol{r}'|$ is the unit vector directed from the scattering point $\boldsymbol{r}'$ to the center of the receiving aperture $\boldsymbol{r}$ (Fig. 4.4).

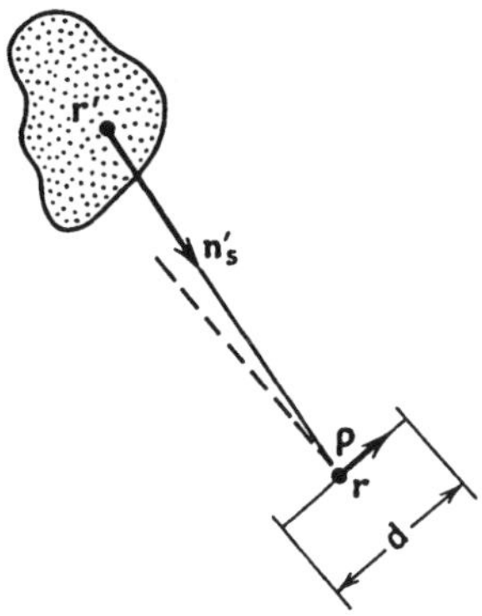

Fig. 4.4. An antenna receiving scattered radiation

Let d be the largest size of the receiving antenna, so that within its aperture $\varrho \lesssim d$, and let the condition $kd^2 \ll |\boldsymbol{r}-\boldsymbol{r}'|$ be met (point $\boldsymbol{r}'$ lies in the Fraunhofer zone of the antenna, where the radiation pattern has already been established). Then, in the exponent of (4.56) we may keep only the first two terms of the expansion (4.57), and in the denominator of the integrand, the first term of this expansion. Integrating with respect to $\boldsymbol{\varrho}$, we find

$$v = \frac{k_0^2}{4\pi} \int \frac{f_1(\boldsymbol{n}_s')e^{ik|\boldsymbol{r}-\boldsymbol{r}'|}}{|\boldsymbol{r}-\boldsymbol{r}'|} M(\boldsymbol{r}')u_0(\boldsymbol{r}')\tilde{\varepsilon}(\boldsymbol{r}')\, d^3r', \quad \text{where} \tag{4.58}$$

$$f_1(\boldsymbol{n}_s') = \int_{\Sigma} b(\boldsymbol{\varrho}) \exp(ik\boldsymbol{n}_s'\cdot\boldsymbol{\varrho})\, d^2\varrho \tag{4.59}$$

is the receiving antenna pattern. The maximum of $|f_1(\boldsymbol{n}_s')|$ is conveniently normalized to unity. If the weight function $b(\boldsymbol{\varrho})$ is symmetrical about $\boldsymbol{\varrho}=0$, then $|f_1(\boldsymbol{n}_s')|$ will be maximal at $\boldsymbol{n}_s'\boldsymbol{\varrho}=0$, i.e., when $\boldsymbol{n}_s'$ is normal to the antenna plane.

Expression (4.58) differs from the initial formula (4.21) in that the integrand now contains the pattern factor $f_1(\boldsymbol{n}_s')$. For receiving antennas, small as compared to the wavelength ($kd \ll 1$), this factor is independent of the direction of $\boldsymbol{n}_s'$, and we then return to (4.21).

If we take as the primary wave the directed spherical wave (4.44), we obtain

$$v = \frac{k_0^2 A}{4\pi} \int_{-\infty}^{\infty} M(\boldsymbol{r}') \frac{e^{ik(|\boldsymbol{r}-\boldsymbol{r}'| + |\boldsymbol{r}'-\boldsymbol{r}_0|)}}{|\boldsymbol{r}-\boldsymbol{r}'||\boldsymbol{r}'-\boldsymbol{r}_0|} f_0(\boldsymbol{n}_i') f_1(\boldsymbol{n}_s') \tilde{\varepsilon}(\boldsymbol{r}') d^3 r' \ . \tag{4.60}$$

If we introduce the notation

$$\mathscr{P}(\boldsymbol{r}') = \frac{k_0^2 A}{4\pi} M(\boldsymbol{r}') \frac{f_0(\boldsymbol{n}_i') f_1(\boldsymbol{n}_s')}{|\boldsymbol{r}-\boldsymbol{r}'||\boldsymbol{r}'-\boldsymbol{r}_0|} \ , \tag{4.61}$$

we can then write

$$v = \int_{-\infty}^{\infty} \mathscr{P}(\boldsymbol{r}') e^{ik(|\boldsymbol{r}-\boldsymbol{r}'| + |\boldsymbol{r}'-\boldsymbol{r}_0|)} \tilde{\varepsilon}(\boldsymbol{r}') d^3 r' \ . \tag{4.62}$$

It should be noted that in this case, where both emission and reception are carried out using directed antennas, the integration region in (4.60) or (4.62) remains finite even when the inhomogeneities $\tilde{\varepsilon}(\boldsymbol{r})$ fill the *entire* space, i.e., $M(\boldsymbol{r}')$ is unity over the whole space. Now we can dispense with constraints on the volume occupied by the inhomogeneities, because the product of the pattern factors falls off to around zero outside the region where the main lobes of both patterns overlap (Fig. 4.5).

Knowing (4.60), we can readily work out the mean intensity of signal at the antenna output $\bar{I}_v = \langle |v^2| \rangle$. As opposed to (4.61), the integrand will now contain the factor $|f_0(\boldsymbol{n}_i) f_1(\boldsymbol{n}_s')|^2$

$$\bar{I}_v = \frac{\pi k_0^4 |A|^2}{2} \int_{-\infty}^{\infty} \frac{M(\boldsymbol{R}) \Phi_\varepsilon(\boldsymbol{q}, \boldsymbol{R})}{|\boldsymbol{r}-\boldsymbol{R}|^2 |\boldsymbol{R}-\boldsymbol{r}_0|^2} |f_0(\boldsymbol{n}_i)|^2 |f_1(\boldsymbol{n}_s)|^2 d^3 R \ . \tag{4.63}$$

For narrow-beam receiving and transmitting antennas, the intensity of the scattered field is conveniently characterized by the *effective scattering volume*. We will denote by $\boldsymbol{q}_0$ the scattering vector $k(\boldsymbol{n}_{i0} - \boldsymbol{n}_{s0})$ corresponding to the intersection point of the pattern maxima. It is only natural to place the origin of the coordinates at that point. We now write (4.63) in the form analogous to (4.48) ($I_0 = |A|^2/r_0^2$)

$$\bar{I}_v = \frac{\pi k_0^4 \Phi_\varepsilon(\boldsymbol{q}_0, 0) I_0}{2r^2} V_{\text{eff}} \ . \tag{4.64}$$

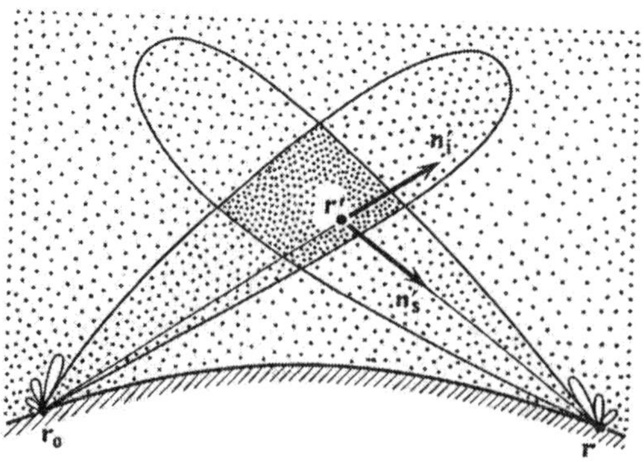

Fig. 4.5. A scattering volume formed by the overlapping of the radiation patterns of a transmitting and a receiving antennas

It is the factor

$$V_{\text{eff}} = \int M(\boldsymbol{R}) \left[\frac{r r_0}{|\boldsymbol{r} - \boldsymbol{R}||\boldsymbol{R} - \boldsymbol{r}_0|} \right]^2 |f_0(\boldsymbol{n}_i) f_1(\boldsymbol{n}_s)|^2 \frac{\Phi_\varepsilon(\boldsymbol{q}, \boldsymbol{R})}{\Phi_\varepsilon(\boldsymbol{q}_0, 0)} d^3R \ , \quad (4.65)$$

obtained by comparing (4.63) and (4.48), that is the effective scattering volume.

As is seen from (4.65), V_{eff} is influenced by the distances to the source and receiver, the pattern shape of both antennas and the form of the spectrum $\Phi_\varepsilon(\boldsymbol{q}, \boldsymbol{R})$. For narrow-beam antennas, if the pattern overlapping region belongs wholly to the scattering volume, $M(\boldsymbol{R})$ and $r r_0 / |\boldsymbol{R} - \boldsymbol{r}_0||\boldsymbol{r} - \boldsymbol{R}|$ may, to a good approximation, be taken to be unity and then

$$V_{\text{eff}} = \int_{-\infty}^{\infty} |f_0(\boldsymbol{n}_i) f_1(\boldsymbol{n}_s)|^2 \frac{\Phi_\varepsilon(\boldsymbol{q}, \boldsymbol{R})}{\Phi_\varepsilon(\boldsymbol{q}_0, 0)} d^3R \ . \quad (4.66)$$

This relation is widely used to compute the intensity of radio waves scattered in the troposphere and lower ionosphere [4.1].

4.3 Effective Scattering Cross-Section. Applicability of Single Scattering Approximation

Consider a scalar monochromatic field $u(\boldsymbol{r})$ satisfying the Helmholtz equation (4.1) in a transparent medium, i.e., *real* permittivity $\varepsilon(\boldsymbol{r})$. We have, directly from (4.1),

$$u^* \Delta u - u \Delta u^* = 0 \quad \text{or} \quad \operatorname{div}(u^* \nabla u - u \nabla u^*) = 0 \ . \quad (4.67)$$

According to (4.67), the quantity

$$\mathscr{S} = \frac{a}{2ik}(u^* \nabla u - u \nabla u^*) = \frac{a}{k} \operatorname{Im}\{u^* \nabla u\} \quad (4.68)$$

that, in a transparent medium, satisfies the conservation law

$$\operatorname{div} \mathscr{S} = 0 \ , \quad (4.69)$$

can be thought of as the energy flux density, provided the normalizing factor a is adequately selected[6].

Let us find the mean energy flux density for the single-scattered field

$$\bar{\mathscr{S}}_1 = \frac{a}{2ik} \langle u_1^* \nabla u_1 - u_1 \nabla u_1^* \rangle \ . \quad (4.70)$$

[6] For sound waves, if u is the velocity potential, the factor a equals $\varrho \omega^2 / 2c$ (ϱ is the liquid density, c the velocity of sound), and for electromagnetic fields $a = c/8\pi$, where c is the velocity of the electromagnetic wave, if by u we understand the strength of the electric field.

Differentiating (4.21) gives

$$\nabla u_1(\boldsymbol{r}) = \frac{k_0^2}{4\pi}\int\limits_{-\infty}^{\infty} M(\boldsymbol{r}')\tilde{\varepsilon}(\boldsymbol{r}')u_0(\boldsymbol{r}')\left(ik - \frac{1}{|\boldsymbol{r}-\boldsymbol{r}'|}\right)\frac{\boldsymbol{r}-\boldsymbol{r}'}{|\boldsymbol{r}-\boldsymbol{r}'|}\frac{e^{ik|\boldsymbol{r}-\boldsymbol{r}'|}}{|\boldsymbol{r}-\boldsymbol{r}'|}d^3r' . \tag{4.71}$$

Using (4.25), we can transform the vector $\boldsymbol{n}_s' = (\boldsymbol{r}-\boldsymbol{r}')/|\boldsymbol{r}-\boldsymbol{r}'|$ directed from $\boldsymbol{r}'$ to $\boldsymbol{r}$ to obtain

$$\boldsymbol{n}_s' = \frac{\boldsymbol{r}-\boldsymbol{r}'}{|\boldsymbol{r}-\boldsymbol{r}'|} \approx \frac{\boldsymbol{r}-\boldsymbol{r}'}{r-(\boldsymbol{n}\cdot\boldsymbol{r}')} \approx \boldsymbol{n}_{s0} - \frac{\boldsymbol{r}'_\perp}{r} , \quad \boldsymbol{r}'_\perp = \boldsymbol{r}' - \boldsymbol{n}_{s0}(\boldsymbol{n}_{s0}\cdot\boldsymbol{r}') , \tag{4.72}$$

where, as before, $\boldsymbol{n}_{s0} = \boldsymbol{r}/r$. Assuming that the observation point $\boldsymbol{r}$ is sufficiently distant from the scattering region ($r \gg L$), we have $|\boldsymbol{r}'_\perp| \ll r$ and $\boldsymbol{n}_s' \lesssim \boldsymbol{n}_{s0}$. Furthermore, when $r \gg L$ the inequality $k \gg |\boldsymbol{r}-\boldsymbol{r}'|^{-1}$ is met *a fortiori*. As a result, (4.71) is greatly simplified to yield

$$\nabla u_1(\boldsymbol{r}) \approx ik\boldsymbol{n}_{s0}u_1(\boldsymbol{r}) ,$$

and (4.70) gives

$$\bar{\mathscr{S}}_1 \approx a\boldsymbol{n}_{s0}\langle |u_1|^2\rangle = a\boldsymbol{n}_{s0}\bar{I}_1 . \tag{4.73}$$

Knowing $\bar{\mathscr{S}}_1$, we can easily find the *effective differential scattering cross-section* σ of a unit volume into a unit solid angle in the direction $\boldsymbol{n}_s$[7].

By definition

$$\sigma(\boldsymbol{n}_s) = \sigma(\theta, \varphi) = \frac{1}{|\mathscr{S}_0|}\frac{d\bar{P}_1}{do\,dV} , \tag{4.74}$$

where $d\bar{P}_1$ is the mean power scattered into the solid angle do in the direction $\boldsymbol{n}_s$ from the small scattering volume dV, and $|\mathscr{S}_0|$ the modulus of the energy flux density in the primary wave. By θ and φ we denote respectively the co-latitude and longitude of $\boldsymbol{n}_s$ (Fig. 4.6).

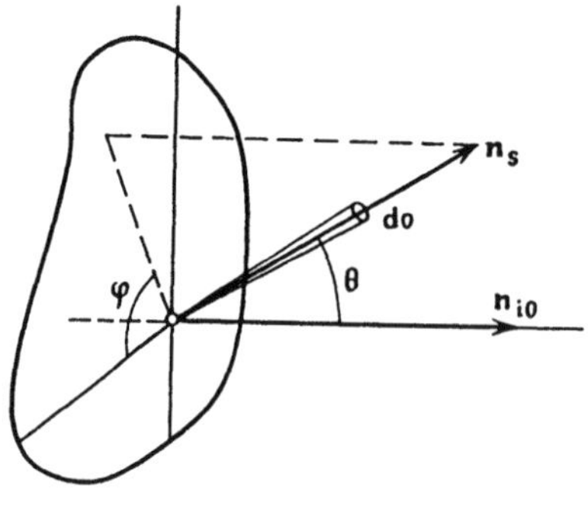

Fig. 4.6. Scattering in the direction (θ, φ)

[7] The effective scattering cross-section is often abbreviated as ESC.

If $d\bar{\mathscr{S}}_1$ and $d\bar{I}_1$ correspond to volume dV, then

$$d\bar{P}_1 = |d\mathscr{S}_1| r^2 do = a d\bar{I}_1 r^2 do \ . \tag{4.75}$$

For the small volume dV the mean intensity $d\bar{I}_1$ may be determined from (4.48), and so

$$d\bar{P}_1 = \frac{\pi a k_0^4}{2} I_0(0) \Phi_\varepsilon(\boldsymbol{q}) do dV \ . \tag{4.76}$$

But the absolute value of the energy flux density in the primary wave (be it a plane, spherical or directed spherical wave) is, by (4.68), $|\mathscr{S}_0| = aI_0(0)$, so that (4.74) gives

$$\sigma(\theta, \varphi) = \tfrac{1}{2}\pi k_0^4 \Phi_\varepsilon(\boldsymbol{q}) \ . \tag{4.77}$$

Owing to the properties of the Fourier transform, the spatial spectrum (4.37) is essentially independent of $\boldsymbol{q}$, if the covariance $\psi_\varepsilon(\boldsymbol{\varrho})$ is nonzero only in a small region $\varrho \lesssim l_\varepsilon \ll \lambda$. In this case of *small-scale* inhomogeneities, the scattering is isotropic

$$\sigma(\theta, \varphi) \approx \text{const} = \tfrac{1}{2}\pi k_0^4 \Phi_\varepsilon(0) \ . \tag{4.78}$$

On the other hand, in the case of *large-scale* inhomogeneities ($l_\varepsilon \gg \lambda$), the spectral density $\Phi_\varepsilon(\boldsymbol{q})$ falls off quickly with q, i.e., with θ, thereby favouring forward scattering. The angle θ within which the radiation is confined can be deduced from the condition $ql_\varepsilon \lesssim 1$ or $\theta \lesssim 1/kl_\varepsilon$.

The *total* scattering cross-section

$$\sigma_0 = \oint \sigma(\theta, \varphi) do = \tfrac{1}{2}\pi k_0^4 \oint \Phi_\varepsilon(\boldsymbol{q}) do \ , \tag{4.79}$$

is by definition the ratio of the mean total scattered power $d\bar{P}_{1\,\text{tot}} = r^2 \int |d\bar{\mathscr{S}}_1| do$ to the energy flux density in the primary wave per unit volume

$$\sigma_0 = \frac{1}{|\mathscr{S}_0|} \frac{d\bar{P}_{1\,\text{tot}}}{dV} = \frac{r^2 \oint |d\mathscr{S}_1| do}{|\mathscr{S}_0| dV} \ . \tag{4.80}$$

Both σ and σ_0 have dimensions of inverse length, whereas for the finite volume V the scattering cross-section σV is measured in units of area.

For the small but finite volume V [when the condition (4.52) is met] the mean density of the scattered energy flux $\bar{\mathscr{S}}_1$ and the mean intensity $\bar{I}_1$ will be

$$\bar{\mathscr{S}}_1 = \frac{\boldsymbol{n}_{s0}}{r^2} |\mathscr{S}_0| V\sigma(\theta, \varphi) \quad \text{and} \quad \bar{I}_1 = \frac{1}{r^2} I_0 V \sigma(\theta, \varphi) \ . \tag{4.81}$$

In optics, $\sigma(\theta, \varphi)$ is normally termed the *scattering indicatrix*, the term being most often applied to $f(\theta, \varphi)$, the value of $\sigma(\theta, \varphi)$ corresponding either to the maximum value of the cross-section σ_{max} or to the total scattering cross-section σ_0:

$$f(\theta, \varphi) = \frac{\sigma(\theta, \varphi)}{\sigma_{max}} \quad \text{or} \quad f(\theta, \varphi) = \frac{\sigma(\theta, \varphi)}{\sigma_0} \ . \tag{4.82}$$

The total scattering cross-section σ_0 is directly related to the attenuation of the primary wave *due to scattering* (the so-called *extinction*). Imagine a small cylinder with the axis along the wave vector of the primary wave $\boldsymbol{k}_i = k\boldsymbol{n}_i$. The volume of the cylinder is $dV = dz\, d\Sigma$, where $d\Sigma$ is the cross-sectional area of the cylinder, and dz is its length. The power brought by the primary wave to the front end face of the cylinder is $P = |\mathscr{S}_0| d\Sigma$, and at the rear end face it will be $P + dP$, where $dP = -d\bar{P}_{1\,tot}$, and $d\bar{P}_{1\,tot}$ is the power scattered by the volume dV, which, by (4.80), is given by

$$d\bar{P}_{1\,tot} = \sigma_0 |\mathscr{S}_0| dV = \sigma_0 |\mathscr{S}_0| d\Sigma\, dz = P\sigma_0\, dz \ . \tag{4.83}$$

Hence,

$$dP = -d\bar{P}_{1\,tot} = -P\sigma_0\, dz \ ,$$

i.e., the power of the primary wave decays exponentially: $P \sim \exp(-\sigma_0 z)$. The energy flux density falls off following the same law

$$|\mathscr{S}_0| \sim e^{-\sigma_0 z} \ . \tag{4.84}$$

The e-fold attenuation occurs over the path $d_{ex} = 1/\sigma_0$, called the *extinction length* (σ_0 is also referred to as the *extinction coefficient*). The product $\tau = \sigma_0 z$, where z is the path covered by the wave in a scattering medium, is called the *optical thickness* ($\tau = 1$ at $z = d_{ex}$).

Single scattering theory is constructed on the assumption that the amplitude of the primary wave u_0, and hence the energy flux density $\mathscr{S}_0$, are essentially constant within the scattering volume. According to (4.84), this will be the case if the optical thickness corresponding to the scattering volume ($L \sim V^{1/3}$) is small as compared to unity

$$\sigma_0 L \ll 1 \ . \tag{4.85}$$

Condition (4.85) can be written as

$$\sigma_0 V \approx \sigma_0 L^3 \ll L^2 \ , \tag{4.86}$$

i.e., the total scattering cross-section of the entire volume $\Sigma_0 \equiv \sigma_0 V$ must be small as compared to the area of order L^2 "cut" by V from the primary wave front.

Let us consider some examples of calculations of the scattering cross-section $\sigma(\theta, \varphi)$.

1) Inhomogeneities with Isotropic Gaussian Covariance:

$$\psi_\varepsilon(r) = \sigma_\varepsilon^2 e^{-r^2/2l_\varepsilon^2} \ . \tag{4.87}$$

In this case, the spatial spectral density of the fluctuations is [see (1.55)]

$$\Phi_\varepsilon(q) = \frac{\sigma_\varepsilon^2 l_\varepsilon^3}{(2\pi)^{3/2}} \exp\left(-\frac{q^2 l_\varepsilon^2}{2}\right) = \frac{\sigma_\varepsilon^2 l_\varepsilon^3}{(2\pi)^{3/2}} \exp\left[-2k^2 l_\varepsilon^2 \sin^2(\theta/2)\right] , \tag{4.88}$$

so that, by (4.77), the effective scattering cross-section of a unit volume will be

$$\sigma(\theta) = \frac{\sigma_\varepsilon^2 k_0^4 l_\varepsilon^3}{4\sqrt{2\pi}} \exp\left[-2k^2 l_\varepsilon^2 \sin^2(\theta/2)\right] . \tag{4.89}$$

The total scattering cross-section σ_0 is obtained by integrating $\sigma(\theta)$ over the unit sphere ($do = \sin\theta\, d\theta\, d\varphi$). It will be

$$\sigma_0 = \frac{\sqrt{2\pi}}{4} \frac{k_0^2 l_\varepsilon \sigma_\varepsilon^2}{\bar{\varepsilon}} (1 - \mathrm{e}^{-2k^2 l_\varepsilon^2}) . \tag{4.90}$$

In the case (4.87), condition (4.85) for the applicability of the single scattering approximation takes the form

$$\sigma_\varepsilon^2 k_0 L \ll \frac{2\sqrt{2}}{\pi k_0 l_\varepsilon [1 - \exp(-2k_0^2 l_\varepsilon^2)]} \equiv \mathscr{F}(k_0 l_\varepsilon) . \tag{4.91}$$

(For simplicity we put $\bar{\varepsilon} = 1$, so that $k = k_0$).

This condition limits the value of $\sigma_\varepsilon^2 k_0 L$. As σ_ε^2 or L increases, the inequality (4.91) is sooner or later violated. As $k_0 l_\varepsilon \to 0$, the function $\mathscr{F}(k_0 l_\varepsilon)$ increases as $(k_0 l_\varepsilon)^{-3} \sqrt{2/\pi}$, and as $k_0 l_\varepsilon \to \infty$ it decreases as $2\sqrt{2}/\pi k_0 l_\varepsilon$. In consequence, the larger the correlation radius l_ε, the more stringent are the constraints on $\sigma_\varepsilon^2 k_0 L$. Therefore, scattering by large inhomogeneities is described by the Born approximation only for small distances. For larger L, multiple scattering should be taken into account. Appropriate methods will be considered in Volume IV.

2) Backscattering by Anisotropic (Anisomeric) Fluctuations with Gaussian Covariance:

$$\psi_\varepsilon(\boldsymbol{r}) = \psi_\varepsilon(x, y, z) = \sigma_\varepsilon^2 \exp\left(-\frac{x^2 + y^2}{2a^2} - \frac{z^2}{2b^2}\right) . \tag{4.92}$$

The spectral density of such fluctuations is given by (1.62). Suppose that $b > a$, i.e., the major axis of the ellipsoidal inhomogeneities is directed along the z-axis.

Consider a primary wave such that its wave vector $\boldsymbol{k}_\mathrm{i}$ lies in the plane (x, z) and forms an angle ψ with the x-axis (Fig. 4.7), so that $\boldsymbol{k}_\mathrm{i} = (k\cos\psi,\, 0,\, k\sin\psi)$. For backscattering $\boldsymbol{q} = -2\boldsymbol{k}_\mathrm{i}$, and hence

$$\sigma_{\mathrm{back}}(\psi) = \frac{\pi}{2} k_0^4 \Phi_\varepsilon(-2\boldsymbol{k}_\mathrm{i}) = \frac{\pi}{2} k_0^4 \frac{\sigma_\varepsilon^2 a^2 b}{(2\pi)^{3/2}} \exp\left[-2k^2(a^2\cos^2\psi + b^2\sin^2\psi)\right] . \tag{4.93}$$

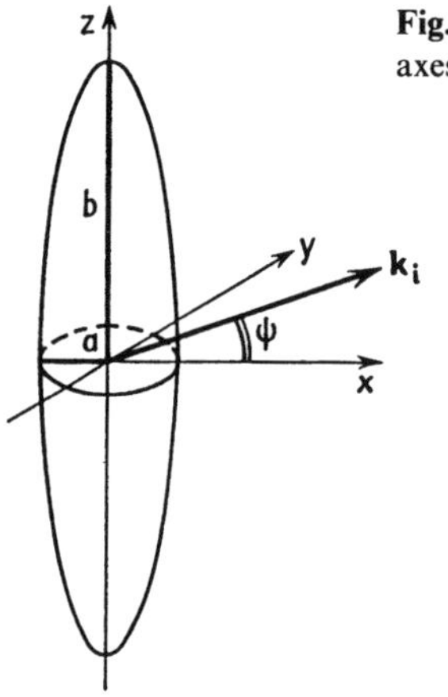

Fig. 4.7. Orientation of the wave vector of the primary field relative to the axes of anisotropic inhomogeneities

We will now consider the variation of the backscattering cross-section σ_{back} with the angle ψ, which is sometimes called the *aspect* angle. At $\psi = 0$ the wave vector of the primary wave, $\boldsymbol{k}_{\text{i}}$, is normal to the major axis of the inhomogeneities. The scattering indicatrix is then given by

$$f(\psi) = \frac{\sigma_{\text{back}}(\psi)}{\sigma_{\text{back}}(0)} = \exp[-2k^2(b^2-a^2)\sin^2\psi] \ . \tag{4.94}$$

It shows a maximum at $\psi = 0$ (Fig. 4.8) and a minimum at $\psi = \pm\pi/2$ ($\boldsymbol{k}_{\text{i}}$ is parallel to the major axis).

This effect (the so-called *aspect sensitivity*) is especially pronounced for markedly elongated inhomogeneities, for which $k\sqrt{b^2-a^2} \gg 1$. In that case, the indicatrix (4.94) may be approximated by

$$f(\psi) \approx \exp[-2k^2(b^2-a^2)\psi^2] \ , \tag{4.95}$$

i.e., by a narrow Gaussian curve of width

$$\Delta\psi \sim 1/k\sqrt{b^2-a^2} \ll 1 \ . \tag{4.96}$$

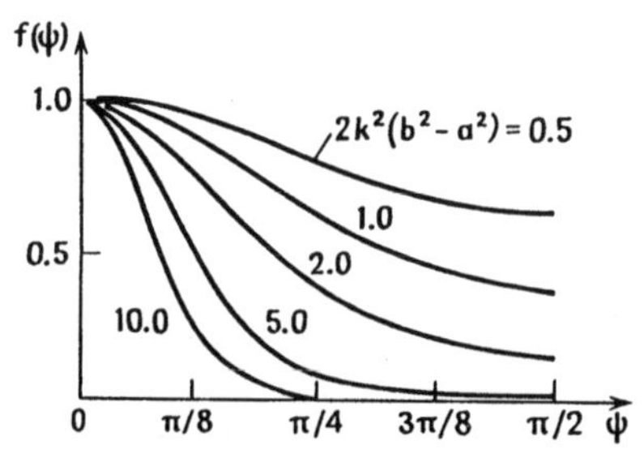

Fig. 4.8. Variation of the scattering indicatrix (f) with the aspect angle (Ψ) for large inhomogeneities

3) Scattering in Turbulent Atmosphere. In the interval of wave numbers $2\pi/L_0 \ll q \ll 2\pi/l_0$, described by the Kolmogorov two-thirds law (Sect. 1.4), the spectral density of turbulent fluctuations is defined by the power law

$$\Phi_\varepsilon(q) = \mathscr{A} C_\varepsilon^2 q^{-11/3}, \quad \mathscr{A} = 0.033 \ . \tag{4.97}$$

For such a spectrum, the scattering cross-section of a unit volume into a unit solid angle is (we put $\bar{\varepsilon} = 1$)

$$\sigma(\theta) = 2^{-8/3} \pi \mathscr{A} C_\varepsilon^2 k_0^{-1/3} [\sin(\theta/2)]^{-11/3} = 0.016 C_\varepsilon^2 k_0^{-1/3} [\sin(\theta/2)]^{-11/3} \ . \tag{4.98}$$

The total scattering cross-section in this case appears to be infinite, since the integral

$$\sigma_0 = \oint \sigma \, do = 2\pi \int_0^\pi \sigma(\theta) \sin\theta \, d\theta$$

diverges at small scattering angles θ. This is accounted for by the fact that the expression for the spectral density (4.97) corresponding to the interval $2\pi/L_0 \ll q \ll 2\pi/l_0$ is unsuitable for small $q = 2k\sin(\theta/2)$. The total scattering cross-section for another model of spectrum (1.82), which assumes finite values as $q \to 0$, is calculated in Exercise 4.9.2.

4.4 Spatial Correlation and Probability Distributions of Scattered Fields

Returning to expression (4.38) for the spatial covariance of the scattered field, we will note its resemblance to (2.128), a relation describing the spatial correlation of the field of a system of independent radiators. In both equations, the dependence on the coordinates of the observation points $\boldsymbol{r}_1$ and $\boldsymbol{r}_2$ is accounted for in the integrand by the same factor $\exp(iks)/|\boldsymbol{r}_1 - \boldsymbol{R}|\,|\boldsymbol{r}_2 - \boldsymbol{R}|$, where

$$s = |\boldsymbol{r}_1 - \boldsymbol{R}| - |\boldsymbol{r}_2 - \boldsymbol{R}| \ . \tag{4.99}$$

This is no chance coincidence. It has already been noted (Sect. 4.1) that finding the *single*-scattered field — a problem coming under scheme (2), i.e., the propagation of waves in randomly inhomogeneous media — actually reduces to a problem of scheme (1), i.e., to the excitation of fields by definite random sources. According to (4.11), these sources are continuously distributed over V. However, as the correlation radius l_ε of the inhomogeneities $\tilde{\varepsilon}$ is finite, the continuous distribution of sources is equivalent to the finite number (of order V/l_ε^3) of discrete uncorrelated sources. This accounts for the similarity of equations (4.38) and (2.128).

This similarity enables us to draw a number of qualitative and quantitative conclusions as to the character of the spatial correlation of the scattered field. So it may be argued that within and near the scattering volume (i.e., for $r \lesssim L$), the correlation radius of the field, l_u, is of order λ for small inhomogeneities ($l_\varepsilon \ll \lambda$), and of order l_ε for $l_\varepsilon \gg \lambda$. Both statements follow from the estimate

$$l_u \sim \lambda/\gamma , \tag{4.100}$$

where γ is the apparent angular size of the region occupied by the sources. In fact, in the case of small-scale inhomogeneities, which scatter in an isotropic way, the angle γ is comparable to π, so that $l_u \sim \lambda/\pi \sim \lambda$. But large-scale inhomogeneities have a narrow indicatrix ($\gamma \sim \lambda/l_\varepsilon$), whose width determines the effective angular size of the region. As a result, here $l_u \sim \lambda/\gamma \sim l_\varepsilon$.

As the distance from the scattering volume is increased, a measure of ordering is observed in the field, thereby increasing the scale of spatial correlation. When $r \gg L$, the angular size of the scattering volume γ becomes of the order of $L/r \ll 1$, with the result that the transverse (relative to the direction of scattering $\boldsymbol{n}_s$) correlation radius exceeds the wavelength by r/L times

$$l_\perp \sim \lambda r/L . \tag{4.101}$$

Near the Fraunhofer zone ($r \sim kL^2$), the transverse correlation radius becomes as large as the diameter of the scattering volume $l_\perp \sim L$, and for $r \gg kL^2$ it exceeds L.

As to the longitudinal correlation radius $l_\parallel$, this can be calculated from the equation of exercise 13, Chap. 2. Within the confines of the near zone of the scattering volume ($r \ll kL^2$), we have

$$l_\parallel \sim \frac{\lambda}{\gamma^2} \sim \frac{\lambda r^2}{L^2} \quad (r \ll kL^2) . \tag{4.102}$$

Beginning with $r \sim kL^2$, the correlation extends to infinity

$$l_\parallel = \infty \quad (r \gg kL^2) . \tag{4.103}$$

The above statements can be partially illustrated by referring to the spatial covariance (4.38), on condition that the observation points lie within the *Fraunhofer* zone ($r \gg kL^2$), and that the scattering volume is filled with statistically homogeneous fluctuations. Given these assumptions, we can replace ks by the approximate value

$$ks \approx kr_1 - kr_2 + k\Delta \boldsymbol{n}_s \cdot \boldsymbol{R} , \tag{4.104}$$

where $\Delta \boldsymbol{n}_s = \boldsymbol{n}_{s1} - \boldsymbol{n}_{s2} = \boldsymbol{r}_1/r_1 - \boldsymbol{r}_2/r_2$, and place all the factors outside the integration sign in (4.23), with the exception of $M(\boldsymbol{R})$ and $\exp(ik\Delta\boldsymbol{n}_s\boldsymbol{R})$

$$\psi_u(\boldsymbol{r}_1, \boldsymbol{r}_2) = \frac{\pi k_0^4 |A|^2}{2} \frac{\Phi_\varepsilon(\boldsymbol{q}_0)}{r^2 r_0^2} \mathrm{e}^{ik(r_1 - r_2)} \int_{-\infty}^{\infty} M(\boldsymbol{R}) \mathrm{e}^{ik\Delta\boldsymbol{n}_s \cdot \boldsymbol{R}} d^3 R . \tag{4.105}$$

We have here substituted $r^2 = (r_1 + r_2)^2/4$ for $|\boldsymbol{r}_1 - \boldsymbol{R}|\,|\boldsymbol{r}_2 - \boldsymbol{R}|$ by additionally assuming that the distance between the observation points $|\boldsymbol{r}_1 - \boldsymbol{r}_2|$ is small as compared to the distance r from the center of the scattering volume to the "center of mass" of the observation points.

The integral in (4.105) is a delta-type function. If we introduce the notation

$$\delta_V(\kappa) = \frac{1}{8\pi^3} \int_{-\infty}^{+\infty} M(\boldsymbol{R}) \exp(\mathrm{i}\boldsymbol{\kappa}\cdot\boldsymbol{R})\, d^3 R \ , \tag{4.106}$$

then (4.105) will become

$$\psi_u(\boldsymbol{r}_1, \boldsymbol{r}_2) = \frac{\pi k_0^4 |A|^2 \Phi_\varepsilon(\boldsymbol{q}_0)\, \mathrm{e}^{\mathrm{i}k(r_1 - r_2)}}{2r^2 r_0^2} 2\pi^3 \delta_V(k\,\Delta\boldsymbol{n}_\mathrm{s}) \ . \tag{4.107}$$

The function $\delta_V(\kappa)$ is equal to $V/8\pi^3$ at $\kappa = 0$ and, according to the properties of the Fourier transform, it falls off to small values for $\kappa \gtrsim 2\pi/L$, where L is the linear size of V. As $V \to \infty$, it becomes $\delta(\boldsymbol{\kappa})$.

Dividing the covariance (4.107) by the mean intensity (4.48) gives the correlation coefficient

$$K_u(\boldsymbol{r}_1, \boldsymbol{r}_2) \equiv \frac{\psi_u(\boldsymbol{r}_1, \boldsymbol{r}_2)}{\sqrt{\bar{I}_1(\boldsymbol{r}_1)\bar{I}_1(\boldsymbol{r}_2)}} = \frac{8\pi^3 \mathrm{e}^{\mathrm{i}k(r_1 - r_2)}}{V} \delta_V(k\,\Delta\boldsymbol{n}_\mathrm{s}) \ . \tag{4.108}$$

Since $\delta_V(k\Delta\boldsymbol{n}_\mathrm{s})$ differs markedly from zero only when $k|\Delta\boldsymbol{n}_\mathrm{s}| \lesssim 2\pi/L$, the values of the scattered field at $\boldsymbol{r}_1$ and $\boldsymbol{r}_2$ become uncorrelated for $\Delta\boldsymbol{n}_\mathrm{s} \gtrsim 2\pi/kL = \lambda/L$.

Now $|\Delta\boldsymbol{n}_\mathrm{s}|$ is almost equal to the angle $\Delta\theta$ between the unit vectors $\boldsymbol{n}_{\mathrm{s}1}$ and $\boldsymbol{n}_{\mathrm{s}2}$ (Fig. 4.1), so that the estimate derived determines the "correlation angle"

$$\Delta\theta_\mathrm{c} \sim \frac{\lambda}{L} \ . \tag{4.109}$$

But at the same time $\Delta\theta_\mathrm{c} \sim l_\perp/r$, so that from (4.109) we find that the transverse radius is still given by (4.101).

As to longitudinal correlation, when the observation points appear to lie on the same straight line and $\Delta\boldsymbol{n}_\mathrm{s} = 0$, the correlation coefficient is

$$K_u(\boldsymbol{r}_1, \boldsymbol{r}_2) = \mathrm{e}^{\mathrm{i}k(r_1 - r_2)} \ ,$$

i.e., its absolute value is unity at any r_1 and r_2 lying in the far zone. This suggests that, by (4.103), for the amplitude of the scattered field $A_1 \equiv |u_1|$ the longitudinal correlation radius is infinite, $l_{\|} = \infty$, and the difference of the field phases at points $\boldsymbol{r}_1$ and $\boldsymbol{r}_2$ lying on the straight line equals the optical path difference $k(r_1 - r_2)$.

It should be noted that (4.108) contains only the *geometric* parameters of the problem (r and L) and the wavelength of the primary field, λ; it does not include any statistical characteristics of the fluctuations $\tilde{\varepsilon}$. This strongly suggests that the correlation behavior of the scattered field is conditioned by purely dynamic considerations, and is only indirectly influenced by the statistical picture.

Actually, a sinusoidal diffraction grating of finite length L and period Λ_q produces a wave beam of finite angular width $\Delta\theta \sim \lambda/L$ in the direction θ [see (4.47)]. Since the fluctuation spectrum contains various spatial harmonics, scattering at a given angle is determined not only by the volume grating meeting the Bragg condition (4.47), but also by gratings of similar size and orientation, for which the direction lies within the main diffraction maxima of width $\Delta\theta \sim \lambda/L$. Thus, as long as the angular distance between observation points is less than $\Delta\theta \sim \lambda/L$, scattering is controlled by quite a definite group of diffraction gratings. By contrast, at $\Delta\theta \sim \lambda/L$ the fields scattered in the direction $\boldsymbol{n}_{s1} = \boldsymbol{r}_1/r_1$ and $\boldsymbol{n}_{s2} = \boldsymbol{r}_2/r_2$ are already determined by *different* groups of gratings (with nonoverlapping main maxima), and this destroys spatial correlation.

We will now turn to probability distributions of the scattered field. In the single scattering approximation, the field $u_1(\boldsymbol{r})$ is expressed by the integral (4.11) of the product of a deterministic function with the random function $\tilde{\varepsilon}(\boldsymbol{r}')$. Because the linear size L of the scattering volume is assumed to be large as compared with the correlation radius l_ε of the fluctuations $\tilde{\varepsilon}$, it may be argued that by the Central Limit Theorem of probability theory the distribution of the scattered field is close to the normal one.

Let us now find the parameters characterizing the joint distribution functions of the real and imaginary parts of the complex field

$$u_1(\boldsymbol{r}) = U(\boldsymbol{r}) + \mathrm{i}V(\boldsymbol{r}) \ . \tag{4.110}$$

Since $U_j \equiv U(\boldsymbol{r}_j)$ and $V_k \equiv V(\boldsymbol{r}_k)$ have a normal distribution, it suffices to derive the matrix of the second moments for these quantities, which coincides with the correlation matrix (since the mean scattered field, and hence the means of $\langle U_j\rangle$ and $\langle V_k\rangle$ are zero). The bilinear means $\langle U_j V_k\rangle$ can then be expressed in terms of the covariances of a complex field

$$\begin{aligned} \psi_u(\boldsymbol{r}_j, \boldsymbol{r}_k) &= \langle u_1(\boldsymbol{r}_j) u_1^*(\boldsymbol{r}_k)\rangle \ , \\ \tilde{\psi}_u(\boldsymbol{r}_j, \boldsymbol{r}_k) &= \langle u_1(\boldsymbol{r}_j) u_1(\boldsymbol{r}_k)\rangle \ . \end{aligned} \tag{4.111}$$

The first of these functions was computed in Sect. 4.2 and considered above. We will show that the second covariance $\tilde{\psi}_u$ is almost everywhere small as compared to ψ_u. To this end, we will estimate $\langle u_1^2(\boldsymbol{r})\rangle = \tilde{\psi}_u(\boldsymbol{r}, \boldsymbol{r})$ in the Fraunhofer zone. By (4.53),

$$\begin{aligned} \langle u_1^2(\boldsymbol{r})\rangle &= \frac{k_0^4 A^2 \mathrm{e}^{2ikr}}{(4\pi r r_0)^2} \int_{-\infty}^{\infty} M(\boldsymbol{r}')\, d^3r' \int_{-\infty}^{\infty} M(\boldsymbol{r}'')\, d^3r''\, \psi_\varepsilon(\boldsymbol{r}' - \boldsymbol{r}'')\mathrm{e}^{-\mathrm{i}\boldsymbol{q}(\boldsymbol{r}' + \boldsymbol{r}'')} \\ &= \frac{k_0^4 A^2 \mathrm{e}^{2ikr}}{(4\pi r r_0)^2} \iint_{-\infty}^{\infty} d^3\varrho\, d^3R\, M(\boldsymbol{R} + \boldsymbol{\varrho}/2)\, M(\boldsymbol{R} - \boldsymbol{\varrho}/2)\psi_\varepsilon(\boldsymbol{\varrho})\mathrm{e}^{-2\mathrm{i}\boldsymbol{q}\cdot\boldsymbol{R}} \ . \end{aligned} \tag{4.112}$$

Substituting $M(\boldsymbol{R})$ for $M(\boldsymbol{R} + \varrho/2)M(\boldsymbol{R} - \varrho/2)$ gives

$$\langle u_1^2(\boldsymbol{r})\rangle = \frac{k_0^4 A^2 \mathrm{e}^{2ikr}}{(4\pi r r_0)^2} (2\pi)^6\, \Phi_\varepsilon(0)\delta_V(2\boldsymbol{q}) \ , \tag{4.113}$$

where $\delta_V(\kappa)$ is defined by (4.106).

With forward scattering ($\boldsymbol{q} = 0$), the mean square of the field $\langle u_1^2\rangle$ coincides up to phase with the mean intensity $\bar{I}_1 = \langle |u_1|^2\rangle$ since $\delta_V(0) = V/8\pi^3$. When

$|2\boldsymbol{q}| \gtrsim 2\pi/L$, however, $\delta_V(2\boldsymbol{q})$ drops sharply as compared to $\delta_V(0)$, i.e.,

$$|\langle u_1^2(\boldsymbol{r})\rangle| \ll \bar{I}_1 = \langle |u_1^2(\boldsymbol{r})|\rangle, \quad q \gtrsim \pi/L \ . \tag{4.114}$$

Since $q = 2k\sin(\theta/2)$, the inequality $q \gtrsim \pi/L$ holds outside the narrow cone with angle $\theta \sim \lambda/2L$, whence it follows that for scattering at not-too-small angles ($\theta \gtrsim \lambda/2L$),

$$|\tilde{\psi}_u| \ll |\psi_u| \ . \tag{4.115}$$

To all intents and purposes, it suggests that outside the cone the second covariance $\tilde{\psi}_u$ may be taken to be zero

$$\tilde{\psi}_u(\boldsymbol{r}_j, \boldsymbol{r}_k) \approx 0 \ . \tag{4.116}$$

This result has been derived for the far zone of the scattering volume ($r \gg kL^2$), but it is all the more valid in the near zone ($r \ll kL^2$). In fact, if we break down the scattering volume into individual elements with linear size small as compared to L but large as compared to the correlation radius of the inhomogeneities, and place the observation point at the Fraunhofer zone of each of these elements, then the resultant mean square of the field, $\langle u_1^2\rangle$, can be obtained by summing up expressions of the type (4.113), since the fluctuations in the various elements are statistically independent. Clearly, the summation of quantities containing the oscillating factor $\exp(2ikr)$ may only cause a reduction in $\langle u_1^2\rangle$ as compared to $\bar{I}_1 = \langle |u_1|^2\rangle$, even within the cone with $\theta \sim \lambda/2L$.

We will make use of this, and to derive the moments $\langle U_j V_k\rangle$ we will apply equations (1.25) that, for a zero second covariance, give

$$\begin{aligned} \langle U_1 U_2\rangle &= \langle V_1 V_2\rangle = \tfrac{1}{2}\operatorname{Re}\{\psi_u(\boldsymbol{r}_1, \boldsymbol{r}_2)\} \ , \\ \langle V_1 U_2\rangle &= -\langle U_1 V_2\rangle = \tfrac{1}{2}\operatorname{Im}\{\psi_u(\boldsymbol{r}_1, \boldsymbol{r}_2)\} \ . \end{aligned} \tag{4.117}$$

Putting in (4.117) $\boldsymbol{r}_1 = \boldsymbol{r}_2 = \boldsymbol{r}$, we find, in particular, that

$$\langle U^2\rangle = \langle V^2\rangle = \tfrac{1}{2}\bar{I}_1, \quad \langle U(\boldsymbol{r})\,V(\boldsymbol{r})\rangle = 0 \ . \tag{4.118}$$

The remainder of the procedure is as given in Sects. I.5.5 and II.2.2.

4.5 Scattering by Nonstationary Inhomogeneities

1) Temporal Covariance. The field scattered by time-dependent inhomogeneities $\tilde{\varepsilon}(t, \boldsymbol{r})$ produces a response similar to (4.26) at the antenna output,

$$v(t, \boldsymbol{r}) = \exp(-i\omega t)\int \mathscr{P}(\boldsymbol{r}')\exp[ik(|\boldsymbol{r}-\boldsymbol{r}'| - |\boldsymbol{r}'-\boldsymbol{r}_0|)]\,\tilde{\varepsilon}\left(t - \frac{|\boldsymbol{r}-\boldsymbol{r}'|}{c}, \boldsymbol{r}'\right)d^3r' \ . \tag{4.119}$$

However, we have restored here the factor $\exp(-i\omega t)$, which we omitted for simplicity in (4.62). The wave perturbation which emerges in scattering at $\boldsymbol{r}'$ reaches the observation point $\boldsymbol{r}$ in a finite time $\Delta t = |\boldsymbol{r}-\boldsymbol{r}'|/c$ (for simplicity we put $\bar{\varepsilon} = 1$). Therefore, the integrand includes the values of $\tilde{\varepsilon}$ not at t, but at an earlier moment $t' = t - \Delta t$. The function $\mathscr{P}(\boldsymbol{r}')$ is still given by (4.61).

We compute the temporal covariance for the scattered field on the assumption that the fluctuations $\tilde{\varepsilon}$ are statistically *homogeneous* and *stationary*, so that the space-time covariance of the inhomogeneities is only dependent on the difference (spatial and temporal) variables

$$\psi_\varepsilon(t', \boldsymbol{r}'; t'', \boldsymbol{r}'') = \langle \tilde{\varepsilon}(t', \boldsymbol{r}')\tilde{\varepsilon}(t'', \boldsymbol{r}'')\rangle = \psi_\varepsilon(t'-t'', \boldsymbol{r}'-\boldsymbol{r}'') \ . \tag{4.120}$$

It is easily seen from (4.119) that the temporal covariance of the response at the antenna output

$$\psi_v(t', t'') = \langle v(t')\, v^*(t'')\rangle$$

is also dependent on the difference $\tau = t' - t''$

$$\psi_v(\tau) = \mathrm{e}^{-\mathrm{i}\omega\tau} \iint\limits_{-\infty}^{\infty} \mathscr{P}(\boldsymbol{r}')\, \mathscr{P}^*(\boldsymbol{r}'')\mathrm{e}^{\mathrm{i}\Psi(\boldsymbol{r}', \boldsymbol{r}'')}\psi_\varepsilon(\tau - \Delta\tau, \boldsymbol{r}'-\boldsymbol{r}'')\, d^3r'\, d^3r'' \ , \tag{4.121}$$

where

$$\Psi(\boldsymbol{r}', \boldsymbol{r}'') = k\{|\boldsymbol{r}-\boldsymbol{r}'| + |\boldsymbol{r}'-\boldsymbol{r}_0| - |\boldsymbol{r}-\boldsymbol{r}''| - |\boldsymbol{r}''-\boldsymbol{r}_0|\} \ .$$

Thus, in the case of stationary $\tilde{\varepsilon}$, the response $v(t)$, like the scattered field $u_1(t)$ itself, is a stationary random process.

The quantity

$$\Delta\tau = (|\boldsymbol{r}'-\boldsymbol{r}| - |\boldsymbol{r}''-\boldsymbol{r}|)/c$$

in (4.121) is the difference in the time delays of wave perturbations coming from $\boldsymbol{r}'$ and $\boldsymbol{r}''$. Owing to the fact that the inhomogeneities are spatially correlated over the distances $|\boldsymbol{r}'-\boldsymbol{r}''| \lesssim l_\varepsilon$, the difference $\Delta\tau$ is actually not larger than the time l_ε/c taken by the wave to travel through one inhomogeneity ($\Delta\tau \lesssim l_\varepsilon/c$). Generally, $\Delta\tau$ is much smaller than τ_ε — the correlation time of $\tilde{\varepsilon}$

$$\Delta\tau \sim l_\varepsilon/c \ll \tau_\varepsilon \ . \tag{4.122}$$

Therefore, we can, with reasonable accuracy, put $\Delta\tau = 0$ in (4.121), and then

$$\psi_v(\tau) = \mathrm{e}^{-\mathrm{i}\omega\tau} \iint\limits_{-\infty}^{\infty} \mathscr{P}(\boldsymbol{r}')\, \mathscr{P}^*(\boldsymbol{r}'')\mathrm{e}^{\mathrm{i}\Psi(\boldsymbol{r}', \boldsymbol{r}'')}\psi_\varepsilon(\tau, \boldsymbol{r}'-\boldsymbol{r}'')\, d^3r'\, d^3r'' \ . \tag{4.123}$$

Expression (4.123) corresponds to the so-called *quasi-stationary* approximation that scattering at individual inhomogeneities occurs as if the inhomogeneities were at rest, and the time dependence of ψ_ε is now restored in the final form (4.123).

It is useful to perform further analysis of temporal fluctuations individually for two cases: (1) a randomly inhomogeneous medium that is on average at rest, and (2) a randomly inhomogeneous medium that on average travels at a uniform velocity $\bar{v}$, generally termed the *drift velocity*.

2) Scattering Without Regular Drift. For a medium that is on average at rest, we may repeat the argument of Sect. I.5.5 almost word for word, i.e., pass over to the integration variables $\boldsymbol{R}=(\boldsymbol{r}_1+\boldsymbol{r}_2)/2$, $\boldsymbol{\varrho}=\boldsymbol{r}_1-\boldsymbol{r}_2$ and expand the exponent $\Psi(\boldsymbol{r}',\boldsymbol{r}'')$ into a series in $\boldsymbol{\varrho}$

$$\begin{aligned}\psi_v(\tau)&=\mathrm{e}^{-\mathrm{i}\omega\tau}\int_{-\infty}^{\infty}d^3R|\mathscr{P}(\boldsymbol{R})|^2\int_{-\infty}^{\infty}d^3\varrho\,\psi_\varepsilon(\tau,\boldsymbol{\varrho})\exp(\mathrm{i}\,\boldsymbol{q}\cdot\boldsymbol{\varrho})\\&=8\pi^3\mathrm{e}^{-\mathrm{i}\omega\tau}\int_{-\infty}^{\infty}d^3R|\mathscr{P}(\boldsymbol{R})|^2\,\Phi_\varepsilon(\tau,\boldsymbol{q})\ ,\end{aligned}\tag{4.124}$$

where

$$\Phi_\varepsilon(\tau,\boldsymbol{\kappa})=\frac{1}{8\pi^3}\int_{-\infty}^{\infty}\psi_\varepsilon(\tau,\boldsymbol{\varrho})\exp(-\mathrm{i}\boldsymbol{\kappa}\cdot\boldsymbol{\varrho})\,d^3\varrho\tag{4.125}$$

is the spatial Fourier transform of the covariance $\psi_\varepsilon(\tau,\boldsymbol{\varrho})$, and $\boldsymbol{q}=\boldsymbol{q}(\boldsymbol{R})$ is the scattering vector for $\boldsymbol{R}$.

It follows from (4.124) that for the antenna output response, the correlation time is of the same order of magnitude as for the fluctuations $\tilde{\varepsilon}$. The same result can also be represented in a spectral form, through the spectral density of the stationary process $v(t)$

$$\begin{aligned}g_v(\Omega)&=\frac{1}{2\pi}\int_{-\infty}^{\infty}\psi_u(\tau)\mathrm{e}^{\mathrm{i}\Omega\tau}\,d\tau\\&=\frac{1}{2\pi}\int_{-\infty}^{\infty}d^3R|\mathscr{P}(\boldsymbol{R})|^2\int_{-\infty}^{\infty}d^3\varrho\,\psi_\varepsilon(\tau,\boldsymbol{\varrho})\exp[\mathrm{i}(\Omega-\omega)\tau-\mathrm{i}\boldsymbol{q}\cdot\boldsymbol{\varrho}]\ .\end{aligned}\tag{4.126}$$

If we take into account that the space-time spectrum of $\tilde{\varepsilon}$ is given by

$$G_\varepsilon(\Omega,\boldsymbol{\kappa})=\frac{1}{(2\pi)^4}\int_{-\infty}^{\infty}d\tau\int_{-\infty}^{\infty}d^3\varrho\,\psi_\varepsilon(\tau,\boldsymbol{\varrho})\exp(\mathrm{i}\Omega\tau-\mathrm{i}\boldsymbol{\kappa}\cdot\boldsymbol{\varrho}]\ ,$$

we can represent the frequency spectrum of $v(t)$ in the form

$$g_v(\Omega)=8\pi^3\int d^3R\,|\mathscr{P}(\boldsymbol{R})|^2\,G_\varepsilon[\Omega-\omega,\boldsymbol{q}(\boldsymbol{R})]\ .\tag{4.127}$$

The fluctuation spectrum $G_\varepsilon(\Omega,\boldsymbol{\kappa})$ is normally concentrated in the vicinity of the zero frequency $\Omega\approx 0$, whereas the spectrum of the scattered signal $g_v(\Omega)$ is concentrated near the frequency of the primary wave ω. It follows from (4.127) that the frequency spectrum of the scattered field can be obtained by summing up the space-time spectrum of the inhomogeneities $G_\varepsilon[\Omega-\omega,\boldsymbol{q}(\boldsymbol{R})]$ over the scattering region, with the weight function

$$8\pi^3|\mathscr{P}(\boldsymbol{R})|^2=\frac{\pi k_0^4|A|^2}{2}\frac{|f_0(\boldsymbol{n}_\mathrm{i})f_1(\boldsymbol{n}_\mathrm{s})|^2}{|\boldsymbol{r}-\boldsymbol{R}|^2|\boldsymbol{R}-\boldsymbol{r}_0|^2}M(\boldsymbol{R})\ .$$

If the nature of the problem is such that we may ignore the variation of the scattering vector $\boldsymbol{q}$ within the confines of the integration region [by (4.52), this is possible when $r \gg r_{\min} = \min(L, Ll_\varepsilon/\lambda)$], then we can place $G_\varepsilon[\Omega-\omega, \boldsymbol{q}(\boldsymbol{R})]$ outside the integration sign as a value at $\boldsymbol{q}=\boldsymbol{q}_0$, where $\boldsymbol{q}_0 = k(\boldsymbol{n}_{s0}-\boldsymbol{n}_{i0})$ is the scattering vector corresponding to the center of the scattering region. As a result, the frequency spectrum of the field appears to be proportional to $G_\varepsilon(\Omega-\omega, \boldsymbol{q}_0)$

$$\begin{aligned} g_v(\Omega) &= 8\pi^3 G_\varepsilon(\Omega-\omega, \boldsymbol{q}_0) \int_{-\infty}^{\infty} d^3R\, |\mathscr{P}(\boldsymbol{R})|^2 \\ &= G_\varepsilon(\Omega-\omega, \boldsymbol{q}_0) \frac{\pi}{2} \frac{k_0^4 |A|^2}{r_0^2 r^2} V_{\text{eff}} , \end{aligned} \tag{4.128}$$

where V_{eff} is the effective scattering volume. For $r \gg r_{\min}$, we can thus find the spectrum shape of ε directly from the measured values of $g_v(\Omega)$. But in the case of extended scattering regions, where $g_v(\Omega)$ and G_ε are related by the integral relationship (4.127), finding the shape of $G_\varepsilon(\Omega)$ from the measured $g_v(\Omega)$ presents more of a problem.

The spectrum of the complex envelope $\mathfrak{A}(t)$ of $v(t) = \mathfrak{A}(t)\exp(-\mathrm{i}\omega t)$ is concentrated near the zero frequency, since $g_{\mathfrak{A}}(\Omega) = g_v(\Omega+\omega)$. In particular, for $r \gg r_{\min}$, when $g_v(\Omega)$ is given by (4.128), we have

$$g_{\mathfrak{A}}(\Omega) = G_\varepsilon(\Omega, \boldsymbol{q}_0) \frac{\pi}{2} \frac{k_0^4 |A|^2}{r^2 r_0^2} V_{\text{eff}} .$$

3) Scattering with Regular Drift. It is somewhat more difficult to derive the covariance of the field scattered by drifting inhomogeneities than it is at $\bar{\boldsymbol{v}}=0$. In considering this issue we follow the results of *Gorelik* [4.3, 4] (see also [4.5, 6]), the treatment given in [4.1].

Let $\psi_{dr}(\tau, \boldsymbol{\varrho})$ be the covariance of the inhomogeneities in a comoving reference frame, i.e., the frame that travels steadily with the medium with a velocity $\bar{\boldsymbol{v}}$. In a stationary frame associated with the observer, the covariance will clearly be

$$\psi_\varepsilon(\tau, \boldsymbol{\varrho}) = \psi_{dr}(\tau, \boldsymbol{\varrho} - \bar{\boldsymbol{v}}\tau), \quad \boldsymbol{\varrho} = \boldsymbol{r}' - \boldsymbol{r}'' . \tag{4.129}$$

We substitute this expression in (4.123) and introduce new integration variables $\boldsymbol{R}$ and $\boldsymbol{\varrho}'$

$$\boldsymbol{R} = \tfrac{1}{2}(\boldsymbol{r}' + \boldsymbol{r}''), \quad \boldsymbol{\varrho}' = \boldsymbol{\varrho} - \bar{\boldsymbol{v}}\tau = \boldsymbol{r}' - \boldsymbol{r}'' - \bar{\boldsymbol{v}}\tau . \tag{4.130}$$

In these variables, the covariance of $v(t)$ will be

$$\begin{aligned} \psi_v(\tau) = \mathrm{e}^{-\mathrm{i}\omega\tau} \iint_{-\infty}^{\infty} \mathscr{P}\left(\boldsymbol{R} + \frac{\boldsymbol{\varrho}' + \bar{\boldsymbol{v}}\tau}{2}\right) \mathscr{P}^*\left(\boldsymbol{R} - \frac{\boldsymbol{\varrho}' + \bar{\boldsymbol{v}}\tau}{2}\right) \psi_{dr}(\tau, \boldsymbol{\varrho}') \\ \times \exp[\mathrm{i}\Psi'(\tau, \boldsymbol{\varrho}', \boldsymbol{R})]\, d^3\varrho'\, d^3R , \end{aligned} \tag{4.131}$$

where

$$\Psi'(\tau, \varrho', R) = \Psi\left(R + \frac{\varrho' + \bar{v}\tau}{2}, R - \frac{\varrho' + \bar{v}\tau}{2}\right)$$

$$= k\left(\left|r - R - \frac{\varrho' + \bar{v}\tau}{2}\right| + \left|R + \frac{\varrho' + \bar{v}\tau}{2} - r_0\right|\right.$$

$$\left. - \left|r - R + \frac{\varrho' + \bar{v}\tau}{2}\right| - \left|R - \frac{\varrho' + \bar{v}\tau}{2} - r_0\right|\right).$$

The covariance $\psi_{dr}(\tau, \varrho')$ essentially vanishes when $\varrho' \gtrsim l_\varepsilon$; therefore, we expand Ψ' into a power series in ϱ' and retain only the first two terms

$$\Psi'(\tau, \varrho', R) = \Phi(\tau, R) - q'(\tau, R)\varrho' + \dots . \tag{4.132}$$

Here

$$\Phi(\tau, R) = \Psi'(\tau, 0, R) = k(|r - R - \bar{v}\tau/2|$$
$$+ |R + \bar{v}\tau/2 - r_0| - |r - R + \bar{v}\tau/2| - |R - \bar{v}\tau/2 - r_0|),$$

$$q'(\tau, R) = \frac{k}{2}\left(\frac{r - R - \bar{v}\tau/2}{|r - R - \bar{v}\tau/2|} - \frac{R + \bar{v}\tau/2 - r_0}{|R + \bar{v}\tau/2 - r_0|}\right.$$
$$\left. - \frac{r - R + \bar{v}\tau/2}{|r - R + \bar{v}\tau/2|} + \frac{R - \bar{v}\tau/2 - r_0}{|R - \bar{v}\tau/2 - r_0|}\right).$$

If we substitute (4.132) into (4.131), an approximation which is valid if the fairly weak conditions (4.32) are met, then integrating with respect to ϱ' gives

$$\psi_v(\tau) = 8\pi^3 e^{-i\omega\tau} \int_{-\infty}^{\infty} d^3R\, \mathscr{P}_+ \mathscr{P}_-^* e^{i\Phi(\tau, R)} \Phi_{dr}(\tau, q'), \tag{4.133}$$

where $\mathscr{P}_\pm \equiv \mathscr{P}(R \pm \bar{v}\tau/2)$, and the function

$$\Phi_{dr}(\tau, \kappa) = \frac{1}{8\pi^3} \int_{-\infty}^{\infty} \psi_{dr}(\tau, \varrho) \exp(i\kappa \cdot \varrho)\, d^3\varrho \tag{4.134}$$

is the spatial Fourier transform of the covariance of the inhomogeneities in the comoving reference frame.

Expression (4.133) is more complicated than (4.123). First, the integrand now includes the additional exponential factor $\exp[i\Phi(\tau, R)]$. Second, the scattering vector q' and the product $\mathscr{P}_+ \mathscr{P}_-^*$ that now take the place of $|\mathscr{P}(R)|^2$ vary with τ. If we suppose, however, that the source of the primary field and the receiver are separated from the scattering region by a distance larger than its size ($r \gg L$ and $r_0 \gg L$), then (4.133) becomes simpler. Our further estimates will only contain inequalities referring to r. Similar inequalities must also hold for r_0.

When $R \gtrsim L$, the product $\mathscr{P}_+ \mathscr{P}_-^*$ tends to zero. Accordingly, in the region $R \lesssim L$ that is of importance for integrating in (4.133), we have $R/r \ll 1$ on the assumption that $r \gg L$. For the same reason, in (4.133), only the values $\bar{v}\tau \lesssim L$ are important. Bearing this in mind, we expand the exponent, $\Phi(\tau, \boldsymbol{R})$, into a power series in the small parameters $R/r \lesssim L/r \ll 1$ and $\bar{v}\tau/r \lesssim L/r \ll 1$

$$\Phi(\tau, \boldsymbol{R}) = -\boldsymbol{q}_0 \cdot \bar{\boldsymbol{v}}\tau + k\boldsymbol{v} \cdot \boldsymbol{R}\tau + kO\left[\left(\frac{R}{r}\right)^2 \bar{v}\tau\right] + kO\left[\frac{(\bar{v}\tau)^3}{r^2}\right], \qquad (4.135)$$

where $\boldsymbol{v}$ is a vector that has dimensions of frequency and is proportional to the drift velocity $\bar{\boldsymbol{v}}$

$$\boldsymbol{v} = \frac{1}{r}[\boldsymbol{n}_{s0}[\bar{\boldsymbol{v}}\boldsymbol{n}_{s0}]] + \frac{1}{r_0}[\boldsymbol{n}_{i0}[\bar{\boldsymbol{v}}\boldsymbol{n}_{i0}]] \; . \qquad (4.136)$$

We have also expanded the vector $\boldsymbol{q}'$ into a series, but in τ only

$$\boldsymbol{q}'(\tau, \boldsymbol{R}) = \boldsymbol{q}(\boldsymbol{R}) + kO\left[\left(\frac{\bar{v}\tau}{r}\right)^2\right], \quad \boldsymbol{q}(\boldsymbol{R}) = k(\boldsymbol{n}_s - \boldsymbol{n}_i) \qquad (4.137)$$

(here there is no term linear in τ).

We will see below that the condition $r \gg L$ is sufficient for the first two terms to be retained in (4.135), and only the first term in (4.137). Then, (4.133) takes the form

$$\psi_v(\tau) = 8\pi^3 \exp[-\mathrm{i}(\omega + \boldsymbol{q}_0 \cdot \bar{\boldsymbol{v}})\tau] \int \mathscr{P}_+ \mathscr{P}_-^* \exp(\mathrm{i}k\boldsymbol{v} \cdot \boldsymbol{R}\tau) \Phi_{dr}(\tau, \boldsymbol{q}) \, d^3R \; . \qquad (4.138)$$

Of course, as $\bar{v} \to 0$ this expression becomes (4.123), an expression obtained without drift.

As τ increases, the integral in (4.138) decreases for three reasons. Firstly, the product $\mathscr{P}_+ \mathscr{P}_-^*$ decreases, and tends to zero for $\bar{v}\tau \gtrsim L$. Secondly, the exponential function $\exp(\mathrm{i}k\boldsymbol{v}\boldsymbol{R}\tau)$ begins to oscillate markedly within the region significant for integration. Thirdly, the function $\Phi_{dr}(\tau, \boldsymbol{q})$, which describes the temporal correlation of fluctuations in the comoving reference frame, decreases and tends to zero as $\tau \to \infty$.

From the conditions $\bar{v}\tau \sim L$ and $k\boldsymbol{v}\boldsymbol{R}\tau \lesssim k\bar{v}L\tau/r \sim 1$, we find two characteristic time intervals corresponding to the first two factors

$$\tau_1 \sim \frac{L}{\bar{v}}, \qquad \tau_2 \sim \frac{r}{k\bar{v}L} \sim \frac{\lambda r}{\bar{v}L} \; . \qquad (4.139)$$

But the characteristic time scale of $\Phi_{dr}(\tau, \boldsymbol{q})$, i.e., the time interval of the correlation of the inhomogeneities in the comoving reference frame, will be denoted by τ_3. In essence, this is the "lifetime" of the inhomogeneities. The smallest of the three values of τ mentioned above clearly defines the correlation time τ_c of the signal

$$\tau_c = \min\{\tau_1, \tau_2, \tau_3\} \; . \qquad (4.140)$$

As $\bar{v} \to 0$ (no drift), τ_1 and τ_2 grow indefinitely, and $\tau_c = \tau_3$.

We will apply (4.138) to the particular case of scattering by *frozen* inhomogeneities that are independent of time ($\tau_3 = \infty$) in the comoving frame.

4) Scattering by Frozen Inhomogeneities. Obviously, in the laboratory frame of reference in which time-constant inhomogeneities drift at a velocity $\bar{v}$, the permittivity fluctuations $\tilde{\varepsilon}(t, \boldsymbol{r})$ have the property

$$\tilde{\varepsilon}(t, \boldsymbol{r}) = \tilde{\varepsilon}(t-\tau, \boldsymbol{r}-\bar{v}\tau) \ , \tag{4.141}$$

i.e., a perturbation $\tilde{\varepsilon}(t', \boldsymbol{r}')$, that at an earlier instant $t' = t-\tau$ was at $\boldsymbol{r}' = \boldsymbol{r}-\bar{v}\tau$ and during the time τ was travelling to $\boldsymbol{r}$ at $\bar{v}$, comes to a point $\boldsymbol{r}$ at a time t.

It has already been said that the "lifetime" of the inhomogeneities τ_3 in a comoving reference frame is formally infinite, i.e., $\tau_3 = \infty$. We can consider that, for all practical purposes, inhomogeneities are frozen if

$$\tau_3 \gg \tau_1, \tau_2 \ . \tag{4.142}$$

If this condition is met, we may assume that $\Phi_{dr}(\tau, \boldsymbol{q})$ in (4.138) is $\Phi_{dr}(0, \boldsymbol{q}) = \Phi_{\varepsilon}(\boldsymbol{q})$, because the integral in (4.138) will have become small earlier than $\Phi_{dr}(\tau, \boldsymbol{q})$ began to decrease. For frozen inhomogeneities, we thus have

$$\psi_v(\tau) = 8\pi^3 \exp[-\mathrm{i}(\omega + \boldsymbol{q}_0 \cdot \bar{v})\tau] \int\limits_{-\infty}^{\infty} \mathscr{P}_+ \mathscr{P}_-^* \exp(\mathrm{i}k\boldsymbol{v}\cdot\boldsymbol{R}\tau)\Phi_{\varepsilon}(\boldsymbol{q})d^3R \ . \tag{4.143}$$

The behavior of this covariance is influenced to a considerable degree by the relationship between τ_1 and τ_2. By (4.139), they are comparable at $r \sim kL^2$. In the near zone, where $r \ll kL^2$ and also, as stated, $r \gg L$, the inequality $\tau_2 \ll \tau_1$ holds. Consequently, the correlation time in the near zone is equal to τ_2

$$\tau_c = \tau_2 \sim \frac{r}{k\bar{v}L} \quad (L \ll r \ll kL^2) \ . \tag{4.144}$$

In contrast, in the far (Fraunhofer) zone, where $r \gg kL^2$, the correlation time coincides with τ_1, which is far shorter than τ_2

$$\tau_c = \tau_1 \sim L/v \quad (r \gg kL^2) \ . \tag{4.145}$$

It is easily seen that in both zones we may well discard the "extraneous" terms in (4.135) and (4.137). For example, in the near zone, where $r \ll kL^2$ and $\tau_c = \tau_2$, the fourth term of (4.135) will have the order of magnitude

$$k\frac{(\bar{v}\tau)^3}{r^2} \lesssim \frac{k}{r^2}\left(\frac{r}{kL}\right)^3 = \frac{r}{kL^2}\frac{1}{kL} \ ,$$

i.e., less than the product of two small parameters. But it is also small in the far zone, where $r \gg kL^2$ and $\tau_c = \tau_1$, so that

$$k\frac{(\bar{v}\tau)^3}{r^2} \lesssim \frac{k}{r^2}L^3 = \frac{kL^2}{r}\frac{L}{r} \ll 1 \ .$$

Note that, if the frozen condition (4.142) is violated, i.e., if $\tau_1, \tau_2 \gtrsim \tau_3$, the discarding of "unnecessary" terms is all the more justified the smaller τ_3 is as compared to τ_1 and τ_2. In other words, the only condition for (4.138) to hold is that $r \gg L$.

Let us now look in more detail at the behavior of the covariance (4.143) in the near and far zones.

For the *near* zone, (4.143) can be simplified if we consider that $\bar{v}\tau \lesssim \bar{v}\tau_2 \sim r/kL \ll L$, and that for $\bar{v}\tau \ll L$ the product $\mathscr{P}_+ \mathscr{P}_-^*$ is approximately equal to $|\mathscr{P}(\boldsymbol{R})|^2$. Thus,

$$\psi_v(\tau) = \exp[-\mathrm{i}(\omega + \boldsymbol{q}_0 \bar{\boldsymbol{v}})\tau] \int_{-\infty}^{\infty} |\mathscr{P}(\boldsymbol{R})|^2 \exp(\mathrm{i}k\boldsymbol{v} \cdot \boldsymbol{R}\tau)\Phi_\varepsilon(\boldsymbol{q})\, d^3R \; . \tag{4.146}$$

Clearly, to take the integral we will have to specify $\Phi_\varepsilon(\boldsymbol{q})$ and the weight function $|\mathscr{P}(\boldsymbol{R})|^2$. A spectral analysis of (4.146) can, however, be carried out even without specifying these functions.

We perform the temporal Fourier transformation to find the spectral density of the signal $v(t)$

$$g_v(\Omega) = 8\pi^3 \int_{-\infty}^{\infty} |\mathscr{P}(\boldsymbol{R})|^2 \Phi_\varepsilon(\boldsymbol{q})\delta(\Omega - \omega - \boldsymbol{q}_0 \bar{\boldsymbol{v}} - k\boldsymbol{v} \cdot \boldsymbol{R})\, d^3R \; . \tag{4.147}$$

From (4.147), a spectral line at the frequency $\Omega = \omega + \boldsymbol{q}_0 \bar{\boldsymbol{v}} + k\boldsymbol{v}\boldsymbol{R}$ with "intensity" $8\pi^3 |\mathscr{P}(\boldsymbol{R})|^2 \Phi_\varepsilon(\boldsymbol{q})$ corresponds to each elementary volume d^3R. The frequency spectrum $g_v(\Omega)$ is formed by the superposition of these lines. The line $\Omega = \omega + \boldsymbol{q}_0 \bar{\boldsymbol{v}} + k\boldsymbol{v}\boldsymbol{R}$ shifts relative to the frequency of the primary field ω due to the Doppler effect.

The Doppler shift

$$\Omega_\mathrm{d} = \bar{\boldsymbol{v}} \cdot \boldsymbol{q}_0 + k\boldsymbol{v} \cdot \boldsymbol{R}$$

contains two terms, the first of which is the same for all the points in the scattering region, whereas the second varies from point to point as the vector $\boldsymbol{q}(\boldsymbol{R})$ varies within the scattering volume[8].

The frequency spectrum width $\Delta\Omega$ is obviously defined by the difference of the Doppler shifts at the boundaries of the scattering volume $R \sim L$

$$\Delta\Omega \approx \Delta\Omega_\mathrm{d} \sim kvL \sim k\bar{v}L/r \; , \tag{4.148}$$

which, as expected, agrees with the estimate for the spectrum width $\Delta\Omega \sim 1/\tau_\mathrm{c} \sim 1/\tau_2$. Estimate (4.144) also permits of another interpretation. As was shown in Sect. 4.4, for stationary inhomogeneities the quantity r/kL is the transverse correlation radius of the field, $l_\perp$, i.e., the space scale of the interference pattern produced by the superposition of the scattered waves. When the frozen inhomogeneities drift,

[8] A more accurate expression for the Doppler shift is $\Omega_\mathrm{d} = \boldsymbol{v} \cdot \boldsymbol{q}(\boldsymbol{R})$. The expression given above only contains the first terms of the Taylor expansion in $\boldsymbol{R}$.

the interference pattern will pass the observation point with velocity $\bar{v}$, so that the characteristic fluctuation time for the field, $l_\perp/\bar{v} \sim r/k\bar{v}L$, coincides with $\tau_c = \tau_2$.

Estimate (4.144) is, strictly speaking, suitable only for $r \gg L$. If, however, we apply it to the case of $r \sim L$, where the observation point lies close to the scattering region, then we find that $\tau_c \sim 1/k\bar{v} \sim \lambda/\bar{v}$. Moreover, this does not contradict the estimate $\tau_c \sim l_\perp/\bar{v}$, since near the scattering volume $l_\perp \sim \lambda$ (see Sect. 4.4).

Different behavior is observed in the *far zone* for which, by (4.145), the correlation time τ_c equals the time $\tau_1 \sim L/\bar{v}$ that an individual inhomogeneity stays within the volume under consideration. In the far zone, the exponent in (4.143) is small as compared to unity, and $\Phi_\varepsilon(\boldsymbol{q}) \approx \Phi_\varepsilon(\boldsymbol{q}_0)$, so that the covariance of the field is given by

$$\psi_v(\tau) = \mathrm{e}^{-\mathrm{i}(\omega + \boldsymbol{q}_0 \cdot \bar{\boldsymbol{v}})\tau} 8\pi^3 \Phi_\varepsilon(\boldsymbol{q}_0) \int_{-\infty}^{\infty} \mathscr{P}_+ \mathscr{P}_-^* \, d^3R \ . \tag{4.149}$$

To take the integral it is necessary, as in (4.146), to specify the form of $\mathscr{P}(\boldsymbol{R})$. In the idealized case where the function $\mathscr{P}(\boldsymbol{R}) = \mathscr{P}_0 = \text{const}$ within the cube with side L, and zero elsewhere, and the drift velocity $\bar{\boldsymbol{v}}$ is normal to one of the cube sides, we have

$$\int_{-\infty}^{\infty} \mathscr{P}_+ \mathscr{P}_- d^3R = \begin{cases} |\mathscr{P}_0|^2 L^3 \left(1 - \dfrac{|\bar{\boldsymbol{v}}\tau|}{L}\right) & \text{for} \quad |\tau| \leqslant L/|\bar{\boldsymbol{v}}| \ , \\ 0 & \text{for} \quad |\tau| > L/|\bar{\boldsymbol{v}}| \ . \end{cases}$$

The laws governing scattering by wandering inhomogeneities, when the velocity field consists of the common drift velocity $\bar{\boldsymbol{v}}$ and the random component $\tilde{\boldsymbol{v}}$, i.e., $\boldsymbol{v} = \bar{\boldsymbol{v}} + \tilde{\boldsymbol{v}}$ are more complicated. Some of the aspects of this case are considered in [4.1] and [4.3–6].

5) Intensity Fluctuations. As was noted in Sect. 4.4, the single-scattered field has a Gaussian distribution with zero mean. For such a field, the intensity covariance

$$\psi_I(\tau) = \langle I_v(t) I_v(t-\tau)\rangle - \bar{I}_v(t)\bar{I}_v(t-\tau), \quad I_v = |v|^2 \ ,$$

will be (see Exercise 12, Chap. 1)

$$\psi_I(\tau) = |\psi_v(\tau)|^2 \ . \tag{4.150}$$

It follows that the characteristic time of the intensity pulsations is rather shorter [by 1.5–2 times, depending on the form of $\psi_v(\tau)$] than the correlation time of the signal $v(t)$ itself. Accordingly, the frequency spectrum of the intensity fluctuations

$$g_I(\Omega) = \frac{1}{2\pi} \int_{-\infty}^{\infty} |\psi_v(\tau)|^2 \mathrm{e}^{\mathrm{i}\Omega\tau} d\tau = \int_{-\infty}^{\infty} g_v(\Omega_1) g_v(\Omega_1 - \Omega)\, d\Omega_1 \tag{4.151}$$

is somewhat wider than $g_v(\Omega)$.

4.6 Scattering of Pulsed and Modulated Signals

The scattering of polychromatic waves is interesting in a number of ways. We will restrict ourselves to the particular, but important, case of *quasi-monochromatic* signals often used in optics, acoustics, and radiophysics. If we represent such a signal as the product of a slowly varying (generally speaking, complex) amplitude $\mathfrak{A}(t)$ by $\exp(-\mathrm{i}\omega t)$ (ω is the carrier frequency), we clearly must write down the signal at the antenna output in the form of (4.60) derived for the monochromatic wave where the constant amplitude A is replaced by the variable amplitude $\mathfrak{A}(t-b'/c)$ with a delayed argument $b'=|\boldsymbol{r}'-\boldsymbol{r}_0|+|\boldsymbol{r}-\boldsymbol{r}'|$. If, in addition, we restore the temporal factor $\exp(-\mathrm{i}\omega t)$ omitted in (4.60), then for the quasi-monochromatic signal we will have

$$v(t)=\frac{k_0^2\mathrm{e}^{-\mathrm{i}\omega t}}{4\pi}\int\limits_{-\infty}^{\infty} M(\boldsymbol{r}')\frac{f_0(\boldsymbol{n}_\mathrm{i}')f_1(\boldsymbol{n}_\mathrm{s}')}{|\boldsymbol{r}-\boldsymbol{r}'||\boldsymbol{r}'-\boldsymbol{r}_0|}\mathfrak{A}\left(t-\frac{b'}{c}\right)\mathrm{e}^{\mathrm{i}k_0b'}\tilde{\varepsilon}(\boldsymbol{r}')d^3r' \; . \tag{4.152}$$

To simplify the argument, we will introduce a dimensionless amplitude $a(t)=\mathfrak{A}(t)/\mathfrak{A}_\mathrm{m}$, where $\mathfrak{A}_\mathrm{m}=\max|\mathfrak{A}(t)|$, and we will denote by $\mathscr{P}(\boldsymbol{r}')$ the product of the nonoscillating factors which differ from those in (4.61) in that $\mathfrak{A}_\mathrm{m}$ is substituted for A. In this notation, (4.152) will become

$$v(t)=\mathrm{e}^{-\mathrm{i}\omega t}\int\limits_{-\infty}^{\infty}\mathscr{P}(\boldsymbol{r}')a\left(t-\frac{b'}{c}\right)\mathrm{e}^{\mathrm{i}k_0b'}\tilde{\varepsilon}(\boldsymbol{r}')d^3r' \; , \tag{4.153}$$

and the temporal covariance will be given by

$$\begin{aligned}\psi_v(t_1,t_2)&=\langle v(t_1)v^*(t_2)\rangle\\&=\mathrm{e}^{-\mathrm{i}\omega\tau}\iint\limits_{-\infty}^{\infty}\mathscr{P}(\boldsymbol{r}')\mathscr{P}^*(\boldsymbol{r}'')a\left(t_1-\frac{b'}{c}\right)a^*\left(t_2-\frac{b''}{c}\right)\mathrm{e}^{\mathrm{i}k_0(b'-b'')}\\&\quad\times\psi_\varepsilon(\tau,\boldsymbol{r}'-\boldsymbol{r}'')d^3r'd^3r'' \; ,\end{aligned} \tag{4.154}$$

where $\tau=t_1-t_2$. For the sake of generality, we take into consideration here the possible variation of the fluctuations $\tilde{\varepsilon}$ with time, for which purpose we introduced the temporal argument τ into the covariance ψ_ε. The process $v(t)$ is, of course, nonstationary, ψ_v varying individually with t_1 and t_2. Let us estimate the correlation interval for $v(t)$ in the case of a single pulse (the periodic sequence of pulses is considered in Problem 5).

Let $a(t)$ be a square pulse of length T [$a(t)=1$ for $|t|\leqslant T/2$ and $a(t)=0$ for $|t|>T/2$]. It is clear that with T larger than the correlation time for the inhomogeneities, τ_ε, the correlation time for $v(t)$ coincides with τ_ε. In the opposite limiting case of $T\ll\tau_\varepsilon$ the correlation time will be of the order of T, as the product $a(t_1-b'/c)a^*(t_2-b''/c)$ vanishes when $|t_1-t_2|>T$ at any value of $\boldsymbol{r}'$ and $\boldsymbol{r}''$.

A more detailed examination of (4.154) is possible on the assumption that the space length of the pulse, cT, is large as compared to the correlation radius of the inhomogeneities, l_ε. In that case, we can carry out the integration with respect to $\boldsymbol{\varrho} = \boldsymbol{r}' - \boldsymbol{r}''$, as was done several times in earlier sections. As a result

$$\psi_v(t_1, t_2) = 8\pi^3 e^{-i\omega\tau} \int_{-\infty}^{\infty} |\mathscr{P}(\boldsymbol{R})|^2 a\left(t_1 - \frac{B}{c}\right) a^*\left(t_2 - \frac{B}{c}\right) \Phi_\varepsilon(\tau, \boldsymbol{q})\, d^3R \ , \tag{4.155}$$

where $B = |\boldsymbol{r} - \boldsymbol{R}| + |\boldsymbol{R} - \boldsymbol{r}_0|$.

In contrast to the case of the monochromatic primary wave, the mean intensity $\bar{I}_v = \langle |v|^2 \rangle$ is now dependent on time

$$\bar{I}_v(t) = \psi_v(t, t) = 8\pi^3 \int_{-\infty}^{\infty} |\mathscr{P}(R)|^2 \left| a\left(t - \frac{B}{c}\right)\right|^2 \Phi_\varepsilon(0, \boldsymbol{q})\, d^3R \ . \tag{4.156}$$

The quantity $\bar{I}_v$ is smaller than for the monochromatic signal of the same amplitude because the integrand now contains the factor $|a|^2 \leqslant 1$. This factor is nonzero within the ellipsoidal layer, such that its internal points satisfy the relation $|t - B/c| \leqslant T/2$, or if we write it in full

$$c(t - T/2) \leqslant B(\boldsymbol{R}) \equiv |\boldsymbol{r} - \boldsymbol{R}| + |\boldsymbol{R} - \boldsymbol{r}_0| \leqslant c(t + T/2) \ . \tag{4.157}$$

The foci of the ellipsoids bounding the layer (Fig. 4.9) coincide with the observation points ($\boldsymbol{r}$) and radiation points ($\boldsymbol{r}_0$), and the ellipsoids grow with time. The ellipsoidal layer (4.157) can be called the *pulse volume*. It is a region that, at a given instant of time, belongs simultaneously to the pulse volume (4.157) and the effectively scattering volume V_{eff} that determines the value of the intensity $\bar{I}_v$.

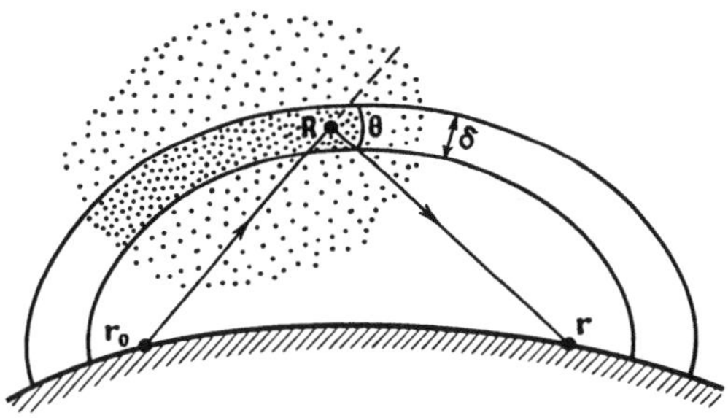

Fig. 4.9. A pulse volume overlapping with a scattering volume

The thickness of the pulse volume δ is expressed in the simplest way for backscattering, where $\boldsymbol{r} = \boldsymbol{r}_0$ and the ellipsoidal layer (4.157) degenerates into a spherical one. As is clear from (4.157), $\delta = cT/2$, i.e., the thickness of the pulse volume is equal to half the spatial length of the pulse. In the general case the expressions for δ are fairly complex, but for not-very-long pulses ($cT \ll |\boldsymbol{r} - \boldsymbol{r}_0|$), δ is given by the simple expression [4.1]

$$\delta = cT/2\sin(\theta/2) \ , \tag{4.158}$$

where θ is the scattering angle (Fig. 4.9). With backscattering ($\theta = \pi$) we obtain, from (4.158), the earlier result $\delta = cT/2$. Note that the ellipsoid $|t - B/c| = \text{const}$ expands with the normal component of the velocity $v_n = (c/2)\sin(\theta/2)$.

The temporal behavior of $\bar{I}_v(t)$ is markedly affected by the relationship between the thickness of the pulse volume δ and the size of the scattering region L. For *long* pulses [$\delta \gtrsim L$ or $T \gtrsim (2L/c)\sin(\theta/2)$] $\bar{I}_v(t)$ grows monotonically until the whole of the scattering volume appears within the pulse volume (4.158). The time $\bar{I}_v(t)$ taken to reach the maximum value $\bar{I}_{\max}$ has the order of magnitude $L/v_n = (2L/c)\sin(\theta/2)$. During this time, $\bar{I}_v(t)$ decays to zero as the trailing edge of the pulse volume leaves the scattering region. Between the increase and decrease stages, $\bar{I}_v(t)$ is constant and equal to the mean intensity of the monochromatic signal of infinite length. If

$$\frac{2L}{c}\sin\frac{\theta}{2} \ll T \, , \tag{4.159}$$

i.e., the increase and decrease times are small as compared to the signal length T, then the scattered pulse repeats almost exactly the shape of the emitted pulse.

The inequality (4.159) may also be represented in a different way, by introducing for the signal the frequency band width

$$\Delta\omega \approx \frac{2\pi}{T} \ll \frac{\pi c}{L\sin(\theta/2)} \, . \tag{4.160}$$

This condition restricts the bandwidth of channels which rely on scattering. In the transmission and reception of microwave signals using narrow-beam antennas, when the size L of a scattering volume in the atmosphere is in the region of several dozen kilometers, the band-width $\Delta\omega$ appears to be sufficient to transform television signals ($\Delta f = \Delta\omega/2\pi \sim 10^6$ Hz).

In the opposite limiting case of $\delta \ll L$ or $T \ll (2L/c)\sin(\theta/2)$ (*short* pulse), the "time of entrance", and also the "time of exit", of a pulse into and out of the scattering region are of the order of the pulse duration: $\delta/v_n \sim T$. But the time of residence of the pulse volume inside the scattering region is about $L/v_n \approx (2L/c)\sin(\theta/2)$. For a short pulse, this time is much longer than the duration of the signal T, i.e., the short primary pulse elongates considerably in scattering—generally, by about $(2L/cT)\sin(\theta/2)$ times, and in backscattering, by $2L/cT$ times.

It is to be noted that although the scattered signal is significantly extended as compared to the primary pulse, the correlation time is still a quantity of order T. This follows both from the above considerations and from the expression

$$\psi_v(t_1, t_2) = \bar{I}_v\left(\frac{t_1 + t_2}{2}\right) K_{\mathfrak{A}}(\tau)\mathrm{e}^{-\mathrm{i}\omega\tau} \quad (\tau = t_1 - t_2) \, , \tag{4.161}$$

which, as may be shown using (4.155), connects for short pulses the covariance of the field with the mean intensity $\bar{I}_v$ and the "correlation coefficient" of the primary pulse envelope

$$K_{\mathfrak{A}}(\tau) = \frac{\int\limits_{-\infty}^{\infty} \mathfrak{A}(t)\mathfrak{A}^*(t-\tau)\,dt}{\int\limits_{-\infty}^{\infty} |\mathfrak{A}(t)|^2\,dt} . \tag{4.162}$$

It is clear from (4.161) that for short pulses, the scattered field is a quasi-stationary random process with variable variance $\bar{I}_v(t)$ and correlation coefficient $K_{\mathfrak{A}}(\tau)\exp(-i\omega\tau)$. The correlation coefficient of the envelope, $K_{\mathfrak{A}}(\tau)$, vanishes for $\tau > T$, thus suggesting that the correlation time of the scattered field is about T.

4.7 Scattering of Electromagnetic Waves

The scattering of electromagnetic waves (like the scattering of transverse elastic waves) produces two new effects which do not feature in the scalar problem. First, scattering changes the wave polarization; second, in an anisotropic medium some types of polarization can transform into others. We will only consider the single scattering of electromagnetic waves in an *isotropic nondispersive* medium, where waves of one type only can propagate, namely transverse electromagnetic waves.

1) Scattering of Monochromatic Electromagnetic Waves by Stationary Inhomogeneities in an Isotropic Medium. In such a case, Maxwell's equations

$$\operatorname{curl}\boldsymbol{H} = \frac{1}{c}\frac{\partial \boldsymbol{D}}{\partial t}, \quad \operatorname{curl}\boldsymbol{E} = -\frac{1}{c}\frac{\partial \boldsymbol{H}}{\partial t} \tag{4.163}$$

should be solved together with the material equation for the isotropic nondispersive medium

$$\boldsymbol{D} = \varepsilon\boldsymbol{E} . \tag{4.164}$$

The material equation (4.164) excludes from consideration not only scattering in anisotropic media (this scattering is considered in more detail, for example, in [4.7]), but also scattering by *anisotropic fluctuations* in the isotropic medium (the so-called anisotropy fluctuations). Such fluctuations occur, for instance, in viscous liquids, due to random rotations of the molecules if the latter are optically anisotropic (see [4.8] and the review [4.9]).

Let a monochromatic wave propagate in a medium with constant mean electric permittivity $\bar{\varepsilon}$ and stationary inhomogeneities, i.e.,

$$\varepsilon = \bar{\varepsilon} + \tilde{\varepsilon}(\boldsymbol{r}), \quad \bar{\varepsilon} = \text{const}, \quad \langle\tilde{\varepsilon}\rangle = 0 . \tag{4.165}$$

Maxwell's equations will take the form ($k_0 \equiv \omega/c$)

$$\operatorname{curl}\boldsymbol{H} + \mathrm{i}k_0\bar{\varepsilon}\boldsymbol{E} = -ik_0\tilde{\varepsilon}\boldsymbol{E}, \quad \operatorname{curl}\boldsymbol{E} - \mathrm{i}k_0\boldsymbol{H} = 0 , \tag{4.166}$$

and for weak fluctuations $\langle(\tilde{\varepsilon})^2\rangle \ll (\bar{\varepsilon})^2$, their solution can be represented by the expansion in perturbations

$$\boldsymbol{E} = \boldsymbol{E}_0 + \boldsymbol{E}_1 + \boldsymbol{E}_2 + \ldots, \quad \boldsymbol{H} = \boldsymbol{H}_0 + \boldsymbol{H}_1 + \boldsymbol{H}_2 + \ldots. \tag{4.167}$$

Here the primary field $\boldsymbol{E}_0$, $\boldsymbol{H}_0$ obeys the homogeneous equations

$$\operatorname{curl} \boldsymbol{H}_0 + \mathrm{i}k_0\bar{\varepsilon}\boldsymbol{E}_0 = 0, \quad \operatorname{curl} \boldsymbol{E}_0 - \mathrm{i}k_0\boldsymbol{H}_0 = 0 \tag{4.168}$$

and the single-scattered field obeys the inhomogeneous equations

$$\operatorname{curl} \boldsymbol{H}_1 + \mathrm{i}k_0\bar{\varepsilon}\boldsymbol{E}_1 = -\mathrm{i}k_0\tilde{\varepsilon}\boldsymbol{E}_0, \quad \operatorname{curl} \boldsymbol{E}_1 - \mathrm{i}k_0\boldsymbol{H}_1 = 0\ . \tag{4.169}$$

For the higher-order approximations, equations are derived from (4.169) with successive substitutions $\boldsymbol{E}_m \to \boldsymbol{E}_{m+1}$, $\boldsymbol{H}_m \to \boldsymbol{H}_{m+1}$.

The scattered fields $\boldsymbol{E}_1$ and $\boldsymbol{H}_1$ can be found using known Green's functions, i.e., the solution to the problem of the excitation of electromagnetic waves by a point electric source $\boldsymbol{j}_e = \boldsymbol{l}_1\delta(\boldsymbol{r}-\boldsymbol{r}')$. In the wave zone, i.e., for $R = |\boldsymbol{r}-\boldsymbol{r}'| \gg \lambda$, the elementary source produces the fields

$$\begin{aligned} \boldsymbol{E}_{0e}(\boldsymbol{r},\boldsymbol{r}') &= \frac{\mathrm{i}k_0^2}{\omega}\,[\boldsymbol{n}_s'[\boldsymbol{l}_1\boldsymbol{n}_s']]\,\frac{\mathrm{e}^{\mathrm{i}kR}}{R}\ , \\ \boldsymbol{H}_{0e}(\boldsymbol{r},\boldsymbol{r}') &= \frac{\mathrm{i}k_0^2}{\omega}\,[\boldsymbol{n}_s'\boldsymbol{l}_1]\,\frac{\mathrm{e}^{\mathrm{i}kR}}{R}\ , \end{aligned} \tag{4.170}$$

where $\boldsymbol{n}_s' = (\boldsymbol{r}-\boldsymbol{r}')/|\boldsymbol{r}-\boldsymbol{r}'| = \boldsymbol{R}/R$ is the unit vector directed from $\boldsymbol{r}'$ to $\boldsymbol{r}$.

Using (4.170), we can represent the solution to the inhomogeneous equations (4.169) as

$$\begin{aligned} \boldsymbol{E}_1(\boldsymbol{r}) &= \frac{k_0^2}{4\pi}\int_{-\infty}^{\infty} M(\boldsymbol{r}')\frac{\mathrm{e}^{\mathrm{i}kR}}{R}\,\tilde{\varepsilon}(\boldsymbol{r}')\,[\boldsymbol{n}_s'[\boldsymbol{E}_0(\boldsymbol{r}')\boldsymbol{n}_s']]\,d^3r'\ , \\ \boldsymbol{H}_1(\boldsymbol{r}) &= \frac{k_0^2\sqrt{\bar{\varepsilon}}}{4\pi}\int_{-\infty}^{\infty} M(\boldsymbol{r}')\frac{\mathrm{e}^{\mathrm{i}kR}}{R}\,\tilde{\varepsilon}(\boldsymbol{r}')\,[\boldsymbol{n}_s'\boldsymbol{E}_0(\boldsymbol{r}')]\,d^3r'\ . \end{aligned} \tag{4.171}$$

Since the only new element here as compared to the scalar problem concerns only the polarization of the scattered field, we will consider the simplest case where the observation point is located within the Fraunhofer zone of the scattering volume ($r \gg kL^2$), and the primary field is the plane wave

$$\boldsymbol{E}_0(\boldsymbol{r}) = A\boldsymbol{e}\mathrm{e}^{\mathrm{i}\boldsymbol{k}_i\cdot\boldsymbol{r}}, \quad \boldsymbol{k}_i = k\boldsymbol{n}_i = k_0\sqrt{\bar{\varepsilon}}\boldsymbol{n}_i\ , \tag{4.172}$$

where $\boldsymbol{e}$ is the polarization vector of the primary wave, which is generally complex. We will take it to be normalized to unity by the condition $\boldsymbol{e}\cdot\boldsymbol{e}^* = 1$. The waves in the isotropic medium being transverse, the polarization vector $\boldsymbol{e}$ is normal to the direction of propagation $\boldsymbol{n}_i$. A simple calculation, using expansions of the form

(4.25), gives for the single-scattered fields E_1 and H_1 the expressions

$$\begin{aligned} E_1(r) &= \frac{k_0^2 A e^{ikr}}{4\pi r} [n_{s0}[en_{s0}]] \int_{-\infty}^{\infty} M(r')\tilde{\varepsilon}(r') e^{-iq_0 \cdot r'} d^3r' , \\ H_1(r) &= \frac{k_0^2 \sqrt{\bar{\varepsilon}} A e^{ikr}}{4\pi r} [n_{s0} e] \int_{-\infty}^{\infty} M(r')\tilde{\varepsilon}(r') e^{-iq_0 \cdot r'} d^3r' , \end{aligned} \tag{4.173}$$

where $n_{s0} = r/r$ is the unit vector directed at a point r from the center of the scattering region, and $q_0 = k(n_{s0} - n_{i0})$ is the scattering vector for the center of this region. We will, for simplicity, omit below the subscript 0 on the vectors n_{s0}, n_{i0}, and q_0.

It is easily seen that the fields E_1 and H_1 only differ from the scalar field

$$u_1(r) = \frac{k_0^2 A e^{ikr}}{4\pi r} \int_{-\infty}^{\infty} M(r')\,\tilde{\varepsilon}(r') e^{-iq \cdot r'} d^3r' , \tag{4.174}$$

calculated with the same assumptions, by the vector factors

$$E_1(r) = [n_s[en_s]]u_1(r), \quad H_1(r) = \sqrt{\bar{\varepsilon}}[n_s e]u_1(r) . \tag{4.175}$$

In consequence, in the single scattering approximation for the deterministic polarization e of the primary wave, any mean bilinear quantities composed of the components of E_1 and H_1 can be expressed in terms of the covariance of the scalar field $\psi_u(r_1, r_2) = \langle u_1(r_1) u_1^*(r_2)\rangle$. If the primary field polarization is random, then averaging over the directions e and the ensemble of inhomogeneities $\tilde{\varepsilon}$ is performed separately, since these quantities are statistically independent. In other words, the means of the bilinear combinations of the components of E_1 and H_1 can also be expressed through ψ_u or $\bar{I}_1$ at the random polarization of a primary wave. Let us consider some examples.

2) Mean Poynting Vector and Effective Scattering Cross-Section. Of the bilinear combinations of the components of E_1 and H_1, the Poynting vector is of paramount importance

$$\mathscr{S}_1 = \frac{c}{8\pi} \mathrm{Re}\{[E_1 H_1^*]\} = \frac{c}{16\pi}([E_1 H_1^*] + [E_1^* H_1]) . \tag{4.176}$$

If we now take into account that $H_1 = \sqrt{\bar{\varepsilon}}[n_s E_1]$, and use (4.48) for $\bar{I}_1 = \langle |u_i|^2\rangle$, we find

$$\begin{aligned} \mathscr{S}_1 &= \frac{c\sqrt{\bar{\varepsilon}}}{8\pi} n_s \langle |E_1|^2\rangle = \frac{c\sqrt{\bar{\varepsilon}} n_s}{8\pi} |[n_s[en_s]]|^2 \bar{I}_1 \\ &= \frac{c\sqrt{\bar{\varepsilon}} n_s}{8\pi} |[n_s[en_s]]|^2 \frac{\pi k_0^4 |A|^2}{2r^2} \Phi_\varepsilon(q) . \end{aligned} \tag{4.177}$$

Note that, in contrast to the initial relationships (4.173), this expression is applicable not only in the Fraunhofer zone $r \gg kL^2$, but also at shorter distances $r \gg r_{\min} = \min\{L, Ll/\lambda\}$ [see (4.52)].

As in Sect. 4.3, we will introduce the effective scattering cross-section of a unit volume into a unit solid angle $\sigma = |\mathscr{S}_1|r^2/|\mathscr{S}_0|V$, where $|\mathscr{S}_0| = c\sqrt{\bar{\varepsilon}}|A|^2/8\pi$ is the modulus of the energy flux density of the primary wave. Using (4.177), we find

$$\sigma = \frac{|\bar{\mathscr{S}}_1|r^2}{|\mathscr{S}_0|V} = \gamma\sigma_{\mathrm{sc}} = \gamma\frac{\pi}{2}k_0^4\Phi_\varepsilon(\boldsymbol{q}) \; , \tag{4.178}$$

where σ_{sc} is the "scalar" scattering cross-section (4.78). The factor

$$\gamma = |[\boldsymbol{n}_{\mathrm{s}}[\boldsymbol{e}\boldsymbol{n}_{\mathrm{s}}]]|^2 = |\boldsymbol{e} - \boldsymbol{n}_{\mathrm{s}}(\boldsymbol{e}\cdot\boldsymbol{n}_{\mathrm{s}})|^2 = 1 - (\boldsymbol{n}_{\mathrm{s}}\cdot\boldsymbol{e})^2 \tag{4.179}$$

is determined by the specific features of the scattering of electromagnetic waves as compared to the scalar problem, and may be referred to as the *polarization factor*. We will now calculate γ for certain special cases.

For the *linearly polarized* primary wave, $\boldsymbol{e}$ is a real unit vector. If χ is the angle between $\boldsymbol{e}$ and $\boldsymbol{n}_{\mathrm{s}}$ (Fig. 4.10), then $\gamma = \sin^2\chi$. Accordingly, for the linearly polarized primary wave,

$$\sigma = \sigma_{\mathrm{sc}}\sin^2\chi \; . \tag{4.180}$$

Owing to the dipole character of the scattering of electromagnetic waves in each element of the scattering volume, the factor $\sin^2\chi$ vanishes with $\boldsymbol{e}$ in the colinear directions.

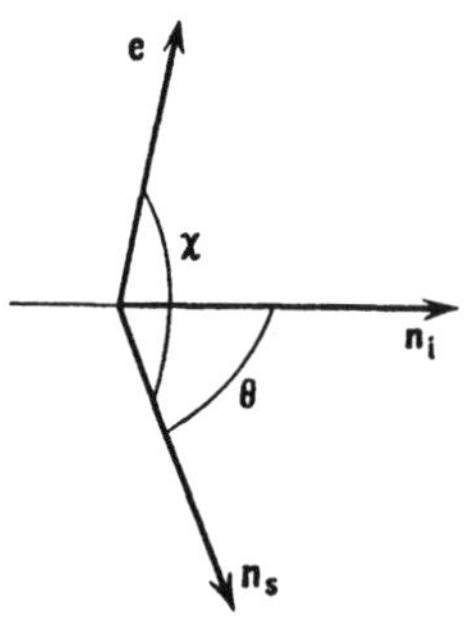

Fig. 4.10. Scattering of a linearly polarized wave

We know that an arbitrary *elliptically polarized* wave can be represented as a sum of two linearly polarized oscillations with a phase difference of $\pi/2$. Let the major axis of the polarization ellipse be directed along the unit vector $\boldsymbol{e}_1$, and the minor one along the unit vector $\boldsymbol{e}_2$. The vectors $\boldsymbol{e}_1$ and $\boldsymbol{e}_2$ are normal to the direction of propagation $\boldsymbol{n}_{\mathrm{i}}$, so that $\boldsymbol{e}_1$, $\boldsymbol{e}_2$ and $\boldsymbol{n}_{\mathrm{i}}$ are orthogonal. For the elliptical polarization

$$\boldsymbol{e} = p_1\boldsymbol{e}_1 + \mathrm{i}p_2\boldsymbol{e}_2 \; , \tag{4.181}$$

where p_1 and p_2 are real coefficients satisfying the condition $p_1^2 + p_2^2 = 1$ (by virtue of the normalization $\boldsymbol{e} \cdot \boldsymbol{e}^* = 1$ we have adopted).

The polarization factor γ for the waves of elliptical polarization is

$$\gamma = 1 - p_1^2(\boldsymbol{n}_s \cdot \boldsymbol{e}_1)^2 - p_2^2(\boldsymbol{n}_s \cdot \boldsymbol{e}_2)^2 = 1 - p_1^2 \cos^2 \chi_1 - p_2^2 \cos^2 \chi_2 \ , \tag{4.182}$$

where χ_1 and χ_2 are the angles between $\boldsymbol{n}_s$ and the unit vectors $\boldsymbol{e}_1$ and $\boldsymbol{e}_2$ (Fig. 4.11). Substituting (4.182) into (4.178), and considering that $p_1^2 + p_2^2 = 1$ gives

$$\sigma = (p_1^2 \sin^2 \chi_1 + p_2^2 \sin^2 \chi_2)\sigma_{sc} \ . \tag{4.183}$$

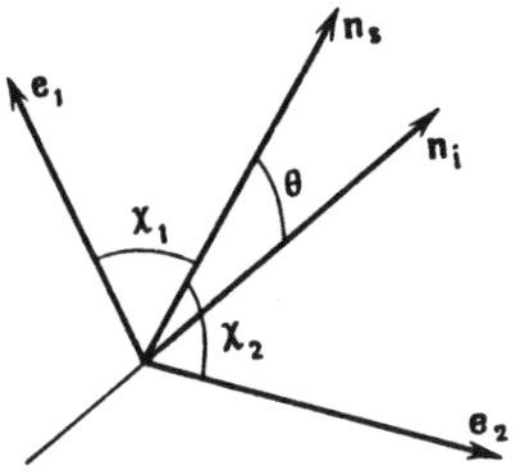

Fig. 4.11. Scattering of an elliptically polarized wave

In contrast to the linearly polarized primary wave, (4.183) never vanishes since $\sin^2 \chi_1$ and $\sin^2 \chi_2$ cannot vanish simultaneously. It follows from (4.175) that the scattered field is polarized linearly in the directions $\boldsymbol{n}_s = \mp \boldsymbol{e}_1$ and $\boldsymbol{n}_s = \mp \boldsymbol{e}_2$, owing to the above-mentioned dipole nature of the scattering. In other directions, the scattered field is elliptically polarized. The parameters of the polarization ellipse are obtainable from expressions (4.175) when (4.181) is substituted into them.

In the special case of the *circular polarization* of a wave where $p_1 = 1/\sqrt{2}$, $p_2 = \mp 1/\sqrt{2}$ (the plus sign corresponds to the right and the minus to the left polarization), from (4.182) we find

$$\gamma = 1 - \tfrac{1}{2}[(\boldsymbol{n}_s \cdot \boldsymbol{e}_1)^2 + (\boldsymbol{n}_s \cdot \boldsymbol{e}_2)^2] = 1 - \tfrac{1}{2}(\cos^2 \chi_1 + \cos^2 \chi_2) \ .$$

By virtue of the fact that the vectors $\boldsymbol{e}_1$, $\boldsymbol{e}_2$ and $\boldsymbol{n}_i$ are mutually orthogonal we have $\cos^2 \chi_1 + \cos^2 \chi_2 + \cos^2 \theta = 1$, so that for circularly polarized waves

$$\gamma = 1 - \tfrac{1}{2}(1 - \cos^2 \theta) = \tfrac{1}{2}(1 + \cos^2 \theta) \ . \tag{4.184}$$

Thus, γ here is dependent only on the scattering angle θ, i.e., it has azimuthal symmetry.

If we measure the intensity of the scattered waves of *one* polarization at the observation point $\boldsymbol{r}$, then the energy flux density will obviously be less than the total density (4.177) and, accordingly, the effective scattering cross-section σ will decrease.

Suppose we have to measure the flux of the energy transferred by the component $E_N = (\boldsymbol{N}\boldsymbol{E}_1)$, where $\boldsymbol{N}$ is the unit (generally speaking, complex) vector normal to the scattering direction $\boldsymbol{n}_s$: $(\boldsymbol{N}\boldsymbol{n}_s) = 0$. The vector $\boldsymbol{N}$ describes the polarization behavior of the receiving antenna (for radio waves) or the polarization analyzer (for light waves). It follows from (4.175) that

$$E_N = \boldsymbol{N}\cdot \boldsymbol{E}_1 = (\boldsymbol{N}(\boldsymbol{n}_s[\boldsymbol{e}\boldsymbol{n}_s]])u_1 \ ,$$

so that the role of the polarization factor γ is played here by

$$\gamma_N = |(\boldsymbol{N}[\boldsymbol{n}_s[\boldsymbol{e}\boldsymbol{n}_s]])|^2 = |\boldsymbol{N}\cdot\boldsymbol{e} - \boldsymbol{N}\cdot\boldsymbol{n}_s\boldsymbol{e}\cdot\boldsymbol{n}_s|^2 = |\boldsymbol{N}\cdot\boldsymbol{e}|^2 \ .$$

It is easily seen that $\gamma_N = \gamma$, if the polarization characteristics of the receiving antenna are "matched" to the polarization of the scattered radiation, i.e., the antenna receives the wave (4.175). Otherwise, $\gamma_N < \gamma$.

3) Scattering of Nonpolarized (Natural) Light. The results derived above are applicable, with only slight modifications, to *partially polarized* fields (Sect. II.2.7) when the components of the primary field $\boldsymbol{E}_0$ are random. It has already been noted that we can average over these quantities independently of the averaging over realizations of the fluctuations $\tilde{\varepsilon}$. We will consider the simple, but important, case of the *nonpolarized* wave.

We will decompose the electric field of the primary wave into orthogonal components along unit vectors $\boldsymbol{e}_1$ and $\boldsymbol{e}_2$

$$\boldsymbol{E}_0 = A_1\boldsymbol{e}_1 + A_2\boldsymbol{e}_2 \ .$$

For the nonpolarized wave, the amplitudes A_1 and A_2 are statistically independent, having zero means and equal variances

$$\bar{A}_1 = \bar{A}_2 = 0, \quad \langle A_1 A_2\rangle = 0, \quad \langle |A_1|^2\rangle = \langle |A_2|^2\rangle = \tfrac{1}{2}\langle |E_0|^2\rangle \ , \tag{4.185}$$

where $|E_0|^2 = |A_1|^2 + |A_2|^2$ is the total intensity.

The scattering cross-section of a nonpolarized wave can be derived using the following simple argument. If the primary wave were linearly polarized along $\boldsymbol{e}_1$, then, by (4.183), the scattering cross-section would be $\sigma = \sigma_1 = \sigma_{sc}\sin^2\chi_1$, even with a fluctuating amplitude A_1. But with linear polarization along $\boldsymbol{e}_2$, we would have $\sigma = \sigma_2 = \sigma_{sc}\sin^2\chi_2$. In the nonpolarized wave, the energy flux splits equally between two orthogonal components, so that the total scattering cross-section must be equal to the half-sum of σ_1 and σ_2

$$\sigma_{nat} = \tfrac{1}{2}(\sigma_1 + \sigma_2) = \tfrac{1}{2}(\sin^2\chi_1 + \sin^2\chi_2)\sigma_{sc} \ .$$

But the coefficient at σ_{sc} is the same here as in the case of the scattering of the circularly polarized wave, and hence it is $(1+\cos^2\theta)/2$ [see (4.184)]. Thus,

$$\sigma_{nat} = \frac{1+\cos^2\theta}{2}\sigma_{sc} \ . \tag{4.186}$$

The factor $g(\theta) = (1+\cos^2\theta)/2$ is called the *Rayleigh scattering indicatrix*. It decreases to 1/2 at $\theta = \mp\pi/2$, which is accounted for by the fact that in this direction, only one component of the primary wave contributes to the scattered field, i.e., the scattered field appears to be linearly polarized.

The degree of polarization $P(\theta)$ for the scattered field is obtainable from (II.2.148)

$$P=\sqrt{1-\frac{4\operatorname{Det}J}{(J_{11}+J_{22})^2}}\,, \tag{4.187}$$

where $J_{\alpha\beta}=\langle E_\alpha E_\beta^*\rangle$ are elements of the polarization matrix in a plane perpendicular to the direction of propagation. A simple calculation gives the expression

$$P(\theta)=\sqrt{1-\frac{4\cos^2\theta}{(1+\cos^2\theta)^2}}=\frac{1-\cos^2\theta}{1+\cos^2\theta}=\frac{\sin^2\theta}{1+\cos^2\theta}\,, \tag{4.188}$$

which is due to Rayleigh.

It is seen from (4.188) that $P=0$ at $\theta=0$ or π, i.e., the field $\boldsymbol{E}_1$ retains its natural polarization with forward and backscattering. For scattering at the right angle ($\theta=\mp\pi/2$), the degree of polarization is unity, and in other directions the scattered field is partially polarized ($0<P<1$).

4) Frequency Spectrum of a Scattered Field. Incoherent Scattering of Electromagnetic Waves in Isotropic Plasma. For time-dependent inhomogeneities $\tilde{\varepsilon}$, we can also take into account the differences between electromagnetic wave scattering and a similar scalar picture by introducing into the results of the scalar theory the polarization factor $\gamma=1-(\boldsymbol{n}_s\boldsymbol{e})^2$. So, for example, when there is no regular drift of inhomogeneities, and when $r\gg r_{\min}$, the frequency spectrum of the scattered field

$$g_E(\Omega)=\frac{1}{2\pi}\int_{-\infty}^{\infty}\langle \boldsymbol{E}_1(t)\boldsymbol{E}_1^*(t-\tau)\rangle e^{i\Omega\tau}\,d\tau \tag{4.189}$$

for the spherical primary wave can be obtained by multiplying the spectrum (4.128) by γ:

$$g_E(\Omega)=\gamma\frac{\pi k_0^4|A|^2V_{\rm eff}}{2r^2r_0^2}\,G_\varepsilon(\omega-\Omega,\boldsymbol{q})\,, \tag{4.190}$$

where $G_\varepsilon(\omega,\boldsymbol{\kappa})$ is the $\omega\boldsymbol{\kappa}$-density of the fluctuations of the electric permittivity.

We will apply (4.190) to the so-called *incoherent scattering* of electromagnetic waves in a plasma. It was initially supposed that in a plasma, waves were scattered by free electrons involved in thermal motion, whereby the intensities of the fields scattered by individual electrons add up in an incoherent manner (whence the term "incoherent scattering"). However, it became clear later that this phenomenon is dominated by collective processes[9]. We will restrict ourselves to the calculation of

[9] In 1958 Gordon suggested that the incoherent scattering of radio waves may be observed in ionospheric plasma, and in that same year the phenomenon was first observed by *Bowls* [4.10]. Collective phenomena in plasma have been treated in many works (see, e.g., [4.11]).

the intensity of the high-frequency field scattered in an isotropic plasma. With this end in view, we will make use of the macroscopic theory of thermal fluctuations (Chap. 3). In Sect. 4.8, we will return to the question of incoherent scattering in the context of the theory of scattering by individual particles.

In macroscopic electrodynamics, the behavior of a cold anisotropic plasma at high frequencies ($\omega \gg \omega_e$) is described by the electric permittivity

$$\varepsilon = 1 - \frac{\omega_e^2}{\omega^2} = 1 - \frac{4\pi e^2 N}{m\omega^2} ,$$

where e, m, and N are the charge, mass, and concentration of the electrons, respectively. Thus,

$$\bar{\varepsilon} = 1 - \frac{4\pi e^2 \bar{N}}{m\omega^2} \quad \text{and} \quad \tilde{\varepsilon} = \varepsilon - \bar{\varepsilon} = -\frac{4\pi e^2 \tilde{N}}{m\omega^2} , \tag{4.191}$$

hence the $\omega\kappa$-densities of $\tilde{\varepsilon}$ and $\tilde{N}$ are related by

$$G_\varepsilon(\omega, \boldsymbol{\kappa}) = \frac{16\pi^2 e^4}{m^2\omega^4} G_N(\omega, \boldsymbol{\kappa}) . \tag{4.192}$$

Substituting this into (4.190) gives

$$g_E(\Omega) = 8\pi^3 r_e^2 \gamma \frac{|A|^2 V_{\text{eff}}}{r^2} G_N(\Omega - \omega, \boldsymbol{q}) , \tag{4.193}$$

where $r_e = e^2/mc^2$ is the classical electron radius. As was shown in exercise 7, Chap. 3, the fluctuation spectrum for the electron concentration G_N varies with the temperature T and mean electron concentration $\bar{N}$. The spectrum of the scattered field g_E is thus proportional to the total number of electrons $N_\Sigma = \bar{N}V$ in the volume V.

It follows from (4.193) that the spectrum of the scattered field g_E is concentrated in the vicinity of the frequency of the primary wave, ω, its shape being completely conditioned by the form of the fluctuation spectrum $G_N(\omega, \kappa)$. Therefore, the frequency spectrum of the scattered signal $g_E(\Omega)$ enables us to make a judgement regarding the properties and state of the plasma. A diagnostic technique based on this is widely employed not only in the analysis of ionospheric, but also of laboratory plasma (the "incoherent scattering method").

4.8 Discrete Scatterers

1) Fields Scattered by a Single Particle. Apart from scattering by volume inhomogeneities, of great interest is scattering by a multitude of bodies (particles, etc.) randomly distributed and/or randomly oriented in space. This is a general problem that has a variety of applications in many branches of physics.

Strictly speaking, the problem of scattering by a system of many bodies should be classified under statistical scheme (3) of Sect. 2.1. But if discrete scatterers are relatively small and sparse, the diffraction problem can be solved in the single scattering approximation based on the assumption that each scatterer treats the incident wave as if there were no other scatterers.

As with continuous inhomogeneities, the single scattering approximation reduces the problem of discrete scatterers to scheme (1), i.e., to fields radiated by random sources. In this case, we have a multitude of discrete sources whose statistics are given by those of the position (and orientation) of the scatterers. We will see that the "external" laws governing single scattering by discrete scatterers and by continuous inhomogeneities of the medium (spatial correlation, frequency spectrum, etc.) have a great deal in common.

Once again, we will discuss a simplified problem, namely the scattering of the monochromatic wave by a multitude of *stationary* particles whose dimensions are *small as compared to the wavelength*, the observation point lying in the *Fraunhofer zone of the volume* V occupied by the particles.

For small particles the scattered field can be derived in the dipole approximation. Let $\boldsymbol{p}$ be the electric dipole moment of a particle produced by an electric field $\boldsymbol{E}$

$$\boldsymbol{p} = \hat{\alpha}\boldsymbol{E} , \tag{4.194}$$

where $\hat{\alpha}$ is the polarizability tensor of the particle. For particles small as compared to the wavelength, $\hat{\alpha}$ can be calculated assuming $\boldsymbol{E}$ to be statistically homogeneous. The treatments for certain symmetrical bodies are, for example, given in the monograph [4.12] devoted to wave scattering by small particles.

For the spherical particles of isotropic materials, and also for particles of arbitrary shape and electric permittivity close to unity, the polarizability α is a scalar

$$\boldsymbol{p} = \alpha\boldsymbol{E} . \tag{4.195}$$

We assume that the electric field in (4.195) is the so-called *acting* field $\boldsymbol{E}_{\text{act}}$, which differs from the field of the primary wave $\boldsymbol{E}_0$ in that it also includes the total field of neighboring scatterers. In the single scattering approximation (Born approximation), $\boldsymbol{E}_{\text{act}}$ essentially coincides with the primary wave field $\boldsymbol{E}_0$, which for simplicity will be taken to be the field of the plane wave $E_0 = A\boldsymbol{e}\exp(\mathrm{i}\boldsymbol{k}_\mathrm{i}\boldsymbol{r} - \mathrm{i}\omega t)$. Then,

$$\boldsymbol{p} = \alpha\boldsymbol{E}_{\text{act}} \approx \alpha\boldsymbol{E}_0 = \alpha A\boldsymbol{e}\mathrm{e}^{\mathrm{i}\boldsymbol{k}_\mathrm{i}\cdot\boldsymbol{r}' - \mathrm{i}\omega t} , \tag{4.196}$$

where $\boldsymbol{r}'$ is the radius vector of a particle, and $\boldsymbol{e}$ is the polarization vector of the primary wave.

The wave field radiated in a vacuum by an individual dipole with the moment (4.196) can be obtained from (4.170), if we substitute $-\mathrm{i}\omega\boldsymbol{p}$ for $\boldsymbol{l}_1$ [as usual, the factor $\exp(-\mathrm{i}\omega t)$ is omitted]

$$\boldsymbol{E}_1(r) = k_0^2[\boldsymbol{n}_\mathrm{s}'[\boldsymbol{e}\boldsymbol{n}_\mathrm{s}']]\alpha A\,\frac{\exp(\mathrm{i}kR + \mathrm{i}\boldsymbol{k}_\mathrm{i}\cdot\boldsymbol{r}')}{R} , \tag{4.197}$$

where $R = |\boldsymbol{r} - \boldsymbol{r}'|$, $\boldsymbol{n}_s' = (\boldsymbol{r} - \boldsymbol{r}')/|\boldsymbol{r} - \boldsymbol{r}'|$ is the unit vector from a point $\boldsymbol{r}'$ to the observation point $\boldsymbol{r}$. The magnetic field of the scattered wave in the wave zone is given by

$$\boldsymbol{H}_1 = [\boldsymbol{n}_s' \boldsymbol{E}_1] = [\boldsymbol{n}_s' \boldsymbol{e}] k_0^2 \alpha A \frac{\exp(ikR + i\boldsymbol{k}_i \cdot \boldsymbol{r}')}{R} . \tag{4.198}$$

2) Mean Single-Scattered Field. Let there be N identical particles with centers at points $\boldsymbol{r}' = \boldsymbol{r}_m$ ($m = 1, 2, \ldots, N$) in a volume V. In the single scattering approximation, the field at the observation point $\boldsymbol{r}$ is the sum of fields scattered by individual particles

$$\boldsymbol{E}_s(\boldsymbol{r}) = k_0^2 \sum_{m=1}^{N} [\boldsymbol{n}_m[\boldsymbol{e}\boldsymbol{n}_m]] \alpha A \frac{\exp(ikR_m + i\boldsymbol{k}_i \cdot \boldsymbol{r}_m)}{R_m} , \tag{4.199}$$

where $R_m = |\boldsymbol{r} - \boldsymbol{r}_m|$, and $\boldsymbol{n}_m = (\boldsymbol{r} - \boldsymbol{r}_m)/|\boldsymbol{r} - \boldsymbol{r}_m|$ is the unit vector from the mth particle to the observation point $\boldsymbol{r}$. With double scattering, $\boldsymbol{E}_{act}$ is the sum of the primary field $\boldsymbol{E}_0$ and the single-scattered field (4.199), and so on.

Note that when we take double scattering into consideration, neglecting the quasi-stationary field of the induced dipoles nearest to $\boldsymbol{r}$ in $\boldsymbol{E}_{act}$, i.e., neglecting the field $\boldsymbol{E}_{q.st.} \sim \alpha \boldsymbol{E}_0/|r - r'|^3$ as compared to $\boldsymbol{E}_0$, imposes definite constraints on the mean separation between the particles, $\bar{\varrho}$. The inequality $\alpha/\bar{\varrho}^3 \ll 1$ follows from the condition $\boldsymbol{E}_{q.st.} \ll \boldsymbol{E}_0$, or since $\bar{n}\bar{\varrho}^3 \approx 1$, where $\bar{n}$ is the mean concentration of particles, the inequality

$$\bar{n}\alpha \ll 1 . \tag{4.200}$$

If this condition is met, both $\bar{\varrho} > \lambda$ and $\bar{\varrho} < \lambda$ are possible. The cloud of particles satisfying (4.100) can be called *tenuous* (clearly, in the electrodynamic sense) in order to distinguish this from the condensed medium, where $\bar{n}\alpha \gtrsim 1$.

We will place the origin of the coordinates at the center of the scattering volume $V \sim L^3$ filled with particles. For observation points brought to the Fraunhofer zone of $V(r \gg kL^2)$, we can use (4.25)

$$|\boldsymbol{r} - \boldsymbol{r}_m| \approx r - \boldsymbol{n}_s \cdot \boldsymbol{r}_m, \quad \boldsymbol{n}_m = \frac{\boldsymbol{r} - \boldsymbol{r}_m}{|\boldsymbol{r} - \boldsymbol{r}_m|} \approx \frac{\boldsymbol{r}}{r} \equiv \boldsymbol{n}_s .$$

If we substitute this into (4.199), we obtain

$$\boldsymbol{E}_s = \boldsymbol{\mathscr{E}}_1 \sum_{m=1}^{N} e^{-i\boldsymbol{q} \cdot \boldsymbol{r}_m} , \tag{4.201}$$

where $\boldsymbol{q} = k\boldsymbol{n}_s - \boldsymbol{k}_i = k(\boldsymbol{n}_s - \boldsymbol{n}_i)$ is the scattering vector, and

$$\boldsymbol{\mathscr{E}}_1 = k_0^2 [\boldsymbol{n}_s[\boldsymbol{e}\boldsymbol{n}_s]] \alpha A \frac{e^{ikr}}{r} \tag{4.201a}$$

is the field scattered by one particle placed at the origin of the coordinates, i.e., at the center of V.

In order to find the statistical moments of the single-scattered field, we must know the distribution law for the random coordinates $\boldsymbol{r}_m = (x_m, y_m, z_m)$, and if the particles move, then, in addition, the distribution of the velocities $\boldsymbol{v}_m = (v_{mx}, v_{my}, v_{mz})$. To obtain the mean and covariance of the scattered field we only need the single-particle (w_1) and binary (w_2) probability densities. We assume that all the particles and their pairs are equivalent, i.e., $w_1(\boldsymbol{r}_m)$ and $w_2(\boldsymbol{r}_m, \boldsymbol{r}_l)$ are independent of the subscripts of $\boldsymbol{r}_m$ and $\boldsymbol{r}_l$.

On averaging (4.201) over the positions of the particles $\boldsymbol{r}_m$, we obtain the mean field scattered by N particles

$$\bar{E}_s(r) = \mathscr{E}_1 \left\langle \sum_{m=1}^{N} \mathrm{e}^{-\mathrm{i}\boldsymbol{q}\cdot\boldsymbol{r}_m} \right\rangle = \mathscr{E}_1 N \int_V \mathrm{e}^{-\mathrm{i}\boldsymbol{q}\cdot\boldsymbol{r}}\, w_1(\boldsymbol{r}) d^3 r = \mathscr{E}_1 N f_1(\boldsymbol{q}) \;, \tag{4.202}$$

where

$$f_1(\boldsymbol{q}) = \int_V \mathrm{e}^{-\mathrm{i}\boldsymbol{q}\cdot\boldsymbol{r}} w_1(\boldsymbol{r})\, d^3 r \tag{4.203}$$

is the characteristic function of the single-particle distribution $w_1(\boldsymbol{r})$.

The intensity of the mean field $\bar{\boldsymbol{E}}^2$, i.e., the *coherent* component of intensity, is

$$I_{\mathrm{coh}} = |\bar{\boldsymbol{E}}_s|^2 = |\mathscr{E}_1|^2 N^2 |f_1(\boldsymbol{q})|^2 \;. \tag{4.204}$$

The coherent component differs from the intensity $|\mathscr{E}_1|^2$ of the field scattered by one particle by the factor $N^2|f_1(\boldsymbol{q})|^2$, i.e., I_{coh} is proportional to the *square* of the particle number. The quantity $N|f_1(\boldsymbol{q})|$ may be called the effective number of coherently scattering particles

$$N_{\mathrm{coh}} = N|f_1(\boldsymbol{q})| = \left| \left\langle \sum_{m=1}^{N} \mathrm{e}^{-\mathrm{i}\boldsymbol{q}\cdot\boldsymbol{r}_m} \right\rangle \right| \;. \tag{4.205}$$

With forward scattering ($\theta = 0, q = 0$), where $f_1(\boldsymbol{q}) = f_1(0) = 1$, all the particles radiate in a coherent way ($N_{\mathrm{coh}} = N$), so that

$$(\bar{\boldsymbol{E}}_s)_{\mathrm{for.}} = (N\mathscr{E}_1)_{\mathrm{for}} = N k_0^2 e\alpha \frac{A}{r} \mathrm{e}^{\mathrm{i}kr} \;. \tag{4.206}$$

For scattering at an angle, N_{coh} drops markedly, since the characteristic function $f_1(\boldsymbol{q})$ is comparable to unity only when $\boldsymbol{q} \lesssim 2\pi/L$, or rather when the scattering angles lie inside the cone $\theta \lesssim \lambda/L \ll 1$ (the rate of decrease for N_{coh} with increasing θ is estimated in Exercise 7).

3) Mean Intensity. According to (4.201), the mean intensity of the single-scattered field $\bar{I}_s = |\boldsymbol{E}_s|^2$ is

$$\bar{I}_s = |\mathscr{E}_1|^2 \left\langle \sum_{m,l=1}^{N} \mathrm{e}^{-\mathrm{i}\boldsymbol{q}\cdot(\boldsymbol{r}_m - \boldsymbol{r}_l)} \right\rangle \;. \tag{4.207}$$

Of the double sum containing N^2 terms, we separate N terms with $m = l$, each of which is unity, and average the remaining $N^2 - N$ terms with $m \neq l$ using the binary probability density w_2. As a result

$$\bar{I}_s = |\mathscr{E}_1|^2[N + (N^2 - N)\tilde{f}_2(\boldsymbol{q})] \;, \quad \text{where} \tag{4.208}$$

$$\tilde{f}_2(\boldsymbol{q}) \equiv f_2(\boldsymbol{q}, -\boldsymbol{q}) = \iint\limits_V w_2(\boldsymbol{r}', \boldsymbol{r}'')\mathrm{e}^{-\mathrm{i}\boldsymbol{q}\cdot(\boldsymbol{r}'-\boldsymbol{r}'')} d^3r' \, d^3r'' \;, \tag{4.209}$$

$f_2(\boldsymbol{q}_1, \boldsymbol{q}_2)$ being the binary characteristic function. Actually, $f_2(\boldsymbol{q})$ is the characteristic function of the distribution of the *relative* coordinates. In fact, if in (4.209) we substitute $W_2(\boldsymbol{\varrho}, \boldsymbol{R})d^3\varrho d^3R$ for $w_2(\boldsymbol{r}', \boldsymbol{r}'')d^3r'd^3dr''$, where $\boldsymbol{\varrho} = \boldsymbol{r}' - \boldsymbol{r}''$, $\boldsymbol{R} = (\boldsymbol{r}' + \boldsymbol{r}'')/2$, and single out the integration with respect to $\boldsymbol{R}$, then

$$\tilde{f}_2(\boldsymbol{q}) = \int W(\boldsymbol{\varrho})\exp(-\mathrm{i}\boldsymbol{q}\cdot\boldsymbol{\varrho})d^3\varrho, \quad W(\boldsymbol{\varrho}) = \int W_2(\boldsymbol{\varrho}, \boldsymbol{R})d^3R \;.$$

Subtracting from (4.208) the intensity of the coherent component (4.204), we get the mean intensity of the *incoherent* scattering

$$\begin{aligned} \bar{I}_{\text{incoh}} &= \bar{I}_s - I_{\text{coh}} = \langle |E_s|^2 \rangle - |\bar{E}_s|^2 \\ &= |\mathscr{E}_1|^2\{N[1 - |f_1(\boldsymbol{q})|^2] + (N^2 - N)[\tilde{f}_2(\boldsymbol{q}) - |f_1(\boldsymbol{q})|^2]\} \;. \end{aligned} \tag{4.210}$$

If the particle coordinates are statistically independent, i.e., $w_2(\boldsymbol{r}', \boldsymbol{r}'') = w_1(\boldsymbol{r}')w_1(\boldsymbol{r}'')$, then $\tilde{f}_2(\boldsymbol{q}) - |f_1(\boldsymbol{q})|^2 = 0$ and the second term in (4.210), which is proportional to $N^2 - N$, vanishes. Consequently, this term describes scattering by particles whose positions are *intercorrelated*, and it may be regarded as a characteristic of the collective effects in scattering

$$\bar{I}_{\text{col}} = |\mathscr{E}_1|^2(N^2 - N)[\tilde{f}_2(\boldsymbol{q}) - |f_1(\boldsymbol{q})|^2] \;. \tag{4.211}$$

The first term in (4.210) describes scattering by *independent* particles

$$\bar{I}_{\text{ind}} = |\mathscr{E}_1|^2 N[1 - |f_1(\boldsymbol{q})|^2] \;. \tag{4.212}$$

Because $f_1(0) = \tilde{f}_2(0) = 1$, in forward scattering ($\theta = 0$) both terms in (4.210) vanish, thus $\bar{I}_{\text{incoh}}(0) = 0$. But this is only the case in the single scattering approximation. It can be shown that even for double scattering $\bar{I}_{\text{incoh}}(0) \neq 0$.

If we disregard forward scattering, i.e., scattering into a narrow cone with $\theta \lesssim \lambda/L$, we can neglect the term $|f_1(\boldsymbol{q})|^2 \ll 1$ in (4.212) (see exercise 7), which gives

$$\bar{I}_{\text{ind}} \approx |\mathscr{E}_1|^2 N \;, \tag{4.213}$$

i.e., the total intensity of the waves scattered by independent particles. If $N \gg 1$, as is often the case in applications of the theory, then in (4.211) we can ignore N as compared to N^2, and

$$\bar{I}_{\text{col}} \approx |\mathscr{E}_1|^2 N^2[\tilde{f}_2(\boldsymbol{q}) - |f_1(\boldsymbol{q})|^2] \equiv |\mathscr{E}_1|^2 N^2 \nu(\boldsymbol{q}) \;. \tag{4.214}$$

Representing the binary distribution function w_2 in the form

$$w_2(\boldsymbol{r}', \boldsymbol{r}'') = w_1(\boldsymbol{r}')w_1(\boldsymbol{r}'')[1 + \mu(\boldsymbol{r}', \boldsymbol{r}'')] \;, \tag{4.215}$$

we can relate $v = \tilde{f}_2 - |f_1|^2$ to the quantity μ that vanishes for particles with independent coordinates $\boldsymbol{r}'$ and $\boldsymbol{r}''$

$$\begin{aligned} v(\boldsymbol{q}) &= \int\int [w_2(\boldsymbol{r}', \boldsymbol{r}'') - w_1(\boldsymbol{r}')w_1(\boldsymbol{r}'')] \mathrm{e}^{-\mathrm{i}\boldsymbol{q}\cdot(\boldsymbol{r}'-\boldsymbol{r}'')} d^3r' \, d^3r'' \\ &= \int\int w_1(\boldsymbol{r}')w_1(\boldsymbol{r}'')\mu(\boldsymbol{r}', \boldsymbol{r}'') \mathrm{e}^{-\mathrm{i}\boldsymbol{q}\cdot(\boldsymbol{r}'-\boldsymbol{r}'')} d^3r' \, d^3r'' \ . \end{aligned} \tag{4.216}$$

If the correlation radius l_μ, i.e., the characteristic space scale of μ with reference to $\boldsymbol{\varrho} = \boldsymbol{r}' - \boldsymbol{r}''$, is small as compared to the cloud size L [at the same time L is the space scale of $w_1(\boldsymbol{r})$ and μ with reference to $\boldsymbol{R} = (\boldsymbol{r}' + \boldsymbol{r}'')/2$], then within the sphere $\varrho = |\boldsymbol{r}' - \boldsymbol{r}''| \lesssim l_\mu$, the product $w_1(\boldsymbol{r}')w_1(\boldsymbol{r}'')$ can be approximated by $w_1^2(\boldsymbol{R})$, and (4.216) integrated with respect to ϱ. This gives

$$v(\boldsymbol{q}) = 8\pi^3 \int w_1^2(\boldsymbol{R}) \Phi_\mu(\boldsymbol{q}, \boldsymbol{R}) d^3R \ , \quad \text{where} \tag{4.217}$$

$$\Phi_\mu(\boldsymbol{q}, \boldsymbol{R}) = \frac{1}{8\pi^3} \int \mu(\boldsymbol{\varrho}, \boldsymbol{R}) \exp(-\mathrm{i}\boldsymbol{q}\cdot\boldsymbol{\varrho}) d^3\varrho \tag{4.218}$$

is the Fourier transform of $\mu(\boldsymbol{\varrho}, \boldsymbol{R})$ with respect to $\boldsymbol{\varrho}$. Thus

$$\bar{I}_{\text{col}} = N^2 |\mathscr{E}_1|^2 v(\boldsymbol{q}) = 8\pi^3 |\mathscr{E}_1|^2 N^2 \int w_1^2(\boldsymbol{R}) \Phi_\mu(\boldsymbol{q}, \boldsymbol{R}) d^3R \ . \tag{4.219}$$

4) Mean Poynting Vector. Effective Scattering Cross-Section. Since in the Fraunhofer zone

$$\mathscr{S}_{\text{s}} = \frac{c}{8\pi} \operatorname{Re}[\boldsymbol{E}_{\text{s}} \boldsymbol{H}_{\text{s}}^*] = \frac{c}{8\pi} \boldsymbol{n}_{\text{s}} |\boldsymbol{E}_{\text{s}}|^2 = \frac{c}{8\pi} \boldsymbol{n}_{\text{s}} I_{\text{s}} \ , \tag{4.220}$$

the mean Poynting vector $\bar{\mathscr{S}}_{\text{incoh}}$ for the fluctuation (incoherent) component of the field is given by

$$\bar{\mathscr{S}}_{\text{incoh}} = \frac{c}{8\pi} \boldsymbol{n}_{\text{s}} \bar{I}_{\text{incoh}} = \frac{c}{8\pi} \boldsymbol{n}_{\text{s}} (\bar{I}_{\text{ind}} + \bar{I}_{\text{col}}) \ . \tag{4.221}$$

We now denote the scattering cross-section of an individual particle into a unit solid angle by σ_1. From (4.201a)

$$\sigma_1 = \frac{|\mathscr{S}_1| r^2}{|\mathscr{S}_0|} = \frac{|\mathscr{E}_1|^2 r^2}{|\boldsymbol{E}_0|^2} = k_0^4 |[\boldsymbol{n}_{\text{s}}[\boldsymbol{e}\boldsymbol{n}_{\text{s}}]]|^2 \alpha^2 = \gamma k_0^4 \alpha^2 \ , \tag{4.222}$$

where $\mathscr{S}_0 = c \operatorname{Re}\{[\boldsymbol{E}_0 \boldsymbol{H}_0^*]\}/8\pi$ is the Poynting vector of the primary field, $\mathscr{S}_1$ is the energy flux of the field (4.197) scattered by one particle, and γ is the polarization factor (4.179). Substituting (4.213) and (4.219) into (4.221), we obtain for the total scattering cross-section σ_N of the system of N particles into the unit solid angle

$$\begin{aligned} \sigma_N &= \frac{|\bar{\mathscr{S}}_{\text{incoh}}| r^2}{|\mathscr{S}_0|} = [N + N^2 v] \sigma_1 \\ &= \sigma_1 [N + 8\pi^3 N^2 \int w_1^2(\boldsymbol{R}) \Phi_\mu(\boldsymbol{q}, \boldsymbol{R}) d^3R] \ . \end{aligned} \tag{4.223}$$

Introducing the mean concentration of the particles

$$\bar{n}(\boldsymbol{r}) = N w_1(\boldsymbol{r}) \tag{4.224}$$

and considering that $N = \int \bar{n}(\boldsymbol{r}) d^3 r$, we can represent (4.223) in the form

$$\sigma_N = \int \sigma_1 [\bar{n} + 8\pi^3 (\bar{n})^2 \Phi_\mu(\boldsymbol{q}, \boldsymbol{R})] d^3 R \ , \tag{4.225}$$

where the integrand is taken to be the scattering cross-section of the unit volume

$$\sigma = \sigma_{\text{ind}} + \sigma_{\text{col}} \ . \tag{4.226}$$

Here the first term

$$\sigma_{\text{ind}} = \sigma_1 \bar{n} \tag{4.227}$$

describes scattering when the correlation between the particle coordinates is ignored, and the second term

$$\sigma_{\text{col}} = \sigma_1 \cdot 8\pi^3 (\bar{n})^2 \Phi_\mu(\boldsymbol{q}, \boldsymbol{R}) \tag{4.228}$$

describes scattering with pairwise correlation of the particle coordinates.

The contributions of these terms may vary widely. It is clear, for example, that when the mean concentration $\bar{n}$ is small, the term σ_{ind} dominates; and when $\bar{n}$ is high, the term σ_{col}. It appears here that scattering by a system of particles with correlated coordinates may be described as scattering by macroscopic (volume) inhomogeneities (see item 5 below). In electromagnetic scattering in a quasi-neutral plasma, the contributions of collective scattering and scattering by independent particles may be comparable.

5) Transition to Continuous Medium. In a macroscopic treatment, the electric permittivity of a tenuous [in the sense of (4.200)] cloud is

$$\varepsilon = 1 + 4\pi n \alpha \ , \tag{4.229}$$

where α is the polarizability of a particle, and n is the *macroscopic* concentration of the particles, which is generally a random variable. Then,

$$\varepsilon = \bar{\varepsilon} + \tilde{\varepsilon}, \quad \bar{\varepsilon} = 1 + 4\pi\bar{n}\alpha, \quad \tilde{\varepsilon} = 4\pi\tilde{n}\alpha \ . \tag{4.230}$$

Let us now try to determine the statistic characteristics ε (or n) at which the macroscopic treatment of scattering, i.e., a description using the language of the fluctuations ε, gives the same results as the microscopic treatment. In discussing the statistical characteristics ε, we will only refer to the moments $\bar{\varepsilon}$ and ψ_ε.

By virtue of (4.224), it is quite obvious that $\bar{\varepsilon}(\boldsymbol{r})$ must be

$$\bar{\varepsilon}(\boldsymbol{r}) = 1 + 4\pi\alpha N w_1(\boldsymbol{r}) = 1 + 4\pi\alpha\bar{n}(\boldsymbol{r}) \ . \tag{4.231}$$

The mean field computed by (4.231) in the Born approximation for the Fraunhofer zone coincides with (4.202).

To find ψ_ε we identify (4.226), obtained for a system of particles, with the cross-section (4.178), which describes scattering by continuous inhomogeneities. We now equate the two expressions, take (4.222) into account and cancel $\pi k_0^4/2$, to obtain

$$\Phi_\varepsilon(\boldsymbol{q}, \boldsymbol{R}) = \frac{2}{\pi}\bar{n}(\boldsymbol{R})\alpha^2 + [4\pi\alpha\bar{n}(\boldsymbol{R})]^2 \Phi_\mu(\boldsymbol{q}, \boldsymbol{R}) \ . \tag{4.232}$$

The covariance $\psi_\varepsilon(\boldsymbol{\varrho}, \boldsymbol{R})$ is obtained from this by the Fourier transformation in $\boldsymbol{q}$

$$\begin{aligned}\psi_\varepsilon(\boldsymbol{\varrho}, \boldsymbol{R}) &= \int \Phi_\varepsilon(\boldsymbol{q}, \boldsymbol{R}) \exp(i\boldsymbol{q}\cdot\boldsymbol{\varrho})\, d^3q \\ &= \frac{2}{\pi}\bar{n}\alpha^2 \int \exp(i\boldsymbol{q}\cdot\boldsymbol{\varrho})\, d^3q + (4\pi\alpha\bar{n})^2 \int \Phi_\mu \exp(i\boldsymbol{q}\cdot\boldsymbol{\varrho})\, d^3q \ .\end{aligned}$$

The first integral here is $8\pi^3\delta(\boldsymbol{\varrho})$, and the second, by (4.218), is $\mu(\boldsymbol{\varrho}, \boldsymbol{R})$:

$$\psi_\varepsilon(\boldsymbol{\varrho}, \boldsymbol{R}) = (4\pi\alpha)^2 [\bar{n}(\boldsymbol{R})\delta(\boldsymbol{\varrho}) + (\bar{n}(\boldsymbol{R}))^2 \mu(\boldsymbol{\varrho}, \boldsymbol{R})] \ . \tag{4.233}$$

If we take into account that $\tilde{\varepsilon} = 4\pi\bar{n}\alpha$, from (4.233) we obtain the covariance of the particle concentration

$$\psi_n(\boldsymbol{\varrho}, \boldsymbol{R}) = \bar{n}(\boldsymbol{R})\delta(\boldsymbol{\varrho}) + (\bar{n}(\boldsymbol{R}))^2 \mu(\boldsymbol{\varrho}, \boldsymbol{R}) \ . \tag{4.234}$$

The first terms in (4.233, 234) are due to the discrete nature of scatterers. The delta-function $\delta(\boldsymbol{\varrho})$ comes from substituting point dipoles for real particles. Given particles of finite size, instead of $\delta(\boldsymbol{\varrho})$, we would have a delta-type function decaying practically to zero at distances of the order of the particle size. The second term is due to the correlation of the particle positions and, by contrast, describes scattering by a collection of particles. If the concentration of particles $\bar{n}$ is sufficiently large[10], this term comes to dominate, and we can then treat scattering by a system of particles with correlated coordinates as scattering by volume macroscopic inhomogeneities of a continuous medium with the covariance

$$\psi_\varepsilon = (4\pi\alpha\bar{n})^2 \mu(\boldsymbol{\varrho}, \boldsymbol{R}) \ .$$

At *any angles* of scattering, we can neglect the "discrete" term σ_{ind} as compared to the term σ_{col} when

$$8\pi^3\bar{n}(\Phi_\mu)_{\min} \gg 1 \ , \tag{4.235}$$

where $(\Phi_\mu)_{\min}$ is the minimal (in $\boldsymbol{q}$) value of the spectrum. But if the inequality $8\pi^3\bar{n}(\Phi_\mu)_{\max} \ll 1$ is obeyed, we can ignore the collective effects. If neither of these conditions is met in the entire range of the scattering angle $0 < \theta < \pi$, then we will have either the complete cross-section (4.226), or, having determined from the condition $8\pi^3\bar{n}\Phi_\mu \sim 1$ the boundary angle θ_{b}, make use of the various limiting relations when $\theta > \theta_{\text{b}}$ and $\theta < \theta_{\text{b}}$.

---

[10] But it meets the condition $4\pi\bar{n}\alpha \ll 1$, which enables us to ignore the contribution of the quasi-stationary field. Should we include $E_{\text{q.st.}}$, we would end up with the Lorentz–Lorenz formula [4.13].

6) Incoherent Scattering in Plasma. We already touched upon this question in Exercise 7, Chap. 3 and Sect. 4.7. Here, we will consider scattering by free electrons in a plasma from another angle, namely as an example of a phenomenon in which, owing to the peculiarities of pairwise correlation, collective scattering appears to be proportional to the *first* power of the electron concentration N_e, not the second.

In the case of a plasma in equilibrium at a temperature T, the two-dimensional probability density $w_2(\boldsymbol{r}_1, \boldsymbol{r}_2)$ is given by the Boltzmann distribution

$$w_2(\boldsymbol{r}_1, \boldsymbol{r}_2) = \text{const} \cdot \exp\left[-\frac{U_{12}(\boldsymbol{r}_1, \boldsymbol{r}_2)}{\Theta}\right],$$

where $\Theta = kT$ is the temperature in energy units, and $U_{12}(\boldsymbol{r}_1, \boldsymbol{r}_2) = U_{12}(\boldsymbol{r}_1 - \boldsymbol{r}_2)$ is the effective potential energy of an electron at a point $\boldsymbol{r}_2$ in the field of another electron at a point $\boldsymbol{r}_1$.

Suppose that particles are uniformly distributed over a large volume V so that $w_1(\boldsymbol{r}) = \text{const} = 1/V$, and consider a case where the energy of the interaction of two electrons, U_{12}, is small as compared to the energy of the thermal motion $\Theta = kT$. The two-dimensional density will then be approximately

$$w_2(\boldsymbol{r}_1, \boldsymbol{r}_2) \approx \frac{1}{V^2}\left(1 - \frac{U_{12}}{\Theta}\right). \tag{4.236}$$

Comparing (4.236) with (4.216), where it is necessary to put $w_1 = 1/V$, we obtain for μ, which characterizes the correlation of the electron coordinates, the following approximate expression:

$$\mu(\boldsymbol{r}_1 - \boldsymbol{r}_2) \approx -\frac{U_{12}}{\Theta} = -\frac{U(\boldsymbol{r}_1 - \boldsymbol{r}_2)}{\Theta}.$$

In the case of an isotropic plasma $U_{12} = e\varphi_{12}$, where φ_{12} is the Debye potential, which differs from the Coulomb potential $e/|\boldsymbol{r}_1 - \boldsymbol{r}_2|$ by the exponential factor $\exp(-|\boldsymbol{r}_1 - \boldsymbol{r}_2|/d)$, where $d = (\Theta/8\pi e^2 \bar{N}_e)^{1/2}$ is the Debye (screening) distance [4.14], and $\bar{N}_e$ is the mean electron concentration. Thus,

$$\mu(\boldsymbol{r}_1 - \boldsymbol{r}_2) \approx -\frac{e^2}{\Theta|\boldsymbol{r}_1 - \boldsymbol{r}_2|}\mathrm{e}^{-|\boldsymbol{r}_1 - \boldsymbol{r}_2|/d}.$$

The Fourier transform for this function

$$\Phi_\mu(\boldsymbol{q}) = -\frac{1}{8\pi^3 \bar{N}_e}\frac{1}{2 + q^2 d^2}$$

is inversely proportional to $\bar{N}_e$, so that from (4.228) the collective scattering cross-section will be

$$\sigma_{\text{col}} = -\frac{\sigma_1 \bar{N}_e}{2 + q^2 d^2}. \tag{4.237}$$

As a result, the total scattering cross-section (4.226) is

$$\sigma = \sigma_1 \bar{N}_e - \frac{\sigma_1 \bar{N}_e}{2 + q^2 d^2} = \sigma_1 \bar{N}_e \frac{1 + q^2 d^2}{2 + q^2 d^2} ,$$

or, taking account of the fact that for the electron the scattering cross-section of an individual particle σ is $\gamma r_e^2 = \gamma e^4/m^2 c^4$, it will be

$$\sigma = \gamma r_e^2 \bar{N}_e \frac{1 + q^2 d^2}{2 + q^2 d^2} . \tag{4.238}$$

In the limiting case of short waves ($qd \gg 1$), it follows from (4.238) that

$$\sigma = \gamma \bar{N}_e r_e^2 \qquad (qd \gg 1) . \tag{4.239}$$

In consequence, the resultant effect in short wave scattering by thermal fluctuations of the electron concentration appears to be the same as with scattering by N_e free electrons. This is because electrons move independently within the confines of the region $\Delta r \sim 1/q \ll d$ that lies completely within the sphere with a radius equal to the Debye distance d.

In the opposite case, where the wavelength $\Lambda_q = 2\pi/q$ is much larger than the Debye distance, the motion of the electrons can no longer be treated as independent of the motion of the ions. In that case, the scattering cross-section will be

$$\sigma = \tfrac{1}{2} \gamma \bar{N}_e r_e^2 \qquad (qd \ll 1) , \tag{4.240}$$

i.e., half the value given in (4.239).

4.9 Exercises

4.9.1 Show that the total scattering cross-section σ_0 of a unit volume in the case of small-scale fluctuations ($kl_\varepsilon \ll 1$) is

$$\sigma_0 = 2\pi^2 k_0^4 \Phi_\varepsilon(0) = \frac{k_0^4}{4\pi} \int_{-\infty}^{\infty} \psi_\varepsilon(\varrho) d^3\varrho , \tag{4.241}$$

and for large-scale fluctuations ($kl_\varepsilon \ll 1$)

$$\sigma_0 = \tfrac{1}{2}\pi k_0^2 \int_{-\infty}^{\infty} \Phi_\varepsilon(\kappa_1, \kappa_2, 0) d\kappa_1 d\kappa_2 = \tfrac{1}{4} k_0^2 \int_{-\infty}^{\infty} \psi_\varepsilon(0, 0, \varrho_3) d\varrho_3 . \tag{4.242}$$

Solution. Equation (4.241) immediately follows from the definition (4.131), since for $kl_\varepsilon \ll 1$

$$\Phi_\varepsilon(\boldsymbol{q}) \approx \Phi_\varepsilon(0) = \frac{1}{8\pi^3} \int_{-\infty}^{\infty} \psi_\varepsilon(\varrho) d^3\varrho .$$

However, if $kl_\varepsilon \gg 1$, we should take into account the fact that the only contributions of significance to (4.241) come from the small scattering angles $\theta \lesssim 1/kl_\varepsilon \ll 1$. Introducing the integration variables $\kappa_1 = k\theta\cos\varphi$, $\kappa_2 = k\theta\sin\varphi$ and considering that at small scattering angles $\Phi_\varepsilon(\boldsymbol{q}) \approx \Phi_\varepsilon(\kappa_1, \kappa_2, 0)$ and $do = \sin\theta \times d\theta\, d\varphi \approx d\kappa_1\, d\kappa_2/k^2$, we obtain (4.242).

4.9.2 Estimate the total light scattering cross-section σ_0 for turbulent fluctuations with the spectral density $\Phi_\varepsilon(\kappa) = \mathscr{A}\, C_\varepsilon^2(\kappa^2 + \kappa_0^2)^{-11/6}$ having no singularity as $\kappa \to 0$. ($\kappa_0 = 2\pi/L_0$ is the wave number corresponding to the outer scale of turbulence L_0).
Solution. In the optical range, the wavelength λ is small as compared to the inner scale of turbulence l_0. Therefore, to derive σ_0 we can make use of (4.242), which gives (for $\mathscr{A} = 0.033$),

$$\sigma_0 = \tfrac{3}{5}\pi^2 k^2 \mathscr{A}\, C_\varepsilon^2 \kappa_0^{-5/3} \approx 0{,}0091\, k^2 C_\varepsilon^2 L_0^{5/3} \ .$$

In the atmospheric surface layer at $\lambda = 5.10^{-5}$ cm, $L_0 = 1$ m, and $C_\varepsilon^2 = 10^{-15}\,\mathrm{cm}^{-2/3}$ we estimate the extinction length to be $d_{\mathrm{ext}} = 1/\sigma_0 \sim 20$ m. This means that in problems concerned with light propagation in the atmospheric surface layer, the applicability of the Born approximation is limited by distances of such an order.

4.9.3 Using (4.101) estimate the transverse correlation radius of the field at backscattering, given that the size of the scattering volume L is limited by the width of the transmitting antenna pattern (Fig. 4.12).

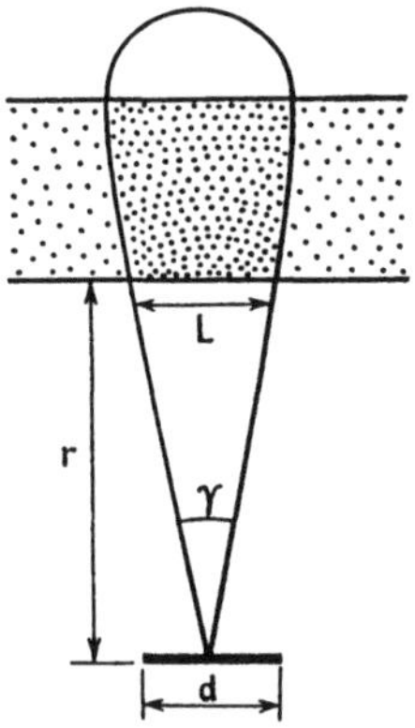

Fig. 4.12. Backscattering at a volume subject to a narrow beam

Solution. The main lobe of the pattern cuts out a region of the layer with the transverse size $L \sim r\gamma$. From (4.101) we have $l_\perp \sim \lambda/\gamma$. However, $\gamma \sim \lambda/d$, where d is the antenna size. Therefore, $l_\perp \sim d$, i.e. at backscattering, the transverse correlation radius of the field is of the order of the antenna diameter d.

4.9.4 A plane wave $A\exp(ik\boldsymbol{m}\cdot\boldsymbol{r})$ is incident on a scattering volume V located near an ideally reflecting plane $z = 0$. Find the mean intensity of the scattered field in the

Fraunhofer zone, assuming that the surface $z = 0$ is absolutely rigid, so that the boundary condition takes the form $\partial u/\partial N|_{z=0} = 0$.

Solution. The total primary field meeting the boundary condition has the form $u_0 = A[\exp(ik\boldsymbol{m}\cdot\boldsymbol{r}) + \exp(ik\boldsymbol{m}^*\cdot\boldsymbol{r})]$, where $\boldsymbol{m}^*$ is the unit normal to the phase front of the reflected wave ($m_z^* = -m_z$). For the absolutely rigid surface, the Green's function is

$$G(\boldsymbol{r}', \boldsymbol{r}) = -\frac{1}{4\pi}\left(\frac{e^{ikR}}{R} + \frac{e^{ikR^*}}{R^*}\right),$$

where $R = |\boldsymbol{r} - \boldsymbol{r}'|$, $R^* = |\boldsymbol{r} - \boldsymbol{r}'^*|$, and $\boldsymbol{r}'^*$ is the radius vector of the mirror-image of the source $\boldsymbol{r}'$, so that $\boldsymbol{r}'^* = \{x', y', -z'\}$.

Substituting u_0 and G into (4.11) and using the expansions of the type (4.26), we find that in the Fraunhofer zone

$$u_1(\boldsymbol{r}) = \frac{Ak_0^2}{4\pi}\frac{e^{ikr}}{r}\int_{-\infty}^{\infty} M(\boldsymbol{r}')\tilde{\varepsilon}(\boldsymbol{r}')\sum_{\alpha=1}^{4} e^{-i\boldsymbol{q}_\alpha\cdot\boldsymbol{r}'}. \tag{4.243}$$

Individual terms in (4.243) correspond to the four possible scattering types shown schematically in Fig. 4.13. The corresponding scattering vectors are

$$\boldsymbol{q}_1 = k(\boldsymbol{n} - \boldsymbol{m}), \quad \boldsymbol{q}_2 = k(\boldsymbol{n} - \boldsymbol{m}^*), \quad \boldsymbol{q}_3 = k(\boldsymbol{n}^* - \boldsymbol{m}), \quad \boldsymbol{q}_4 = k(\boldsymbol{n}^* - \boldsymbol{m}^*),$$

where $\boldsymbol{n}^*$ denotes the vector that is the "mirror-image" relative to the unit vector $\boldsymbol{n} = \boldsymbol{r}/r$, directed from the center of the scattering volume to the observation point ($n_z^* = -n_z$).

The expression for the mean intensity $\bar{I}_1 = \langle |u_1|^2 \rangle$ computed using (4.243) contains 16 terms, but only four of them are of significance [4.15]:

$$\bar{I}_1 = \frac{\pi k_0^4 |A|^2 V}{2r^2}\sum_{\alpha=1}^{4}\Phi_\varepsilon(\boldsymbol{q}_\alpha). \tag{4.244}$$

The remaining terms may be discarded (they are taken into account only for $m_z \to 0$ or $n_z \to 0$, i.e., for small grazing angles of the primary or scattered waves). With large-scale inhomogeneities, the major contributions to (4.244) come only from

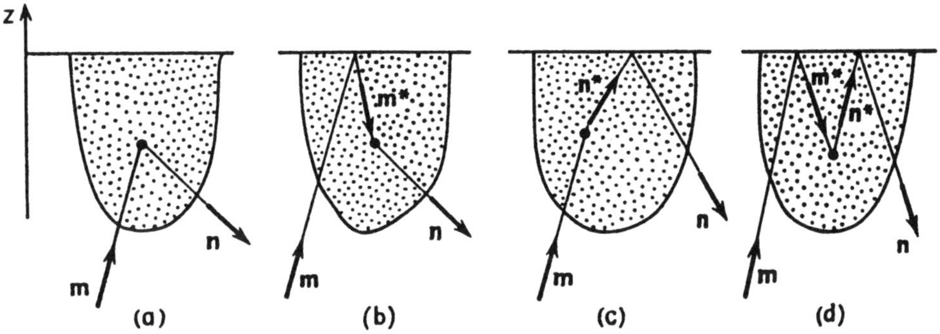

Fig. 4.13a–d. Four possible scattering types near an ideally reflecting plane

terms with $\boldsymbol{q} = \boldsymbol{q}_2$ and $\boldsymbol{q} = \boldsymbol{q}_3$ and, as seen in Fig. 4.13, correspond to scattering at small angles.

4.9.5 Find the covariance of the scattered field if the primary radiation constitutes a periodic sequence of short pulses recurring with period T_r.
Solution. In the case of stationary inhomogeneities, the scattered signal at time $t + T_r$ would be exactly as at t, and the covariance would be periodic in $\tau = t_1 - t_2$ and comprise terms of the form (4.161)

$$\psi_v(t, t-\tau) = \bar{I}_v(t) \sum_{m=-\infty}^{\infty} K_A(\tau - mT_r) \mathrm{e}^{\mathrm{i}\omega(\tau - mT_r)} ,$$

where $K_A(t)$ is the correlation coefficient for the envelope of the individual pulse (4.162).

However, with scattering by time-varying inhomogeneities $\tilde{\varepsilon} = \tilde{\varepsilon}(t, r)$, the correlation of signals at times t and $t + T_r$ will be incomplete, because during the time T_r the scattering medium changes. In that case, from (4.152) and (4.155) we can derive that

$$\psi_v(t, t-\tau) = \sum_{m=-\infty}^{\infty} I_m(t) K_A(t - mT_r) \mathrm{e}^{\mathrm{i}\omega(\tau - mT_r)} , \quad \text{where}$$

$$I_m(t) = 8\pi^3 \int_{-\infty}^{\infty} |\mathscr{P}(\boldsymbol{R})|^2 |a(t - B/c)|^2 \Phi_\varepsilon(mT_r, \boldsymbol{q}) \, d^3R .$$

Clearly, $I_0(t)$ coincides with the mean intensity $\bar{I}_v(t)$ given by (4.156).

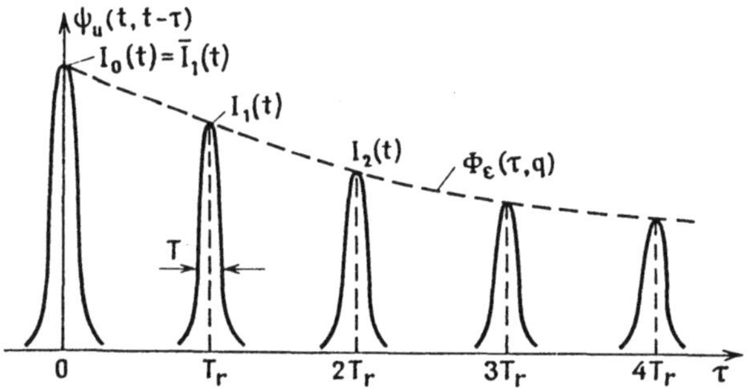

Fig. 4.14. Covariance of a scattered field for a periodic succession of short pulses

Figure 4.14 shows schematically the variation of the covariance $\psi_v(t, t-\tau)$ with τ for a fixed t, on the assumption that $T_r \gg T$. The width of the individual maxima of ψ_v is conditioned by the pulse duration T, and the separation between the maxima is T_r. Individual maxima I_m become lower with increasing m, in accordance with the time behavior of $\Phi_\varepsilon(mT_r, \boldsymbol{q})$. Consequently, the decrease in the height of the maxima with increasing m can give some idea of the correlation time of the inhomogeneities [4.16].

4.9.6 Consider a scattering volume illuminated by two primary waves with different frequencies ω_1 and ω_2. Find the interval of frequency correlation of the scattered field in the Fraunhofer zone.

Solution. We denote by $u_1(\omega_1, \boldsymbol{r})$ and $u_2(\omega_2, \boldsymbol{r})$ the appropriate scattered fields that in the Fraunhofer zone are given by (4.53), and find the frequency correlation coefficient

$$K_u(\Delta\omega) = \frac{\langle u_1(\omega_1, \boldsymbol{r}) u_1^*(\omega_2, \boldsymbol{r})\rangle}{\sqrt{\bar{I}_1(\omega_1)\bar{I}_1(\omega_2)}} = \frac{8\pi^3}{V}\delta_V\left(\frac{\Delta\omega}{\omega}\boldsymbol{q}\right),$$

where $\Delta\omega = \omega_1 - \omega_2$, $\delta_V(\boldsymbol{\kappa})$ is the delta-type function defined by (4.106), and $\boldsymbol{q} = \omega(\boldsymbol{n}_s - \boldsymbol{n}_i)/c$ is the scattering vector for the medium frequency $\omega = (\omega_1 + \omega_2)/2$. The interval of frequency correlation $\Delta\omega$ can be estimated from the inequality $\Delta\omega|q|/\omega \lesssim 2\pi/L$. If it is satisfied, the function $\delta_V(\Delta\omega\boldsymbol{q}/\omega)$ still differs markedly from zero. From (4.46), we obtain

$$\Delta\omega \lesssim \frac{2\pi\omega}{Lq} = \frac{\pi c}{L\sin(\theta/2)},$$

i.e., the correlation interval coincides with the frequency bandwidth (4.160) in which communication channels that use scattering can function without distortion [4.1].

4.9.7 Find the way in which the number of coherently scattering particles varies with the wavelength, in the case of a cloud of particles uniformly distributed over the volume of a sphere, and in the case of a continuous, smooth (in terms of wavelength) distribution.
Solution. For particles uniformly distributed in the volume of a sphere of radius a we have

$$w_1(\boldsymbol{r}) = \text{const} = \frac{1}{V} = (\tfrac{4}{3}\pi a^3)^{-1},$$

so that, by (4.203, 205), we obtain

$$N_{\text{coh}} = \frac{3N}{(aq)^3}|\sin aq - aq\cos aq|, \quad q = 2k\sin(\theta/2).$$

For $qa \gg 1$, i.e., for $\theta \gtrsim \lambda/a$, N_{coh} falls off linearly with $(aq)^{-2}$.

If we consider the smeared boundaries of the cloud of scatterers, N_{coh} falls off even faster (in exponential manner) with increasing qa. Let the probability density $w_1(\boldsymbol{r})$ decay smoothly from the center of the cloud to the periphery, with the characteristic space scale a, equal in order of magnitude to the cloud radius ($a \gg \lambda$). Now, taking into account that $\max w_1 \sim 1/a^3$, and using the known properties of the Fourier integrals [4.17], we obtain

$$|f_1(\boldsymbol{q})| \sim \frac{1}{qa}e^{-qa} = \frac{\lambda}{4\pi a\sin(\theta/2)}\exp\left(-\frac{4\pi a\sin(\theta/2)}{\lambda}\right). \tag{4.245}$$

It follows that for backscattering ($\theta = \pi$, $q = 2k$) N_{coh} accounts for a negligible percentage of N, even when $ka \gtrsim 10$, i.e., when $a \gtrsim 1.6\,\lambda$.

The reason why the number of particles that scatter in a coherent way decreases with increasing ka may easily be seen. Consider a layer of thickness $d \ll \lambda$, in which $\boldsymbol{q}\boldsymbol{r}_k = \text{const}$ up to quantities of order $d/\lambda \ll 1$ (the layer is oriented perpendicularly to the scattering vector $\boldsymbol{q}$). The particles in this layer scatter the incident radiation in an inphase (coherent) way. For each such layer, another thin layer can be chosen that radiates in antiphase (for which purpose it must be separated by an odd number of half-wavelengths Λ_q) and partially cancel the radiation of the first layer. The cancellation would be complete if both layers contained equal numbers of particles. However, owing to the random arrangement of the particles in space, each pair of layers gives a small residue that is uncompensated for. It is the sum of such residues that is given by (4.245).

References

Preface

1. S. A. Akhmanov, A. S. Chirkin: *Statistical Phenomena in Nonlinear Optics* (Moscow University Press, Moscow 1971) (in Russian)
2. A. M. Yaglom: *An Introduction to the Theory of Stationary Random Functions*(Prentice Hall, Englewood Cliffs, New York 1962)
3. A. Blanc-Lapierre, R. Fortet: *Théorie des fonctions aléatoires* (Masson, Paris 1953)
4. J. L. Doob: *Stochastic Processes* (John Wiley, New York 1953)
5. M. S. Barlett: *An Introduction to Stochastic Processes* (Cambridge University Press, Cambridge 1956)
6. Yu. A. Rozanov: *Stationary Random Processes* (Holden-Day, San Fransisco 1967)
7. A. A. Sveshnikov: *Applied Methods of the Theory of Random Functions* (Pergamon, Oxford 1966)
8. H. Cramér, M. R. Leadbetter: *Stationary and Related Stochastic Processes* (John Wiley, New York 1967)
9. S. Karlin: *A First Course in Stochastic Processes* (Academic, New York 1966)
10. V. L. Lebedev: *Random Processes in Electrical and Mechanical Systems* (Gostekhizdat, Moscow 1958) (in Russian)
11. A. van der Ziel: *Noise* (Prentice-Hall, Englewood Cliffs, New York 1954)
12. B. R. Levin: *Fondoments Teorique de la Radio technique Statistique*, (MIR, Moscow) Vols. I, II (1973); Vol. III (1979)
13. W. B. Davenport, Jr., W. L. Root: *An Introduction to the Theory of Random Signals and Noise* (McGraw-Hill, New York 1958)
14. A. van der Ziel: *Fluctuation Phenomena in Semi-Conductors* (Butterworths Scientific Publications, London 1959)
15. R. L. Stratonovich: *Topics in the Theory of Random Noise* Vols. I and II (Gordon and Breach, New York, 1963)
16. D. A. Middleton: *An Introduction to Statistical Communication Theory* (McGraw-Hill, New York 1960)
17. V. I. Tikhonov: *Statistical Radio Engineering* (Sovradio, Moscow 1966) (in Russian)
18. A. N. Malakhov: *Fluctuations in Self-Oscillatory Systems* (Nauka, Moscow 1968) (in Russian)
19. B. R. Levin: *Theoretical Foundations of Statistical Radio Engineering* (Sovradio, Moscow 1974) (in Russian)

Chapter 1

1.1 S. M. Rytov, Yu. A. Kravtsov, V. I. Tatarskii: *Principles of Statistical Radiophysics 4, Wave Propagation Through Random Media* (Springer, Berlin, Heidelberg, New York) in preparation

1.2 S. M. Rytov, Yu. A. Kravtsov, V. I. Tatarskii: *Principles of Statistical Radiophysics 1, Elements of Random Process Theory* (Springer, Berlin, Heidelberg, New York 1987)

1.3 S. M. Rytov, Yu. A. Kravtsov, V. I. Tatarskii: *Principles of Statistical Radiophysics 2, Correlation Theory of Random Processes* (Springer, Berlin, Heidelberg, New York 1988)

1.4 A. N. Kolmogorov: Dokl. Akad. Nauk SSSR **26**, 6 (1940)

1.5 A. N. Kolmogorov: Dokl. Akad. Nauk SSSR **26**, 15 (1940)

1.6 V. I. Tatarskii: *The Effects of the Turbulent Atmosphere on Wave Propagation* (IPTS, Jerusalem 1971)

1.7 A. S. Monin, A. M. Yaglom: *Statistical Hydromechanics Vol. 1* (Nauka, Moscow 1965); *Vol. 2* (Nauka, Moscow 1967)

1.8 V. I. Klyatskin: *Statistical Description of Dynamic Systems with Fluctuating Parameters* (Nauka, Moscow 1975) (in Russian)

1.9 K. Furutsu: J. Res. Nat. Bur. Stand **67**, 303 (1963)

1.10 E. A. Novikov: Zh. Eksp. Teor. Fiz. **47**, 1919 (1964)

1.11 M. D. Donsker: *Proc. Conf. on Theory and Application of Analysis in Function Space* (M.I.T., Cambridge MA 1964) pp. 17–30

1.12 A. G. Vinogradov, Yu. A. Kravtsov, Z. I. Feizulin: Radiotekh. Electron. **19**, 1758 (1974)

Chapter 2

2.1 S. A. Akhmanov, A. S. Chirkin: *Statistical Phenomena in Nonlinear Optics* (Moscow University Press, Moscow 1971) (in Russian);
S. A. Akhmanov, Yu. E. D'yakov, A. S. Chirkin: *Introduction to Statistical Radiophysics and Optics Vols.* 1, 2 (Springer, Berlin, Heidelberg, New York) in preparation

2.2 S. A. Akhmanov: Izv. Vyssh. Uchebn. Zaved, Radiofiz. **17**, 541 (1974)

2.3 V. N. Tsytovich: *A Theory of Turbulent Plasma* (Atomizdat, Moscow 1971)

2.4 B. B. Kadomtsev, V. M. Kantorovich: Izv. Vyssh. Uchebn. Zaved. Radiofiz. **17**, 511 (1974)

2.5 O. M. Phillips: *The Dynamics of the Upper Ocean* (Cambridge University Press, London 1969);

2.6 O. V. Rudenko: Sov. Phys. Usp. **29**, 620 (1986)

2.7 L. Mandel, E. Wolf: Rev. Mod. Phys. **37**, 231 (1965)

2.8 R. J. Glauber: "*Optical Coherence and Photon Statistics*" *in Quantum Optics and Electronics*, Proc. Summer School of Theor. Phys., Grenoble University 1965

2.9 J. Klauder, E. Sudarshan: *Fundamentals of Quantum Optics* (W. A. Benjamin, New York, Amsterdam 1968)

2.10 J. Perina: *Coherence of Light* (van Nostrand Reinhold, New York 1972)

2.11 R. Loudon: *The Quantum Theory of Light* (Oxford University Press, Oxford 1973)

2.12 Yu. A. Ryzhov: Zh. Eksp. Teor. Fiz. **59**, 218 (1970)

2.13 Yu. A. Kravtsov, S. M. Rytov, V. I. Tatarskii: Sov. Phys. Usp. **18**, 118 (1975)

2.14 S. M. Rytov, Yu. A. Kravtsov, V. I. Tatarskii: *Principles of Statistical Radiophysics 4*, Wave Propagation Through Random Media (Springer, Berlin, Heidelberg, New York) in preparation

2.15 S. M. Rytov, Yu. A. Kravtsov, V. I. Tatarskii: *Principles of Statistical Radiophysics 1, Elements of Random Process Theory* (Springer, Berlin, Heidelberg, New York 1987)

2.16 S. M. Rytov, Yu. A. Kravtsov, V. I. Tatarskii: *Principles of Statistical Radiophysics 2, Correlation Theory of Random Processes* (Springer, Berlin, Heidelberg, New York 1988)

2.17 E. O. Neill: *Introduction to Statistical Optics* (Addison–Wesley, Reading, Mass. 1963)

2.18 G. Stroke: *An Introduction to Coherent Optics and Holography* (Academic Press, New York 1966)

2.19 N. S. Shestov: *Detection of Optical Signals in Random Noise* (Sovradio, Moscow 1967) (in Russian)

2.20 V. A. Zverev, E. F. Orlov: *Optical Analyzers* (Sovradio, Moscow 1971) (in Russian)

2.21 A. Van-der-Lugt: Proc. IEEE **62**, 1300 (1974)

2.22 D. D. Klovsky, V. A. Soifer: *Processing of Spatio-Temporal Signals* (Svyaz, Moscow 1976) (in Russian)

2.23 V. I. Siforov, A. V. Prosin (eds.): *Information Transfer along Lines Containing Statistically Inhomogeneous Media* (Nauka, Moscow 1976) (in Russian)

2.24 P. A. Bakut, N. D. Ustinov, I. N. Troitsky, K. N. Sviridov: Zarub. Radioelektron. **7**, 15 (1976)—Part I; **9**, 3 (1976)—Part II; **1**, 3 (1977)—Part III; **3**, 55 (1977)—Part IV

2.25 M. Born, E. Wolf: *Principles of Optics* (Pergamon, Oxford 1964)

2.26 M. J. Beran, G. P. Parrent: *Theory of Partial Coherence* (Prentice-Hall, Englewood Cliffs, NJ 1964)

2.27 N. G. Denisov: Izv. Vyssh. Uchebn. Zaved. Radiofiz. **4**, 630 (1961)

2.28 H. G. Booker, J. A. Ratcliffe, D. H. Schinn: Philos. Trans. R. Soc. London, Ser. A **242**, 579 (1950)

2.29 A. Hewish: Proc. R. Soc. London, Ser. A **209**, 81 (1951)

2.30 J. A. Fejer: Proc. R. Soc. London, Ser. A **220**, 455 (1953)
2.31 Yu. N. Barabanenkov, Yu. A. Kravtsov, S. M. Rytov, V. I. Tatarskii: Sov. Phys. Usp. **13**, 551 (1971)
2.32 L. M. Erukhimov, O. I. Maksimenko: "Studies of Ionospheric Inhomogeneities", in *Drifts and Inhomogeneities in the Ionosphere* (Nauka, Moscow 1973) p. 41 (in Russian)
2.33 B. H. Briggs: Contemp. Phys. **16**, 469 (1975)
2.34 G. S. Gorelik: *Oscillations and Waves*, 2nd ed. (Fizmatgiz, Moscow 1959) (in Russian)
2.35 Yu. E. D'yakov: Radiotekh. Elektron. **8**, 1812 (1963)
2.36 I. S. Vsekhsvyatskaya: *Statistical Behavior of Signals Reflected from the Ionosphere* (Nauka, Moscow 1975) (in Russian)
2.37 Ya. I. Alber, L. M. Erukhimov, V. A. Ryzhov, V. P. Uryadov: Izv. Vyssh. Uchebn. Zaved. Radiofiz. **11**, 1371 (1968)
2.38 E. E. Salpeter: Astrophys. J. **147**, 433 (1967)
2.39 I. G. Yakushkin: Izv. Vyssh. Uchebn. Zaved. Radiofiz. **17**, 1350 (1974)
2.40 V. I. Shishov: Izv. Vyssh. Uchebn. Zaved. Radiofiz. **14**, 85 (1971)
2.41 M. Franson, S. Slansky: *Coherence in Optics* (Nauka, Moscow 1967) (in Russian)
2.42 V. A. Zverev: *Radiooptics* (Sovradio, Moscow 1975) (in Russian)
2.43 Ya. S. Shifrin: *Topics in Statistical Theory of Antennas* (Sovradio, Moscow 1970) (in Russian)
2.44 Yu. A. Kravtsov, V. I. Tatarskii: "Statistics of Wave Diffraction", in *Lectures at IV All-Union School on Diffraction and Wave Propagation* (Ryazan Radiotech. Inst., Ryazan 1976) (in Russian)
2.45 B. F. Kur'yanov: Akust. Zh. **9**, 441 (1964)
2.46 L. S. Dolin: Izv. Vyssh. Uchebn. Zaved. Radiofiz. **7**, 559 (1964)
2.47 S. Chardrasekhar: *Radiative Transfer* (Dover, New York 1960)
2.48 V. V. Zheleznakov: *Radio Emission of the Sun and Planets* (Nauka, Moscow 1964) (in Russian)
2.49 O. V. Rudenko, S. I. Soluyan: *Theoretical Foundations of Nonlinear Acoustics* (Consultant Bureau, New York, London 1977)
2.50 L. A. Apresyan, Yu. A. Kravtsov: Sov. Phys. Usp. **27**, 301 (1984)
2.51 L. A. Apresyan, Yu. A. Kravtsov: *Theory of Radiative Transfer: Statistical and Wave Aspects* (Nauka, Moscow 1983) (in Russian)
2.52 A. Ishimaru: *Wave Propagation and Scattering in Random Media, Vol. 1, 2* (Acad. Press, N.Y. 1978)

Chapter 3

3.1 S. M. Rytov: *A Theory of Electrical Fluctuations and Thermal Radiation* (USSR Academy of Sciences, Moscow 1953) (in Russian)
3.2 L. A. Vainshtein: *Electromagnetic Waves* (Sovradio, Moscow 1957) (in Russian)
3.3 L. D. Landau, E. M. Lifshits: *Electrodynamics of Continuous Media* (Pergamon, Oxford 1960)
3.4 S. M. Rytov: Dokl. Akad. Nauk SSSR **110**, 371 (1956)
3.5 S. M. Rytov: Zh. Eksp. Teor. Fiz. **33**, 166 (1957)
3.6 M. L. Levin, S. M. Rytov: *A Theory of Equilibrium Thermal Fluctuations in Electrodynamics* (Nauka, Moscow 1967) (in Russian)
3.7 M. L. Levin, V. G. Polevoi, S. M. Rytov: Zh. Exp. Teor. Fiz. **79**, 2087 (1980)
3.8 E. M. Lifshits: Zh. Eksp. Teor. Fiz. **29**, 94 (1955)
3.9 V. P. Silin, A. A. Rukhadze: *Electromagnetic Properties of Plasma-Like Bodies* (Gosatomizdat, Moscow 1961) (in Russian)
3.10 G. A. Grinberg: *Selected Topics in the Mathematical Theory of Electric and Magnetic Phenomena* (USSR Academy of Sciences, Moscow 1948) (in Russian)
3.11 E. L. Fainberg: *Propagation of Radio Waves over the Earth's Surface* (USSR Academy of Sciences, Moscow 1961) (in Russian)
3.12 I. E. Dzyaloshinsky, E. M. Lifshits, L. P. Pitaevsky: Zh. Eksp. Teor. Fiz. **37**, 229 (1959); Sov. Phys. Usp., **4**, 153 (1961)
3.13 A. A. Abrikosov, L. P. Gor'kov, I. E. Dzyaloshinsky: *Methods of Quantum Field Theory in Statistical Physics* (Fizmatgiz, Moscow 1962) (in Russian)
3.14 M. A. Leontovich: Zh. Eksp. Teor. Fiz. **40**, 907 (1961)

3.15 V. D. Shafranov: "Electromagnetic Waves in Plasma", in *Topics in Plasma Theory* (Gosatomizdat, Moscow 1963)
3.16 M. L. Levin, S. M. Rytov: Zh. Eksp. Teor. Fiz. **65**, 1382 (1973)
3.17 V. L. Ginzburg: *Propagation of Electromagnetic Waves in Plasma* (Pergamon, Oxford 1970)

Chapter 4

4.1 V. I. Tatarskii: *The Effects of the Turbulent Atmosphere on Wave Propagation* (IPST, Jerusalem 1971)
4.2 Yu. N. Barabanenkov, Yu. A. Kravtsov, S. M. Rytov, V. I. Tatarskii: Sov. Phys. Usp. **13**, 3 (1971) [Sov. Phys. Usp. **13**, 3 (1971)]
4.3 G. S. Gorelik: Radiotekh. Elektron **1**, 695 (1956)
4.4 G. S. Gorelik: Radiotekh. Elektron **2**, 1227 (1957)
4.5 M. I. Rodak, A. V. Frantsesson: Radiotekh. Elektron. **4**, 398 (1959)
4.6 M. I. Rodak: Radiotekh. Elektron. **5**, 1370 (1960)
4.7 N. G. Denisov: Izv. Vyssh. Uchebn. Zaved. Radiofiz. **3**, 393 (1960)
4.8 S. M. Rytov: Zh. Eksp. Teor. Fiz. **33**, 514 (1957)
4.9 V. S. Vikhrenko: Sov. Phys. Usp. **17**, 558 (1975)
4.10 J. V. Evans: Proc. IEEE **63**, 1636 (1975)
4.11 A. I. Akhiezer et al.: *Collective Oscillations in Plasma* (Atomizdat, Moscow 1964) (in Russian)
4.12 H. C. van der Hulst: *Light Scattering by Small Particles* (Wiley, New York 1957)
4.13 M. Born, E. Wolf: *Principles of Optics*, 4th ed. (Pergamon, New York 1970)
4.14 V. L. Ginzburg: *Propagation of Electromagnetic Waves in Plasma*, (Pergamon, Oxford 1970)
4.15 N. G. Denisov: Izv. Vyssh. Uchebn. Zaved. Radiofiz. **7**, 378 (1964)
4.16 L. L. Goryshnik, Yu. A. Kravtsov: Geomagn. Aeron. **9**, 38 (1969)
4.17 A. B. Migdal, V. P. Kraynov: *Approximate Methods in Quantum Mechanics* (Nauka, Moscow 1966) (in Russian)
4.18 A. Ishimaru: *Wave Propagation and Scattering in Random Media, Vol. 1, 2* (Acad. Press, N.Y. 1978)

Note: Some Soviet publications widely cited in the present book are regularly translated into English. We remind the reader that

(1) Izv. Vyssh. Uchebn. Zaved. Radiofiz. appears in English as Radiophys. and Quantum Electron.;
(2) Radiotekh. Elektron. up to 1984 was known as Radioeng. and Electron. Phys. and since 1985 as Sov. J. Communic. Technol. and Electron.;
(3) Usp. Fiz. Nauk as Sov. Phys. Usp.;
(4) Akustich. Zh. as Sov. Phys. Acoust.;
(5) Doklady Akad. Nauk SSSR as Sov. Phys. Dokl.

Subject Index

36310336R00142

Made in the USA
San Bernardino, CA
19 July 2016